Free Radicals
in Organic Chemistry

FREE RADICALS IN ORGANIC CHEMISTRY

Jacques FOSSEY

*CNRS, École polytechnique,
France*

Daniel LEFORT

*CNRS, Groupe des laboratoires de Vitry-Thiais,
France*

Janine SORBA

*CNRS, École polytechnique,
France*

Forewords by
Alwyn DAVIES, *Fellow of the Royal Society (London)*
Guy OURISSON, *Member of the Académie des Sciences (Paris)*

Translated by
John LOMAS, *CNRS, Université de Paris VII, France*

JOHN WILEY & SONS
Chichester • New York • Brisbane • Toronto • Singapore

1995

MASSON
Paris • Milan • Barcelona

Copyright © Masson, Paris, 1995

Original French language edition – *Les radicaux libres en chimie organique*
Copyright © Masson, Paris, 1993

Reprinted July 1997
All rights reserved.

No part of this book may be reproduced by any means,
or transmitted, or translated into a machine language
without the written permission of the publisher

Wiley Editorial Offices

John Wiley & Sons Ltd
Baffins Lane, Chichester
West Sussex PO19 1UD, England

John Wiley & Sons, Inc., 605 Third Avenue,
New York, NY 10158-0012, USA

Jacaranda Wiley Ltd, G.P.O. Box 859, Brisbane,
Queensland 4001, Australia

John Wiley & Sons (Canada) Ltd, 22 Worcester Road,
Rexdale, Ontario M9W 1L1, Canada

John Wiley & Sons (SEA) Pte Ltd, 37 Jalan Pemimpin #05-04,
Block B, Union Industrial Building, Singapore 2057

ISBN : 0-471-95495-0 (cloth) (Wiley)
ISBN : 0-471-95496-9 (paper) (Wiley)
ISBN : 2-225-84730-4 (paper) (Masson)

A catalogue record for this book is available from the British Library

Printed and bound in Great Britain by Bookcraft (Bath) Ltd.

FOREWORD
to the English edition

I did research for my Ph. D. degree in Sir Christopher Ingold's heterolytic school at University College London, where the only mention of radicals was in some intercollegiate lectures by D.Y. Hey from the rival King's College. After graduation, I left and started research into the behaviour of hydrogen peroxide and alkyl hydroperoxides as nucleophiles, which soon inevitably led me into homolysis. When I returned to University College in 1953 as a lecturer, and expressed my intention of moving further into free radical chemistry, my colleagues Charles Vernon and Peter de la Mare feigned alarm and, in the language of those times, advised me that at U.C.L. homolysis, even between consenting adults, was grounds for instant dismissal. Ingold himself, however, was full of support.

The only book from which we could learn the relevant free radical chemistry in those days was that by W.A. Waters. Otherwise we had to rely on M.S. Kharasch's papers in the early volumes of Journal of Organic Chemistry, which were not always models of clarity. Cheves Walling's monumental *Free Radicals in Solution*, published in 1957, was the first book to incorporate the great advances made in the subject in the U.S.A. during the war years, and copies of it are still jealously prized by the people who grew up with it. When, before long, I started to teach a course in free radical chemistry, in the days well before photocopying, I had to hand out cyclostyled notes as there was still no suitable textbook at an appropriate level.

The advances that have been made in the subject since then have been enormous. The selectivity and sensitivity of EPR spectroscopy has made it possible to study radicals at very low concentration. Matrix isolation methods have provided conditions under which these concentrations can be maintained for an adequate time, and the spin trapping technique makes it possible to preserve a transient radical as a long-lived radical. NMR observation of the CIDNP effect provides a complementary way of looking at the products of the radicals, rather than at the radicals themselves. Time-resolved UV, EPR, and IR spectroscopy has made it possible to determine the kinetics of the fast reactions which most free radicals undergo, and to provide standards against which the rates of other reaction can be measured. Chemists who now work with free radicals have access to tabulations of their EPR spectra and their kinetics in many volumes of *Landoldt Börnstein*.

The most rapid developments in the free radical field however have been in organic synthesis: chemists have tamed the radicals which in my student days were regarded as being too unselective in their reactions to be useful in synthesis. This field of radical chemistry lives in symbiosis with the physical studies, and together they are now providing the foundation for the rapidly expanding field of free radicals in biochemistry.

There is still a dearth of text books and monographs in free radical chemistry, particularly at a level suitable for a graduate. All the above topics and more are covered in Jacques Fossey, Daniel Lefort and Janine Sorba's book. It deals with the physical, synthetic, biochemical, and industrial aspects of organic free radicals at a level appropriate for a chemistry graduate, and would provide an excellent grounding in the field, and guide to the literature, for anyone who needs to find their feet in free radical chemistry. I only wish that the equivalent had been available when I was in that position.

Alwyn Davies
Fellow of the Royal Society

FOREWORD
to the French edition

Courses on organic chemistry have no secrets for me, since for 40 years I have followed them, read them and (above all) given them. I know therefore that the last lectures of a basic course are devoted to radical chemistry and that it is customary to say that it is important but that there is not enough time to discuss it in detail. Sometimes a postgraduate course includes a more detailed account, but not always; most organic chemistry researchers have never had a serious introduction to the field. That is my case.

"This work, intended for students at the end of their studies", as the authors say in their introduction, was therefore for me.

And yet, pick up any recent number of a good journal on organic chemistry and read the Contents: it will be very surprising if you don't find several articles devoted to radical cyclization, to one-electron reduction, to oxidative cyclization, etc.

An organic chemist who knows nothing about radical chemistry cannot, at the end of this century, be anything other than a poor organic chemist. But how can one get a reason- able idea of the subject without spending too much time, if one doesn't want to become an expert but simply a potential user ? We need mediators, experts who have taken the trouble to go beyond their narrow field in order to present to others a competent and didactic intro- duction, based on their experience.

This is what Jacques Fossey, Daniel Lefort and Janine Sorba have magnificently succeeded in doing.

I learnt a great deal by reading this book, finding a healthy balance between theoretical bases, mechanistic descriptions and synthetic applications, and I finished reading it with a regret and a desire: the regret that I did not have this book in time to use it in my teaching and the desire to reread certain chapters more attentively.

However, I cannot conclude this preface without recommending that readers bear in mind that an important class of radical reactions appears to be missing from this book, as from most other even much bigger volumes: the most important reaction of organic substances, their combustion in air ... Isn't this organic chemistry ? It is true that this is a particularly non-synthetic and very difficult chapter. We look forward to seeing a good presentation in the next edition.

Guy Ourisson
Member of the French Academy of Sciences

ACKNOWLEDGMENTS

We dedicate this work to the memory of Bianca Tchoubar who directed Research Group #12 of the CNRS *Réactivité et mécanismes en chimie organique* (Reactivity and mechanisms in organic chemistry). Throughout this period she gave us encouragement, stimulus and suggestions.

Our thanks are also addressed to those who have contributed to our scientific activity in radical chemistry and particularly to Drs Michel Gruselle, Jean-Yves Nedelec and Jean Zylber as well as to Ms Andrée Druilhe, Monique Heintz and Nicole Zylber.

We also thank our colleagues, Professors Jean-Marie Surzur, Michel Chanon, Michèle-Paul Bertrand and Paul Tordo, as well as Michel Crozet, Bernard Maillard, Philippe Maitre, Lucien Stella and François Volatron for many informative discussions and for rereading the manuscript. We thank Professor François Francou for constructive criticism of the chapter on biochemistry.

Rhône-Poulenc, Atochem and Interox-International made information available to us on their industrial production by radical processes, in particular on oxidation (hydrogen peroxide, phenol, adipic acid), chlorination and polymerization. We are indebted to them for their kindness and their contributions.

We are grateful to Christiane Lefort for having read this text and corrected various errors and to Anne Sarrobert for having typed most of it.

We thank, our colleague, John Lomas for his precise and meticulous translation which often enabled us to improve upon the original text. We thank also Professor Alwyn Davies for his attentive reading of the English version, which led us to correct a certain number of errors and again to upgrade the first manuscript.

CONTENTS

Foreword to the English edition   V
Foreword to the French edition   VII
Acknowledgments   IX

Chapter 1 Introduction   1

PART I GENERAL CONCEPTS AND BASIC PRINCIPLES   3

Chapter 2 Detection and observation of free radicals   5

2-1   Electron Paramagnetic Resonance (EPR)   5
2-2   Chemically Induced Dynamic Nuclear Polarization (CIDNP)   13
2-3   Electronic Spectroscopy   17

Chapter 3 Radical structure   19

3-1   Definitions   19
3-2   Methods of investigation   21
3-3   Carbon radicals   22
3-4   Radicals centered on an atom other than carbon   26

Chapter 4 Radical stability   31

4-1   Thermodynamic stability   31
4-2   Kinetic stability   37
4-3   Conclusions   38

Chapter 5 Elementary reactions and mechanisms   39

5-1   Elementary reactions   39
5-2   Radical chain mechanisms   42
5-3   Non-chain radical mechanisms   47

Chapter 6 Reactivity of free radicals   49

6-1   Introduction to reactivity   49
6-2   Factors controlling reactivity   50
6-3   Orbital analysis of radical reactivity   57
6-4   Applications of orbital analysis   61
6-5   State correlation diagrams   70

Chapter 7 Radical kinetics   73

7-1   Kinetics of chain reactions   73
7-2   Examples of kinetics   76
7-3   Competitive systems: rate constant determination   79

Chapter 8 Radicals centered on an atom other than carbon — 83

8-1	Si, Ge and Sn-centered radicals	83
8-2	Nitrogen-centered radicals	86
8-3	Phosphorus-centered radicals	91
8-4	Oxygen-centered radicals	93
8-5	Sulfur-centered radicals	97
8-6	Halogen atoms	100

PART II REACTIONS: CLASSIFICATIONS AND MECHANISMS — 103

Chapter 9 Production of free radicals — 105

9-1	General principles	105
9-2	Methodology of radical production	108
9-3	Conclusions	117

Chapter 10 Radical - radical reactions — 119

10-1	Recombination	119
10-2	Disproportionation	123
10-3	Disproportionation/recombination ratio	124

Chapter 11 Substitution reactions — 125

11-1	Different types of S_H2	125
11-2	S_H2 on monovalent atoms	126
11-3	S_H2 on multivalent atoms	133
11-4	Intramolecular substitution, S_Hi	135

Chapter 12 Addition and fragmentation reactions — 139

12-1	Addition to a >C=C< double bond	139
12-2	Addition to various unsaturated systems	147
12-3	Fragmentation	148

Chapter 13 Cyclizations and rearrangements — 151

13-1	Cyclization to a >C=C< double bond	151
13-2	Cyclization on other unsaturated systems	156
13-3	Cyclization by S_Hi	157
13-4	Ring opening	158
13-5	Rearrangements	161

Chapter 14 Aromatic substitution — 166

14-1	Mechanism	166
14-2	Addition step	167
14-3	Ortho, meta, para regioselectivity	172
14-4	Elimination step	176
14-5	Reactions	176
14-6	Conclusions	180

Contents XIII

Chapter 15 Reactions of charged radicals — 181

- *15-1 Radical anions* — 181
- *15-2 Radical cations* — 185

Chapter 16 Free radicals in biochemistry — 191

- *16-1 Biological oxidation: a source of energy* — 191
- *16-2 Dioxygen effect* — 193
- *16-3 Autoxidation of lipids* — 195
- *16-4 Isomerization reactions* — 197
- *16-5 Hydroxylation by cytochrome P450* — 198
- *16-6 Reactions induced by exogenous agents* — 198
- *16-7 Conclusions* — 200

Chapter 17 How to prove that a reaction is a radical process — 201

- *17-1 Transfer reaction* — 201
- *17-2 Coupling with a stable radical* — 203
- *17-3 Isomerization or rearrangement* — 204
- *17-4 Addition to unsaturated compounds* — 206
- *17-5 Inhibitors and kinetics* — 208

PART III APPLICATIONS IN SYNTHESIS — 209

Chapter 18 Functionalization of the C–H bond — 211

- *18-1 Halogenation* — 211
- *18-2 Hydroxylation* — 217
- *18-3 Autoxidation* — 218

Chapter 19 Transformation of functional groups — 221

- *19-1 Halogen reduction* — 221
- *19-2 Replacement of –COOH by another group* — 223
- *19-3 Alcohol deoxygenation* — 225
- *19-4 Deamination* — 227
- *19-5 Denitration* — 228

Chapter 20 Functionalization of multiple bonds — 229

- *20-1 Kinetic aspects* — 229
- *20-2 Addition of carbon radicals: C–C bond formation* — 230
- *20-3 Addition of aminyl groups: C–N bond formation* — 236
- *20-4 Addition of sulfur- and phosphorous- centered radicals* — 238
- *20-5 Addition of group 14 radicals: C–M bond formation* — 240
- *20-6 Addition-fragmentation: allylation reaction* — 240

Chapter 21 Cyclization — 243

- *21-1 Methodology* — 243
- *21-2 Cyclization of carbon radicals* — 244
- *21-3 Cyclization of heteroatomic radicals* — 253
- *21-4 Combination of intermolecular addition and cyclization* — 254

Chapter 22 Aromatic substitution — 257

22-1	Amination reactions	257
22-2	Substitution on heteroaromatic bases by nucleophilic radicals	258
22-3	Nucleophilic radical substitution, $S_{RN}1$	261

Chapter 23 Coupling reactions — 265

23-1	Dehydrodimerization	265
23-2	Electrochemistry	266
23-3	$S_{RN}1$ reaction	267

Chapter 24 Free radical reactions in industry — 269

24-1	Polymers	269
24-2	Halogenation	272
24-3	Nitrosation (lactam)	273
24-4	Oxidation	273
24-5	Oxidation-combustion	275

PART IV REFERENCES AND TABLES — 279

Chapter 25 References — 281

General reviews and books	281
Part I - General concepts and basic principles	281
Part II - Reactions: classifications and mechanisms	284
Part III - Applications in synthesis	287

Chapter 26 Tables — 289

26-1	Rate constants	289
26-2	Ionization potentials	297
26-3	Bond energies	297

Index — 301

1 – INTRODUCTION

Radical processes occur in many fields of chemistry: combustion and autoxidation, the atmosphere and pollution, biological and pathological chain reactions, etc. Radical chemistry has become a valuable tool for the organic chemist, for solving various fundamental problems and also for the construction of complex molecules. The many studies of free radicals (formation, structure, reactivity, etc.) since the beginning of the century now make it possible to integrate this chemistry into the general knowledge of all the reactions of organic chemistry, from both the conceptual and methodological viewpoints.

Historically, it is generally accepted that the first radical was identified by Gomberg in 1900, in the form of a *trivalent compound*, triphenylmethyl. However, the beginnings of synthesis by radical pathways did not appear until 1937 with Hey and Waters and the homolytic phenylation of aromatic substrates, as well as with Kharasch and the regioselectivity of HBr addition to alkenes, called the *peroxide effect* or the Kharasch effect or anti-Markovnikov orientation. The Second World War was an important stimulus for the development of radical chemistry. The Japanese occupation of South-East Asia cut off supplies of latex to the United States; chemists were therefore called upon to find substitutes for natural rubber. Mayo, Walling and Lewis studied the polymerization of vinyl compounds, established the rules of polymerization and copolymerization, and developed kinetic laws. Whereas uncharged free radicals were considered as strictly neutral species, these authors demonstrated their polar character in the reaction process, this character being expressed in the concept of *electrophilic* and *nucleophilic* radicals. We now understand just how important this concept is for explaining the differences in the selectivity and the reactivity of free radicals. Up to a certain period, these were expressed only in relative terms by considering a radical and several substrates or vice versa. Work at that time described above all mechanisms, depending on the types of reaction, and established relative reactivity scales. Nevertheless, certain authors such as Barton, Cadogan, Marc Julia and others had already shown the synthetic interest of radical reactions. For organic chemists, these reactions remained difficult to control and led only to a mixture of products whose formation could not be explained by current theories.

The use of physical methods was all-important. EPR first of all, making it possible to detect radicals, to identify them and to determine their structures; Symons, Kochi, Krusic and Alwyn Davies, amongst others, were particularly active in this field. EPR can also be used to follow the appearance or the disappearance of radicals and, therefore, to determine their kinetics. *Flash photolysis* later provided a valuable additional experimental tool. From the 60s onwards Keith Ingold, Fischer and Beckwith determined the absolute rate constants of the most important radical reactions in solution. The knowledge of these values was a major contribution to mastering the course of a radical reaction and to fixing the experimental conditions in order to attain a given target.

About 1970 onwards sees the beginning of new synthetic methods, using radical reactions, developed by various authors including Barton, Giese, Hart, etc. It is interesting to note that cyclization reactions, all aspects of which were studied in detail by Julia, Surzur and their co-workers in the 60s, were not exploited as synthetic methods, by Stork and Curran, for example, until about 1980 onwards.

The existence of charged radicals, radical cations and radical anions, was recognized very early on. In particular, the reactivity of radical anions and their use by Kornblum (1963) with aliphatics and by Bunnett with aromatics constitutes an attractive synthetic method ($S_{RN}1$) with many applications.

The homolytic substitution of benzenes or polynuclear aromatics aroused little synthetic interest because of the lack of substrate and positional selectivity. By taking advantage of polar effects Minisci was able, from 1965 onwards, to develop, with good yields and very good regioselectivity, a whole series of reactions: amination of aromatic substrates and substitution on heteroaromatic bases by various nucleophilic radicals.

The production of radicals under mild conditions adapted to the reaction medium is still a problem. Amongst recent innovations, the use of the thio function as precursor and reactant, developed by Barton, offers a very wide range of possibilities.

In biological chemistry, although the subject was touched upon in the 50s by radiochemists who examined the effects of ionizing radiations (X-rays and γ-radiation), the creation of and the role played by free radicals in living systems remains highly topical and is only now beginning to be studied seriously, at least in France.

The current output of publications concerning radical processes in synthesis is enormous. These reactions have many advantages compared to heterolytic chemistry and are usually easy to put into practice. They are performed under mild conditions and, unlike ionic reactions, can be used to create complex structures by operating on molecules which have other functional groups without these being modified. Moreover, they can be chemo-, regio- and stereoselective.

In this book, intended for students at the end of their studies and for researchers wishing to learn the basics of radical chemistry, we have attempted in the first two parts to give an overview of the general principles of this chemistry. In the third part we have described what we think at the moment to be the main synthetic applications. We simply give some examples: those interested can consult the many reviews published in the last few years. We hope that reading this book will enable researchers to understand them better; it is divided into four parts:

Part I, *general concepts and basic principles* of free radicals and their reactivity.

Part II, *reactions: classification and mechanisms* of the elementary reactions, substitution, addition, cyclization, fragmentation, etc.

Part III, *applications in synthesis*. We give some examples without attempting to be exhaustive.

Part IV, *references and tables*.

PART I

GENERAL CONCEPTS AND BASIC PRINCIPLES

2 – DETECTION AND OBSERVATION OF FREE RADICALS

2-1 ELECTRON PARAMAGNETIC RESONANCE (EPR)

EPR is a spectroscopic technique which can be used for studying species containing one or more unpaired electrons. This method is very sensitive, being capable of detecting free radicals at concentrations of the order of 10^{-8} M. It provides information about both the nature of the radical and its structure (conformation, configuration, delocalization of the unpaired electron).

2-1-1 Principles of EPR

Historically, EPR was discovered before NMR but its development has been slower because of the difficulty of observing radicals which are usually short-lived. The principle is analogous to that of NMR, the difference being that the resonance phenomenon concerns electron spin and not nuclear spin.

The concept of spin was introduced by Uhlenbeck and Goudsmit in 1925. They suggested that the electron had a magnetic moment (spin) resulting from the rotation of a charged particle about an axis (spinning). The spin of an electron is characterized by the s quantum number. Quantum mechanics shows that the electron has two spin states corresponding to the two orientations in space. The eigenvalues of the projection in a given direction are equal to $m_s h/2\pi$ where $m_s = \pm 1/2$, h being Planck's constant. When $m_s = +1/2$ one says that the electron is in the α spin state and when $m_s = -1/2$ the electron is in the β spin state.

In the same way, the proton has two nuclear spin states, α_N and β_N, the eigenvalues of the projection in a given direction being $m_I h/2\pi$ with $m_I = +1/2$ for α_N and $m_I = -1/2$ for β_N.

Experiment shows that for the electron the spin magnetic moment, μ_s, and its projections, μ_{sz}, are given by:

$$\mu_s = g_e \beta_e \sqrt{s(s+1)} \tag{2.1}$$

$$\mu_{sz} = -g_e \beta_e m_s = \boxed{\gamma_s m_s h/2\pi} \tag{2.2}$$

$g_e \approx 2$ (Landé splitting factor)
$\beta_e = 9.2732 \times 10^{-21}$ erg gauss^{-1} (Bohr magneton)
$|\gamma_s| = 1.7586 \times 10^7$ rad s^{-1}gauss^{-1} (gyromagnetic ratio)
$s = 1/2$, $m_s = \pm 1/2$

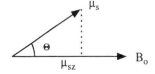

$$E = -\vec{\mu}_s \cdot \vec{B}_o \tag{2.3}$$

$$E = -\mu_s B_o \cos\theta = -\mu_{sz} B_o \tag{2.4}$$

$$E = \boxed{-\gamma_s h/2\pi\, m_s B_o} = g_e \beta_e m_s B_o \tag{2.5}$$

In the absence of an external magnetic field, the electron spin is oriented randomly in space. If a magnetic field, B_o, is applied the free electron acquires a *Zeeman energy* (eqns (2.3) to (2.5)) with two stationary states for $m_s = +1/2$ (α) and $m_s = -1/2$ (β) separated by an energy, ΔE:

$$\Delta E = g_e \beta_e B_o \tag{2.6}$$

The field, B_o, is said to have split the degeneracy of the α and β spin states. This is represented schematically by the diagram in Figure 2.1, where the parallel ($m_s = +1/2$) and antiparallel ($m_s = -1/2$) orientations of the magnetic moment of the electron with respect to the field are denoted by $\uparrow$ and $\downarrow$.

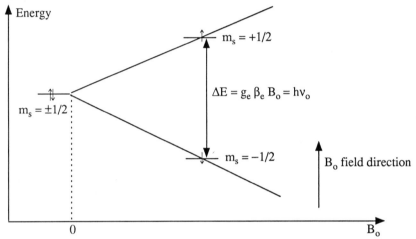

Figure 2.1 – *Energy level of electron spins in a magnetic field, B_o.*

If an oscillating magnetic field, B_1, supplied by an electromagnetic radiation of frequency ω_o, is applied perpendicular to the static magnetic field, with $B_1 \ll B_o$, such that:

$$\frac{h\omega_o}{2\pi} = h\nu_o = g_e \beta_e B_o = \Delta E \tag{2.7}$$

one induces a transition between the α and β levels, called *electron paramagnetic resonance*. Since the probabilities of absorption and induced emission are the same, the populations in each level tend to become equivalent. However, an excess population of about 10^{-3} in the lower level is maintained by relaxation phenomena. For the nuclear spin of the proton this excess population is only about 10^{-5} at room temperature with a ΔE of about 6×10^{-3} cal/mol; in principle, therefore, it is easier to observe a signal in EPR than in NMR.

The populations of the two levels are given by the Boltzmann distribution (eqn. (2.8), Figure 2.2). The population in the β level is increased (n_β/n_α rises) and, therefore, the absorption intensity I is increased either by lowering the temperature or by increasing the field, B_o, which enhances ΔE.

Electron paramagnetic resonance (EPR)

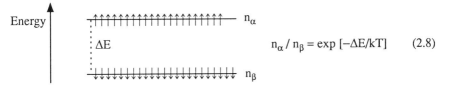

$$n_\alpha / n_\beta = \exp[-\Delta E/kT] \qquad (2.8)$$

Figure 2.2 – *Population of electrons with α and β spins.*

In practice, to satisfy the resonance condition (eqn. (2.7)) the frequency is fixed by means of a microwave source at about 9 GHz and the external field is varied. Unlike NMR apparatuses, EPR spectrometers are usually designed to record the first derivative of the absorption curve (Figure 2.3) in order to obtain higher sensitivity and better definition.

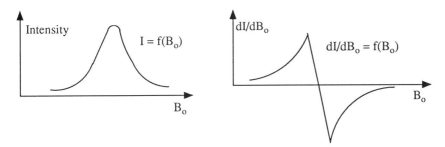

Figure 2.3 – *Absorption curve (NMR) and first derivative of the signal (EPR).*

2-1-2 EPR spectra of free radicals

An EPR spectrum is described by three parameters: the g-factor, the hyperfine coupling constants and the line-widths. We shall briefly examine the first two.

The g-factor
In a free radical the unpaired electron has an orbital kinetic moment due to its movement in space and a spin kinetic moment due to its rotation about itself. The g-factor expresses the coupling between these two kinetic moments; this interaction, called spin-orbit coupling, depends on the electronic environment. Two radicals with different g-values resonate at different fields (eqn. (2.7)), hence the analogy between g and the NMR chemical shift. The variations in g (≈ 2) from one radical to another are small but big enough to provide interesting information about radical structure.

Hyperfine coupling
In presenting the principles of EPR we have considered a single electron, treated as a magnet in a magnetic field. When the electron is close to an atom with a non-zero nuclear spin (^1H, ^{13}C, etc.) there is an interaction between the magnetic moments of the electron and the nucleus; one then observes *hyperfine coupling*, which is expressed by a multiplication of the lines in the EPR signal.

Consider, for example, the simplest case of a hydrogen atom, H•, with spin $I = \pm 1/2$. The nuclear spin has two possible orientations ($m_I = \pm 1/2$) parallel or antiparallel to the electron spin ($m_s = \pm 1/2$); this results in four energy levels (Figure 2.4).

Figure 2.4 – *Energy level resulting from electron-nucleus coupling (I = 1/2).*

Since in EPR the values of the field, B_o, correspond to the resonance of the electron spins, $\alpha_e \leftrightarrow \beta_e$, and not to those of the nuclear spins, m_I does not change during the electronic transition, so only two transitions are observed, (1) and (2) (Figure 2.5). They are equal in energy and intensity but occur at different fields and therefore give two lines in the spectrum. The difference, a(H), is called the *hyperfine coupling constant* and can be compared with the proton/proton coupling constant, J_{HH}, in NMR.

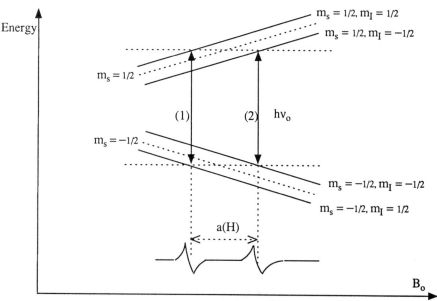

Figure 2.5 – *Transitions due to electron-nucleus hyperfine coupling (I = 1/2).*

In general, *electron-nucleus hyperfine coupling divides an EPR line into 2I + 1 lines*, I being the nuclear spin quantum number (I = 1/2 for ^1H, ^{19}F, ^{15}N, ^{13}C and I = 1 for ^{14}N).

When an unpaired electron is centered on a carbon atom bearing hydrogen atoms (H_α) there is electron $\leftrightarrow H_\alpha$ coupling and the number of lines is equal to n+1, n being the number of protons (I = 1/2). The number and the intensities of the lines are given by Pascal's triangle (Table 2.1). For example, in the case of •CH_3 the EPR spectrum shows 4 lines with relative intensities of 1.3.3.1 (Figure 2.6). In this case the coupling constant, a(H), ($\approx$ 23 Gauss) is small compared with that of electron $\leftrightarrow$ H• nucleus coupling (500 Gauss).

number of nuclei (I = 1/2)	number of lines	relative intensity of lines
0	1	1
1	2	1 1
2	3	1 2 1
3	4	1 3 3 1
4	5	1 4 6 4 1

Table 2.1 – *Pascal's triangle.*

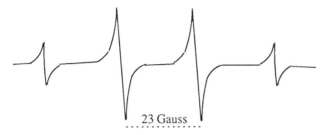

Figure 2.6 – *EPR spectrum of methyl radical, $^\bullet CH_3$.*

For a carbon chain such as that indicated in Scheme 2.1, coupling between the electron and the H_α and H_β protons is observed, with $a(H_\beta) > a(H_\alpha)$. Depending on the conformation of the radical, one may detect long-range couplings, $a(H_\gamma)$ and $a(H_\delta)$. The α couplings give information about the *configuration* of the radical (hybridization of the radical center) and the β couplings about the preferred *conformations* of the radical. This result is not limited to interactions with protons; any nucleus with non-zero spin gives rise to similar α, β and further couplings, depending on the structure.

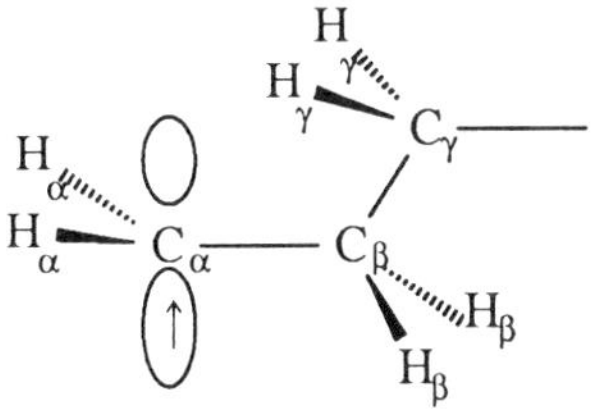

Scheme 2.1

2-1-3 Radical structure and EPR

The electron ↔ proton coupling and the resulting number of lines is a first indication as to the nature of the radical. For example, there are two possibilities for hydrogen abstraction from propionic acid by $^\bullet OH$. The spectrum should consist of 8 lines (quadruplet of doublets) for radical **1** and 9 lines (triplet of triplets) for **2**. In fact, the spectrum corresponds to **2**. This regioselectivity of hydrogen abstraction is explained by the electrophilicity of the $^\bullet OH$ radical.

$$CH_3CH_2CO_2H + HO^{\bullet} \longrightarrow \begin{array}{l} CH_3\overset{\bullet}{C}HCO_2H + H_2O \quad (2.9) \\ \phantom{CH_3\overset{\bullet}{C}}1 \\ \\ {}^{\bullet}CH_2CH_2CO_2H + H_2O \quad (2.10) \\ \phantom{{}^{\bullet}CH_2CH_}2 \end{array}$$

α-Coupling: configuration of radical center

The α-coupling constant, a(N), for a nucleus N with non-zero spin is given by the McConnell equation (2.11). The electron ↔ H_α (N = H) interaction deserves more detailed study. Let us consider the very nearly planar methyl radical, $^{\bullet}CH_3$; the unpaired electron is in an almost pure p orbital.

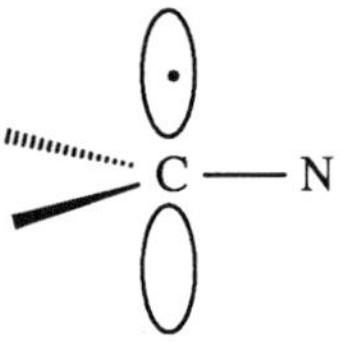

$$a(N) = Q_N \rho \quad (2.11)$$

Figure 2.7 – *Coupling constant, a(N). Q_N: proportionality factor characteristic of a given nucleus. ρ: spin density on the nucleus, i.e. $P_o(\alpha) - P_o(\beta)$, P_o being the probability of either α or β spin density on the nucleus.*

This orbital has a node in the plane of the hydrogen atoms. In this case, the spin density at N should be zero. In fact, a quadruplet with a coupling constant of about 23 Gauss is observed (Figure 2.6). This is due to *spin polarization*. The unpaired electron orients (polarizes) the nearest electron of the σ orbital in the same direction (illustration of Hund's rule) so that electronic configuration **3** is more probable than **4** (Figure 2.8). The spin density on the hydrogen is therefore negative (i.e. excess of β spin over α spin ($P_o(\alpha) - P_o(\beta) < 0$) and a(H) will be negative.

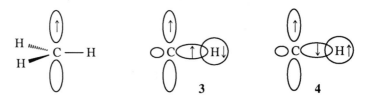

Figure 2.8 – *Spin polarization in $^{\bullet}CH_3$.*

The *electron ↔ ^1H and especially electron ↔ ^{13}C* coupling can be used to determine the configuration (planar or pyramidal) of a radical. For example, for the planar methyl radical, progressive replacement of the hydrogens by fluorine or any other electron-attracting substituent increases the pyramidalization up to virtually sp^3 hybridization for $^{\bullet}CF_3$ (Table 2.2 and Chap. 3). With pyramidalization the 2s C character of the SOMO (spin α) becomes more and more marked, which corresponds to a greater excess of α spin on the central carbon; a(^{13}C) is therefore positive. The α-hydrogens then have an excess of β spin and a(^1H) is negative. Radicals such as cyclopropyl, $^{\bullet}CH_2OH$, $^{\bullet}CH(OH)_2$, and metal-centered radicals (Si, Ge, Sn) have been shown by EPR to have more or less pyramidal structures, while vinyl and phenyl radicals have the σ structure (Chap. 3).

	•CH$_3$	•CH$_2$F	•CHF$_2$	•CF$_3$
a(^{1}H)	−23.04	−21.1	−22.2	-
a(^{13}C)	38.5	54.8	148.8	271.6

Table 2.2 – *Coupling constant (in Gauss) for ^{1}H and ^{13}C for fluoromethyl radicals.*

Coupling with β-hydrogens

Coupling with β-hydrogens can be used to determine whether the conformation of a radical is staggered or eclipsed (Figure 2.9). The staggered structure allows better overlap between the p orbital and the C–H bond, therefore better coupling.

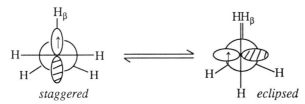

Figure 2.9 – *Staggered and eclipsed conformations of the ethyl radical.*

The coupling, a(H$_\beta$), is given by the Heller-McConnell equation (2.12) which is analogous to that of Karplus in NMR. A and C are constants characteristic of the systems studied but vary little and are about 5 and 40 ~ 50 G, respectively, and Θ is the dihedral angle between the p orbital and the C–H bond.

$$a(H_\beta) = A + C \cos^2\Theta \quad (2.12)$$

It has been shown that the p orbital containing the unpaired electron and a C–X bond are eclipsed when X is a second-row element (Si, P, S, Cl). In particular, calculations indicate that the structure of β-chloro radicals is staggered **5** (Figure 2.10).

Figure 2.10 – *Structure of a β-chloro radical.*

The delocalization of the unpaired electron in conjugated systems and the relative weights of the different mesomeric forms can also be determined by EPR. Thus, for the benzyl radical the unpaired electron density is 1/7 on atoms 2, 4 and 6, 4/7 on atom 7 and practically zero on atoms 3 and 5 (Figure 2.11). The high spin density on atom 7 corresponds to observations of chemical reactivity.

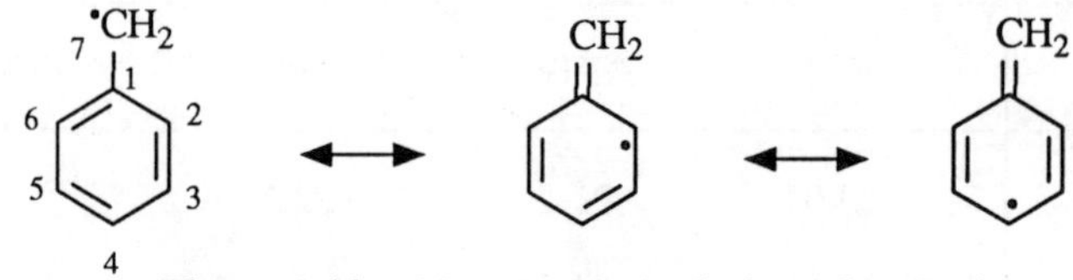

Figure 2.11 – *Mesomerism in the benzyl radical.*

2-1-4 Radical trapping (spin trapping)

On the basis of kinetic criteria (lifetimes) radicals can be classified into three categories (Chap. 4): short-lived (stabilized and destabilized), stable and persistent. The half-life of the first class is less than 10^{-3} s. Radicals in the other two classes can be observed directly by EPR without any major experimental problems, since their instantaneous concentrations are sufficiently high.

Short-lived radicals are associated with reaction systems where the instantaneous concentrations of free radicals are very low, because they are highly reactive; this is the situation in chain reactions, for example. EPR observation then becomes difficult and one has recourse to special experimental techniques such as low-temperature photolysis in flow systems or the technique of trapping radicals, known as *spin-trapping*. In this latter case a nitroso derivative (2.13) or a nitrone (2.14) is included in the mixture being studied; addition of the radical, R•, which is to be investigated, gives a stable, EPR-observable nitroxide radical.

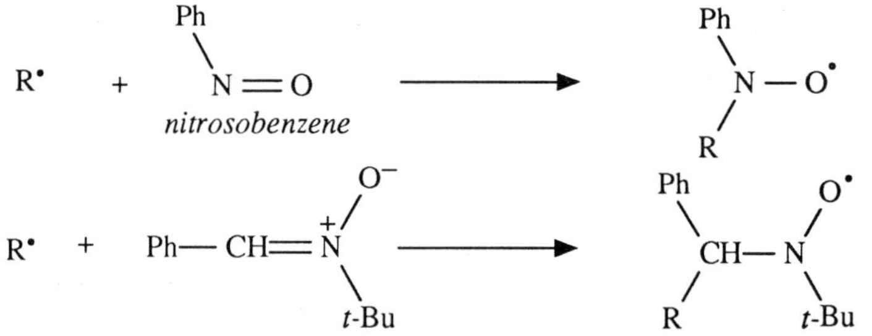

$$R^\bullet + \text{Ph-N=O (nitrosobenzene)} \longrightarrow \text{Ph-N(R)-O}^\bullet \quad (2.13)$$

$$R^\bullet + \text{Ph-CH=N}^+(O^-)(t\text{-Bu}) \longrightarrow \text{Ph-CH(R)-N(O}^\bullet)(t\text{-Bu}) \quad (2.14)$$

N-tert-butyl-α-phenyl nitrone (BPN)

Such reactions are very fast (k = 10^6 to 10^8 M^{-1} s^{-1}) and the nitroso derivatives and nitrones are for this reason called *radical traps*. EPR study of the nitroxide derivatives produced by these reactions gives information about the structure of R•. Reaction (2.13) is the most often used, since R is directly attached to –NO• and the spectrum is the most informative. This technique of *spin-trapping* is much used in biological radical chemistry where the intermediates, in particular those derived from oxygen

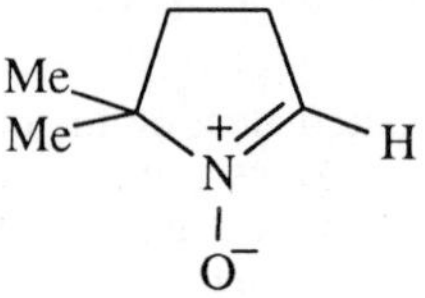

Scheme 2.2

($O_2^{\bullet-}$, $^\bullet OH$, $^\bullet OOH$), enter into reactions at very low concentrations in a complex medium. The most frequently used trap for such studies is a cyclic nitrone, 5,5-dimethyl-1-pyrroline N-oxide (DMPO) **6**.

2-2 CHEMICALLY INDUCED DYNAMIC NUCLEAR POLARIZATION (CIDNP)

When a radical reaction is monitored by recording the NMR spectrum, in certain cases *anomalies* are observed: certain peaks are negative and correspond to radiofrequency *emission* while others show an *absorption increase* compared to normal peaks. This effect is called *chemically induced dynamic nuclear polarization* (CIDNP). The *sole* modification of the NMR spectra concerns the signal intensities.

2-2-1 Explanation of the phenomenon: radical pair theory

Under the influence of the magnetic field, B_0, the nuclear spins of protons are distributed according to Boltzmann's law (eqn. (2.8)) between two energy levels corresponding to the two values of the magnetic quantum number, $m_I = \pm 1/2$. If by any process the relative populations of the levels are modified, return to equilibrium occurs with emission when $m_I = -1/2$ is overpopulated and with absorption in the other case. The formation and the evolution of a radical pair can provoke this population change and therefore induce the emission and absorption phenomena.

The homolytic cleavage of a bond by thermolysis in solution leads to a pair of radicals in the singlet state (antiparallel spins) and by photolysis to a pair in the singlet or triplet (parallel spins) state. If the pair is a singlet the radicals formed can react inside the cage of solvent molecules to recombine and give a *geminate* product. If the pair is a triplet, spin inversion must precede recombination. Furthermore, the radicals can diffuse outside the solvent cage and give rise to encounter (*free pair*) and recombination or react with the solvent to give non-radical products (Scheme 2.3).

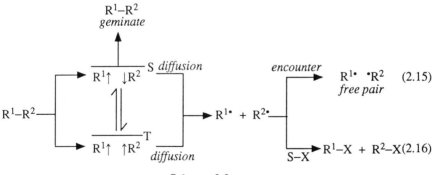

Scheme 2.3

A triplet radical pair in a magnetic field, B_0, has three states: T_{-1}, T_0 and T_1. In Figure 2.12 the energy difference between the singlet state and the triplet states is represented as a function of the distance between the two radicals, $R^{1\bullet}$ and $R^{2\bullet}$.

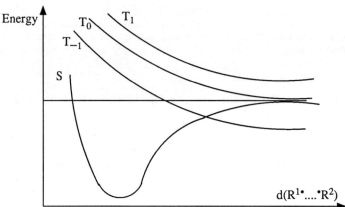

Figure 2.12 – *Energies of the singlet and triplet states of a radical pair in a magnetic field as a function of the separation, d.*

Only $S \leftrightarrows T_0$ exchange is effective. When the singlet pair ($R^{1\bullet}$, $^{\bullet}R^2$) has just been formed, the S state is low whereas the T states have much higher energies. As the two radicals separate, the energy difference decreases and becomes sufficiently small (of the order of magnitude of electron-nucleus hyperfine interactions) for the $S \leftrightarrows T_0$ transformation to be possible without energy absorption; the distance is then about 6 ~ 10 Å. At this distance $S \leftrightarrows T_0$ transformation is faster than radical diffusion. The rate of $S \leftrightarrows T_0$ exchange depends on the relative precession rate, $\Delta\omega$, of each electron spin of the pair; it is therefore characteristic of each of the radicals, $R^{1\bullet}$ and $^{\bullet}R^2$. The first term of the sum concerns the electron and the second the magnetic contributions of all the nuclei on $R^{1\bullet}$ and $^{\bullet}R^2$.

$$\Delta\omega = \omega_1 - \omega_2 = \beta_e h/2\pi \left[(g_1 - g_2) B_o + (\Sigma a_1 m_1 - \Sigma a_2 m_2) \right] \quad (2.17)$$

β_e Bohr magneton
h Planck's constant
g_1 and g_2 Landé factors for electrons 1 and 2
B_o external magnetic field
a_1 and a_2 electron-nucleus (^1H or ^{13}C) hyperfine coupling constants
m_1 and m_2 spin magnetic quantum numbers of the neighboring nuclei.

Let us consider two singlet pairs of radicals with different nuclear spins for $R^{1\bullet}$ and $^{\bullet}R^2$; that which because of its precession rate favors $S \leftrightarrows T_0$ exchange will be in the diffusion product (since there is no recombination in the T_0 state) and the recombination product will be characterized by an excess of the complementary spin state.

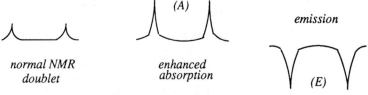

Figure 2.13 – *Net CIDNP effect.*

Chemically induced dynamic nuclear polarization (CIDNP)

In this case, one or other of the nuclear magnetic levels will be overpopulated. Depending on the circumstances, either emission (E) or enhanced absorption (A) will result from relaxation of the energy levels (Figure 2.13). Thus, for a singlet radical pair $(HR^{1\bullet}, {}^{\bullet}R^2)^S$ with $g_1 \neq g_2$, the NMR spectra of the cage recombination products (geminate), HR^1-R^2, and substitution products, HR^1-X, are characterized by *net effects*, i.e. the NMR signal (singlet or multiplet) is either completely in emission or in enhanced absorption. The phases will be necessarily opposed for HR^1-R^2 and HR^1-X.

When the two radicals have the same Landé factor ($g_1 = g_2$) but *several* coupled nuclei, for example, $H^1H^2R^{1\bullet}, {}^{\bullet}R^2$ one observes a *multiplet effect*, i.e. part of the signal of an NMR multiplet will be in E and the other in A, the order of the peaks depending on the coupling constant. The products of cage recombination and diffusion have opposite effects (Figure 2.13).

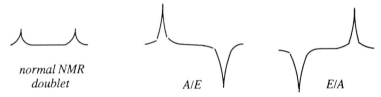

normal NMR doublet   A/E   E/A

Figure 2.14 – *Multiplet CIDNP effect.*

By means of Kaptein's rules the signs of the CIDNP phenomena can be related to the origin of the products: singlet or triplet pair, products of geminate recombination (cage effect) or of encounter (free pair), products of reaction with a non-radical species, SX (Scheme 2.3).

2-2-2 Examples of CIDNP

We shall simply give two examples to illustrate the net and multiplet effects.

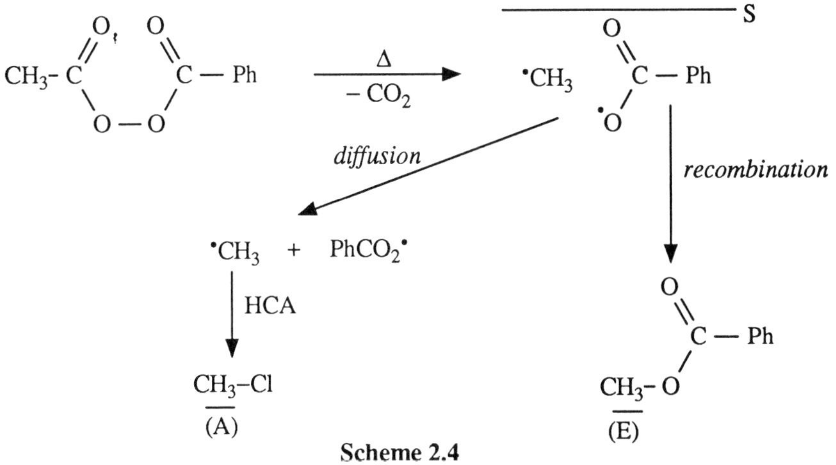

Scheme 2.4

Net effect

The thermal decomposition of benzoyl acetyl peroxide in hexachloroacetone (HCA) gives, by cage recombination, methyl benzoate with an emission effect on the methyl singlet and, by diffusion, methyl chloride with enhanced absorption of the methyl singlet (Scheme 2.4).

Multiplet effect

The thermal decomposition of propionyl peroxide in hexachloroacetone gives, by diffusion, ethyl chloride with an A/E effect on the methyl and methylene protons (Scheme 2.5).

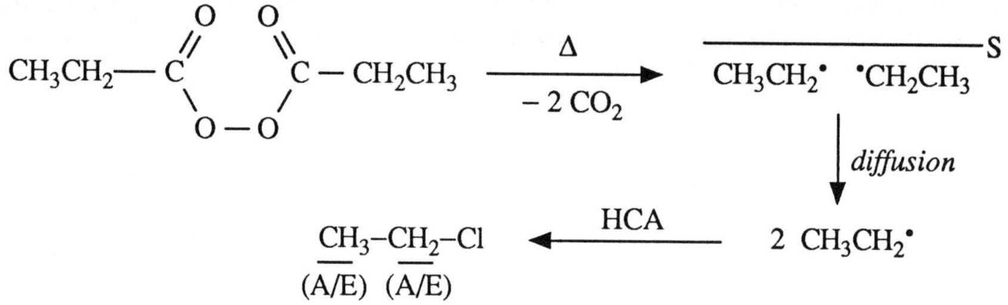

$$CH_3CH_2-C(=O)-O-O-C(=O)-CH_2CH_3 \xrightarrow[-2\,CO_2]{\Delta} \overline{CH_3CH_2^\bullet \quad {}^\bullet CH_2CH_3}\,\text{S}$$

$$\downarrow \text{diffusion}$$

$$\underline{CH_3}-\underline{CH_2}-Cl \xleftarrow{\text{HCA}} 2\,CH_3CH_2^\bullet$$
$$\quad (A/E)\ (A/E)$$

Scheme 2.5

2-2-3 Kinetic studies

The effects of nuclear polarization are produced on a time-scale going from about 10^{-9} to 10^{-7} s (S ⇋ T_0 mixing) depending on the magnitude of the hyperfine interactions and the rate of recombination of the radicals of the pair. A chemical reaction fast enough to lie in this time-scale can therefore change the nature of the radicals of the pair and lead to a new pair with the same electronic state as the original pair. The chemical reaction in question can be bimolecular (2.18) or unimolecular (2.19) (fragmentation, rearrangement).

$$R^{1\bullet} + R-X \longrightarrow R^1-X + R^\bullet \quad (2.18)$$

$$R^{1\bullet} \longrightarrow R^{3\bullet} \quad (2.19)$$

In the latter case the reaction sequence will be (2.20) where k is the rate constant for the replacement of the first pair by the second. The amount of the second product, R^3R^2, and the intensity of its NMR signal will depend mainly on k. It is therefore feasible to situate the rate of the reaction relative to that of the CIDNP process. For this phenomenon to be observed, $k \geq 10^6\,s^{-1}$.

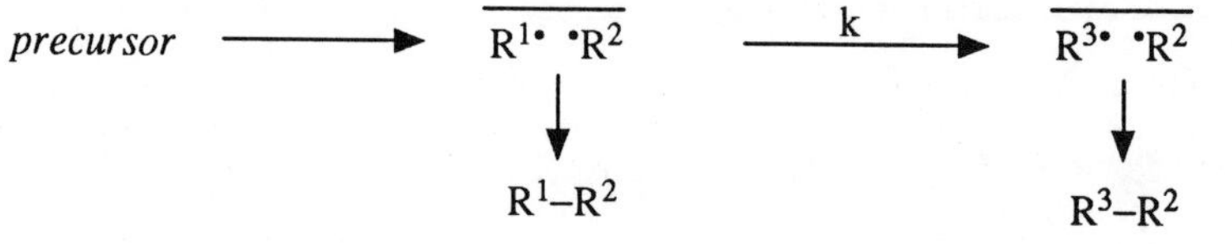

$$precursor \longrightarrow \overline{R^{1\bullet}\ {}^\bullet R^2} \xrightarrow{k} \overline{R^{3\bullet}\ {}^\bullet R^2} \quad (2.20)$$
$$\qquad\qquad\qquad \downarrow \qquad\qquad\qquad \downarrow$$
$$\qquad\qquad\quad R^1-R^2 \qquad\qquad\quad R^3-R^2$$

For example, the rate constants of certain decarboxylation reactions have been estimated: $k \approx 2 \times 10^9\,s^{-1}$ when R = Me, $k > 10^{10}\,s^{-1}$ for R = alkyl and $10^6\,s^{-1}$ for R = Ph.

$$RCO_2^\bullet \longrightarrow R^\bullet + CO_2 \qquad (2.21)$$

2-2-4 Conclusion

The NMR observation of a CIDNP phenomenon indicates unambiguously that the corresponding products are formed by a radical mechanism. Anionic and cationic intermediates do not give this type of effect. This indicates also that the origin is a radical *pair*. Consequently, the products obtained in the propagation steps of a chain do not show CIDNP; strictly speaking, those of initiation and termination, in very small amounts, could be polarized. The signs of polarization (A or E, E/A or A/E) provide information about the type of reaction the pair undergoes (thermal, photochemical, recombination, reaction after diffusion).

However, the quantitative aspect is difficult to evaluate and the NMR CIDNP spectra do not say whether all the molecules or merely a fraction are formed from a radical pair. The CIDNP phenomenon cannot be observed for the products of all radical reactions, since the nuclear polarization may not last long enough; consequently, the absence of CIDNP in an NMR spectrum is not formal proof that free radicals are absent from a reaction path.

2-3 ELECTRONIC SPECTROSCOPY

The UV spectra of free radicals typically show a very marked bathchromic shift (towards high wavelengths) and a molecular extinction coefficient, ε, much greater than that of the corresponding non-radical species (Scheme 2.6).

λ_{max} (ε) 319 (15 000)
          530 (0)

λ_{max} (ε) 332 (19 000)   DPPH
          530 (14 000)

Ph_3C-CPh_3
λ_{max} (ε) 262 (870)

$Ph_3C^\bullet$
λ_{max} (ε) 345 (11 000)

Scheme 2.6

For *stable* radicals like DPPH and galvinoxyl, absorption occurs in the visible and the products are colored; conventional spectrophotometric apparatus can be used. These radicals are used as traps (Chap. 17) for following the appearance or disappearance of *short-lived* radicals by spectrophotometry.

On the other hand, for *short-lived* radicals (half-lives less than about 10^{-3} s) UV spectroscopy requires much more sophisticated equipment. *Pulse photolysis* is used. By means of a light source (laser) very short, intense light pulses are emitted ($\approx$ 4 ns) so as to create a large concentration of radicals in the solution. After each pulse a rapid spectrophotometric system records the spectrum during about 2 ~ 4 µs. In this way one can follow the rate of disappearance or appearance, therefore the reaction rate, of radicals with lifetimes of the order of 0.2 to 1 µs and determine absolute rate constants of about 10^6 to 10^9 s^{-1} at room temperature. We shall give some examples:

$$X-C_6H_4-C(=O)O^{\bullet} \xrightarrow{k/25°C} X-C_6H_4^{\bullet} + CO_2 \quad (2.22)$$

laser irradiation at 308 nm
observation at 720 nm (disappearance of $XC_6H_4CO_2^{\bullet}$)

$k = 2.0 \times 10^6$ s^{-1} for X = H
$\quad\,\, 1.4 \times 10^6$ s^{-1} $\quad\,\,$ X = Cl
$\quad\,\, 0.3 \times 10^6$ s^{-1} $\quad\,\,$ X = OCH$_3$

$$PhCH_2SO_2Cl + R^{\bullet} \xrightarrow{k/30°C} PhCH_2SO_2^{\bullet} + RCl \quad (2.23)$$

$$PhCH_2SO_2^{\bullet} \xrightarrow{very\ fast} PhCH_2^{\bullet} + SO_2 \quad (2.24)$$

laser irradiation at 337.1 or 308 nm
observation at 317 nm (appearance of PhCH$_2^{\bullet}$)

$k = 1.3 \times 10^6$ M^{-1}s^{-1} for R$^{\bullet}$ = n-Bu$^{\bullet}$
$\quad\,\, 5.7 \times 10^9$ M^{-1}s^{-1} $\quad\quad\,\,$ = Et$_3$Si$^{\bullet}$

3 – RADICAL STRUCTURE

The geometric and electronic structures of free radicals interest chemists and physicists both from a fundamental viewpoint and because of the chemical consequences. The very general term, *geometric structure*, describes the spatial form of radicals. We shall emphasize two particular aspects, the *configuration* and the *conformation*. The *electronic structure* or *state* depends on the way in which the orbitals are filled.

3-1 DEFINITIONS

3-1-1 Configuration of the radical center

The *configuration* or geometry of the radical center defines the spatial arrangment of the atoms directly bonded to the center. A radical, $^{\bullet}AX_3$, can be planar or pyramidal and a radical, $^{\bullet}AX_2$, linear or non-linear.

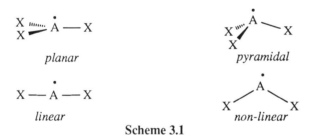

Scheme 3.1

In principle, configurational isomers are interconverted by breaking a covalent

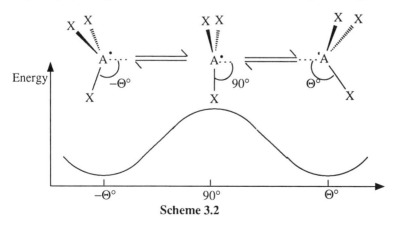

Scheme 3.2

bond, but that is not the case here. No doubt for this reason, various authors use the terms of configuration, conformation and geometry as though they were equivalent. For radicals, a change of configuration implies therefore a change in the hybridization of the atom bearing the unpaired electron. In the case of a pyramidal $^\bullet AX_3$ radical there are two identical configurations interconnected by a planar form, which is the transition structure for this inversion process (Scheme 3.2).

In the case of a planar $^\bullet AX_3$ radical the curve is of the form shown in Scheme 3.3. Similar considerations apply to the linear and non-linear forms of $^\bullet AX_2$.

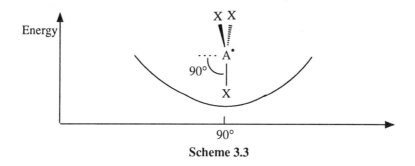

Scheme 3.3

3-1-2 Radical conformation

One *conformation* is converted into another by rotation about a single bond. Thus, for the ethyl radical there are two characteristic conformers (3.1). In the staggered form the orbital with the unpaired electron eclipses one of the β C–H bonds but this is not so in the eclipsed form.

$$\text{staggered} \rightleftharpoons \text{eclipsed} \tag{3.1}$$

3-1-3 Electronic state

A radical can be in various electronic states depending on how the orbitals are filled. Thus, the aminyl radical, $^\bullet NH_2$, has two characteristic electronic states. The more stable is the $^2B_1$ state, where the unpaired electron is in the p orbital perpendicular to the plane of the molecule, quoted b_1 in the group theory notation. For this reason chemists say that this radical is π type. There is also an electronic state, $^2A_1$, with two electrons in the p orbital and one electron in the sp^2 orbital in the plane of the molecule, quoted a_1 in the group theory notation. This is described as being σ type. A change in the geometry or a substituent effect can reverse the relative energies of the electronic states. The $^2A_1$ and $^2B_1$ notations come from group theory.

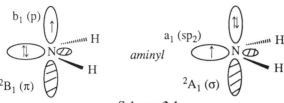

Scheme 3.4

3-2 METHODS OF INVESTIGATION

Information about the structure of radicals is provided by physical methods (IR, EPR, etc.), by calculation and by chemical methods:

– *IR spectra*. The symmetry elements of a molecule and then the structure of the radical can be determined by analyzing *IR spectra*. However, this method is difficult to apply;
– *mass spectrometry* is used to measure the ionization potentials of radicals;
– *EPR* (Chap. 2) is without doubt the most appropriate and the most powerful method for determining radical structures;
– *pulsed radiolysis* and *high-speed electrochemistry* give information about the redox properties of radicals in solution;
– *calculations* can give the various geometries and energies of radicals;
– *chemical methods* are only pertinent provided that the structure of the radical changes sufficiently slowly compared to the rate at which the radical is trapped. With saturated carbon compounds, retention or inversion of configuration is a strong argument in favor of a pyramidal radical. Racemization indicates either a planar radical or a pyramidal radical where inversion is faster than trapping. Thus, retention of configuration in reaction (3.2) argues in favor of a strongly pyramidal cyclopropyl radical. In the same way, reactions (3.3) and (3.4) indicate that the *cis* ⇌ *trans* isomerization of 9-decalyl radicals is very slow.

3-3 CARBON RADICALS

3-3-1 Methyl radical

The most recent EPR data and *ab initio* calculations are consistent with a planar structure for the Me• radical (•CH$_3$). The potential well for Me• is much more open than that for the corresponding cation, Me$^+$. Unlike carbocations, carbon-centered radicals readily tolerate serious deviations from the planar structure.

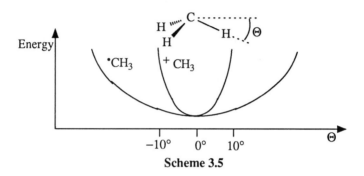

Scheme 3.5

3-3-2 Methyl radicals with electronegative substituents

The replacement of a hydrogen by a σ-attractor, π-donor element or group (halogen, OH, NH$_2$) leads to pyramidalization of the radical. When the hydrogens are successively replaced by fluorine the radical goes from planar to practically tetrahedral, as is shown by the EPR coupling constants (Table 3.1). The •CH$_2$OH radical is pyramidal by about 5 to 10°. With electron-attracting substituents such as –C≡CH and –C≡N the radical center remains planar.

radical	a(H)	a(F)	a(C)	Θ°
•CH$_3$	–23.0		38.5	0
•CH$_2$F	–21.1	64.3	54.8	5
•CHF$_2$	–22.2	84.2	148.8	12.7
•CF$_3$		142.4	271.6	17.8

Table 3.1 – *Pyramidalization of •CH$_n$F$_{n-3}$ radicals according to EPR coupling constants (in Gauss).*

3-3-3 Orbital factors determining radical configuration

The more or less pyramidal character of a radical can be explained in terms of orbital interactions (Chap. 6). In a planar radical the unpaired electron is in the p orbital of the central atom. When the radical is pyramidalized this orbital, called the SOMO

(Singly Occupied Molecular Orbital) interacts with the LUMO (Lowest Unoccupied

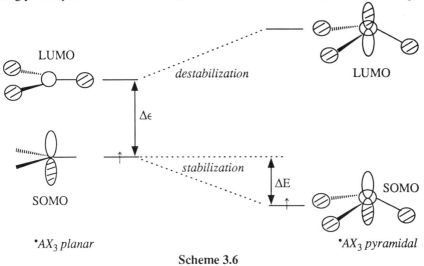

Scheme 3.6

Molecular Orbital) of the planar radical (Scheme 3.6). Because of this interaction the SOMO is stabilized by in-phase mixing with the LUMO, while the LUMO is destabilized by out-of-phase mixing with the SOMO. From the electronic energy viewpoint the only thing which counts is the stabilization of the occupied orbital, i.e. the SOMO. The energetic effect of this interaction (ΔE) is, among other things, inversely proportional to the energy difference between the orbitals before interaction ($\Delta\epsilon$); consequently, any electronic effect which reduces the difference between the SOMO and the LUMO will favor pyramidalization.

Thus, in the case of halogens, the π-donor effect raises the energy of the SOMO and the σ-acceptor effect lowers that of the LUMO. These two effects combine to favor pyramidalization; •CF_3 is practically tetrahedral. The same reasoning applies to all π-donor, σ-acceptor substituents. π-Acceptor groups such as –C≡CH and –C≡N lower the SOMO energy, increasing the energy difference, $\Delta\epsilon$, and consequently enhance the planar character of the radical center.

For Si, Ge and Sn-centered radicals the energy difference, $\Delta\epsilon$, decreases when the principal quantum number of the valence shell increases, which explains why the pyramidal character of the Group 14 hydrides varies in the order:

$$•CH_3 \;<\; •SiH_3 \;<\; •GeH_3 \;<\; •SnH_3 \tag{3.5}$$

3-3-4 Alkyl radicals

The ethyl radical is slightly pyramidal (Scheme 3.7) for three reasons:

– reduction of the torsional interaction between the C_α–H and C_β–H bonds;
– increase in hyperconjugative stabilization between the SOMO and the third β C–H bond;
– low resistance of the radical carbon to pyramidalization.

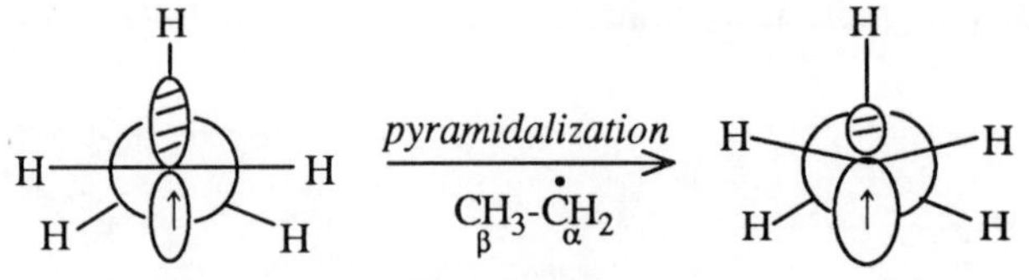

Scheme 3.7

For the same reasons we observe that when the hydrogens of the methyl radical are progressively replaced by methyl groups, to give i-Pr• and t-Bu• radicals, the degree of pyramidalization and the inversion barrier increase. For t-Bu• the deviation from the plane is 10° and the inversion barrier about 0.5 kcal/mol.

3-3-5 Cycloalkyl radicals

The radical carbon of cyclohexyl, cyclopentyl and cyclobutyl radicals is close to planar and inverts very rapidly. Stereochemical results on the 4-*tert*-butylcyclohexyl radical illustrate this statement, the decomposition of the *cis* and *trans* peracids giving the same alcohol mixture; this implies that there is a common intermediate which is planar or whose invertion is fast (Scheme 3.8).

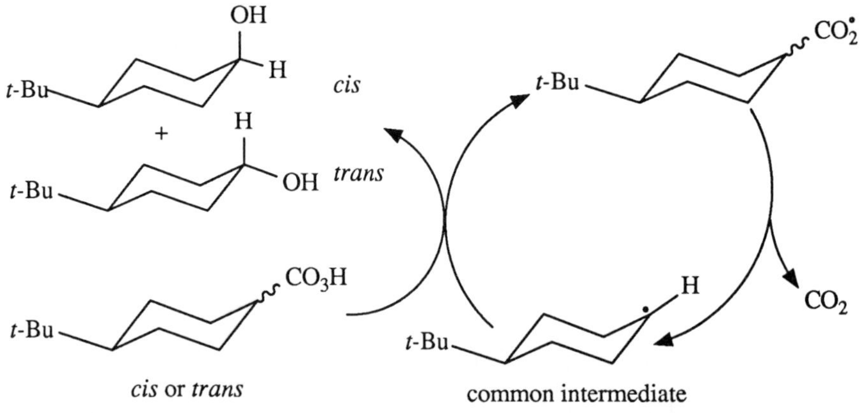

Scheme 3.8

On the other hand, in strained ring systems the radical center is more or less pyramidal depending on the type of structure. This is clearly the case for the species indicated in Scheme 3.9.

Scheme 3.9

3-3-6 Vinyl radicals

The high EPR coupling constant, $a(^{13}C)$, indicates that the vinyl radical is non-linear ($R^1 = R^2 = R^3 = H$) with a very small inversion barrier. Depending on the substituents (steric or conjugative effect) vinyl radicals can adopt a structure:

- either linear, with the electron in the 2p orbital (π radical);
- or non-linear, with the electron in an orbital with a certain s character (σ radical).

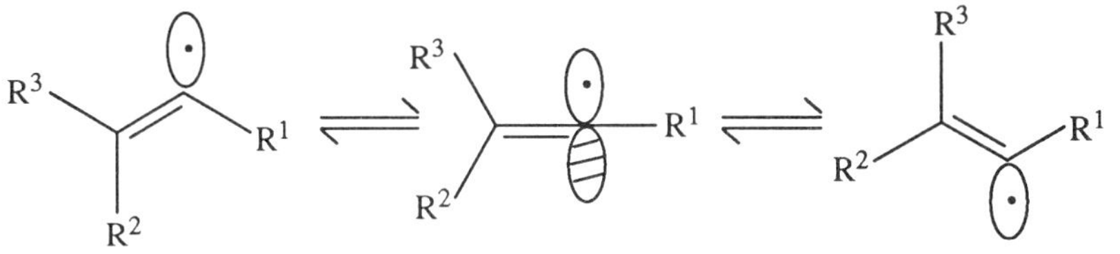

$$(3.6)$$

In most cases vinyl radicals are non-linear but have very low inversion barriers ($\approx$ 2 kcal/mol), which makes the reactions relatively or totally non-stereoselective. The only π type vinyl radicals are those where steric effects are important.

Scheme 3.10

3-3-7 β-Substituted alkyl radicals

Alkyl radicals substituted in the β position have two characteristic conformers: the *staggered* rotamer, where the C–X bond is antiperiplanar to the orbital containing the unpaired electron (SOMO), and the *eclipsed* rotamer, where this bond is in the nodal plane of the SOMO. Analysis of EPR spectra and calculations show that the staggered, pyramidal rotamer is the preferred conformation when X = SR, Cl, SiR$_3$, GeR$_3$ or SnR$_3$, because of the delocalization of the unpaired electron into the C–X bond. The eclipsed rotamer is then the transition structure for rotation. When X = Me, NH$_2$, OH or F the two rotamers have identical energies and there is only a very small rotational barrier.

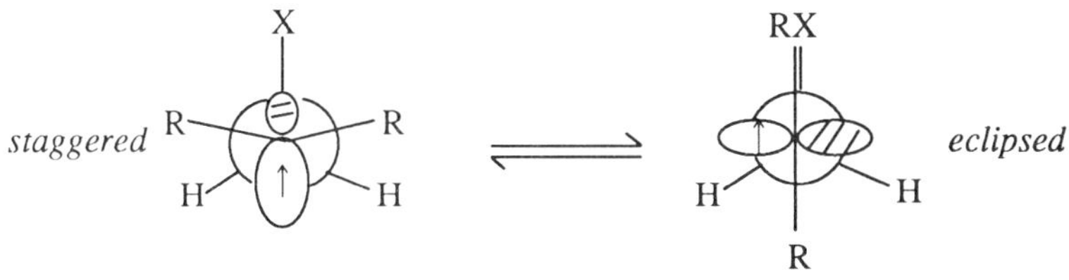

$$(3.7)$$

The EPR data for X = Br, R = Me are consistent with a staggered, pyramidal structure where one of the β C–H bonds is antiperiplanar to the SOMO. β-Halo (Cl or Br) radicals are stereoselective in certain reactions, which indicates that rotation is slower than radical trapping.

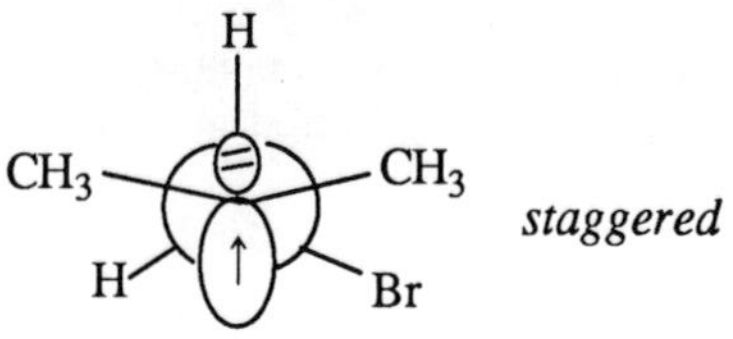

staggered

Scheme 3.11

3-4 RADICALS CENTERED ON AN ATOM OTHER THAN CARBON

3-4-1 Si, Ge and Sn-centered radicals

The pyramidal structure is more pronounced for $R_3M^\bullet$ radicals than for $R_3C^\bullet$; this is due to the greater electronegativity difference between the central atom and the substituents. Experimental evidence for a pyramidal structure is provided by the observation that optically active 1-naphthylphenylmethylsilane reacts with CCl_4 to give the chloride with 90% retention of configuration (eqn (3.8)). The σ orbital containing the unpaired electron in $Me_3Si^\bullet$ has 21% s character, and the out-of-plane deviation is 13 ~ 14° with an inversion barrier of about 6 kcal/mol. Similar data are found for $R_3Ge^\bullet$ and $R_3Sn^\bullet$ radicals.

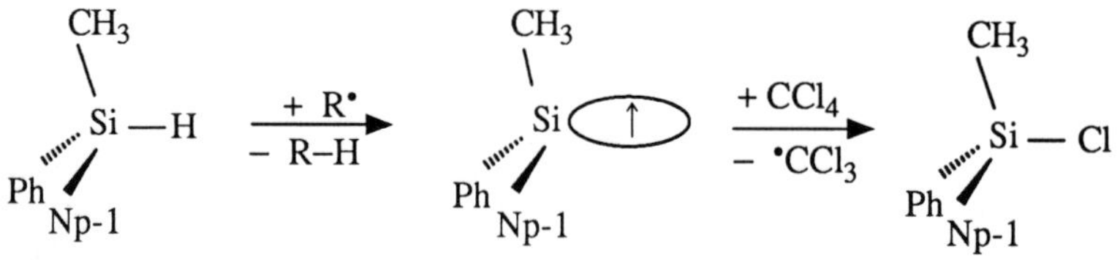

$$\text{Ph(Np-1)(CH}_3\text{)Si-H} \xrightarrow[-\text{R-H}]{+\text{R}^\bullet} \text{Ph(Np-1)(CH}_3\text{)Si}^\bullet \xrightarrow[-{}^\bullet\text{CCl}_3]{+\text{CCl}_4} \text{Ph(Np-1)(CH}_3\text{)Si-Cl} \qquad (3.8)$$

3-4-2 Nitrogen-centered radicals

Both electronic structures, σ and π, can exist for nitrogen-centered radicals (Scheme 3.4).

Aminyl radicals
Whether the unpaired electron is in the p orbital (π structure) or sp² (σ structure) depends on the nature of the substituents. Dialkylaminyl, $R_2N^\bullet$, and protonated dialkylaminyl radicals, $R_2NH^{+\bullet}$, are π type. In arylaminyl radicals the presence of a donor substituent on the aromatic ring favors the π structure and an attractor, the σ structure.

Amidyl radicals
EPR spectra and theoretical calculations on amidyl radicals indicate that the more stable electronic structure is π type with little electron delocalization onto oxygen; the σ structure is higher in energy.

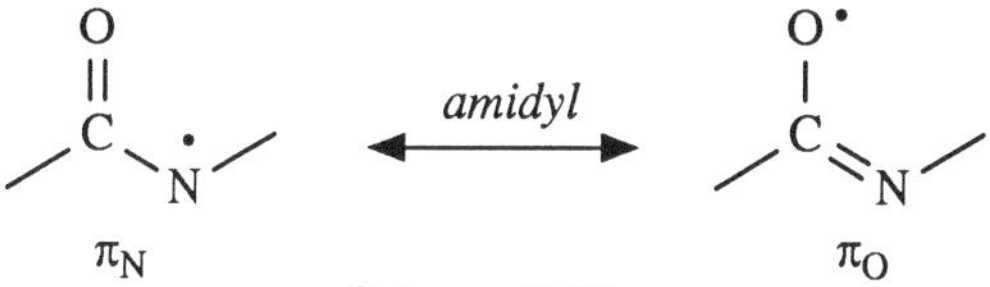

Scheme 3.12

Nitroxide radicals

Nitroxide radicals are represented by two predominant mesomeric forms where the unpaired electron is on either nitrogen or oxygen.

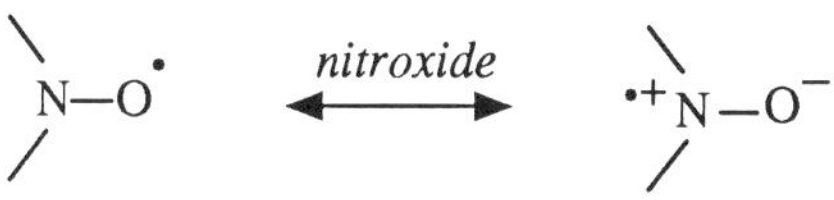

Scheme 3.13

According to the EPR coupling constants the spin density on nitrogen is 0.75 ~ 0.80, indicating that the charged form is the more important and that the system is pyramidal (Scheme 3.14).

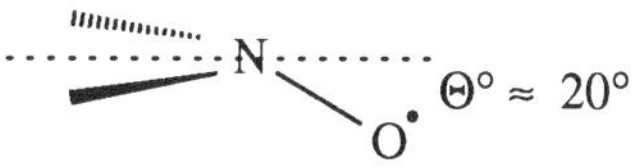

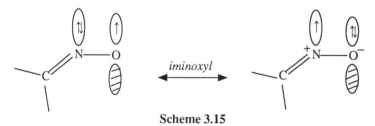

Scheme 3.14

Iminoxyl radicals

According to EPR, iminoxyl radicals have a σ structure in which the unpaired electron is in an orbital of very marked s character in the C=N–O plane and orthogonal to the π system.

Scheme 3.15

3-4-3 Phosphorus-centered radicals

Phosphinyl radicals

As for nitrogen-centered radicals, phosphinyl radicals can have two electronic structures, π and σ (3.9). The π structure, where the unpaired electron is in a phosphorus p orbital perpendicular to the plane of the bonds, is the more stable. The Z–P–Z bond angle is 90° for Z = H and 120° for Z = Cl or Br.

$$\text{²B}_1 (\pi) \xrightleftharpoons{\text{phosphinyl}} \text{²A}_1 (\sigma) \quad (3.9)$$

Phosphonyl radicals

Phosphonyl radicals are pyramidal. Their configurational stability is great enough for retention of configuration to be observed in their reactions with CCl_4 or with an olefin.

$$\text{(i-PrO)(CH}_3\text{)P(=O)H} \xrightarrow{\text{PhS}^\bullet} \text{(i-PrO)(CH}_3\text{)P(=O)}^\bullet \text{ phosphonyl}$$

$$\xrightarrow{CCl_4} \text{(i-PrO)(CH}_3\text{)P(=O)Cl} \quad (3.10)$$

$$\xrightarrow{C_7H_{14}} \text{(i-PrO)(CH}_3\text{)P(=O)C}_7H_{15} \quad (3.11)$$

Phosphoranyl radicals

With four covalent bonds and 9 electrons in the valence shell of phosphorus, phosphoranyl radicals adopt a trigonal bipyramidal structure in which the unpaired electron is in an equatorial rather than an apical orbital (3.12).

$$\text{equatorial} \xrightleftharpoons{\text{phosphoranyl}} \text{apical} \quad (3.12)$$

Phosphorus radicals are relatively stable but little is known about substituent effects on their stability.

3-4-4 Oxygen-centered radicals

Alkoxyl radicals, $RO^\bullet$, have a plane of symmetry and two electronic structures of similar energy. The unpaired electron lies in an orbital which is either symmetric ($^2A'$ state) or antisymmetric ($^2A''$ state) with respect to the plane.

Radicals centered on an atom other than carbon 29

$$\text{(diagram)} \xrightleftharpoons{alkoxyl} \text{(diagram)} \qquad (3.13)$$

$^2A'$ $\qquad$ $^2A''$

In the same way, acyloxyl radicals, $ZCO_2^\bullet$, have two electronic states, π and σ, depending on the position of the singly occupied orbital with respect to the plane of symmetry. It is difficult to distinguish between these structures experimentally, but theoretical calculations indicate that the σ structure is the more stable; this implies that there is no mesomerism.

$$\text{(diagram)} \xrightleftharpoons{acyloxyl} \text{(diagram)} \qquad (3.14)$$

π $\qquad\qquad\qquad\qquad\qquad$ σ

3-4-5 Sulfur-centered radicals

Thiyl radicals
As for alkoxyl radicals, thiyl radicals have a plane of symmetry and two electronic structures, one ($^2A''$) where the unpaired electron is in an orbital antisymmetric with respect to the plane of the molecule, and another ($^2A'$) where this electron is in an orbital symmetric with respect to this plane.

$$\text{(diagram)} \xrightleftharpoons{thiyl} \text{(diagram)} \qquad (3.15)$$

$^2A'$ $\qquad$ $^2A''$

Sulfonyl radicals
Sulfonyl radicals are pyramidal and the degree of pyramidalization increases with the electronegativity of R. It has been shown by EPR that the unpaired electron is very little delocalized onto the aromatic ring when R = Ar.

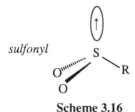

Scheme 3.16

Sulfinyl radicals

Sulfinyl radicals have a $^2A'$ structure (π). Two rotamers have been shown to exist.

$$\text{H}^3-\text{S}=\text{O} \text{ (with H}^1\text{, H}^2\text{)} \xrightleftharpoons{\text{sulfinyl}} \text{S}=\text{O}-\text{H}^2 \text{ (with H}^1\text{, H}^3\text{)} \quad (3.16)$$

4 – RADICAL STABILITY

The idea of *stability* almost always has experimental connotations: the rate of disappearance or of formation of a radical. This notion is therefore often ambiguous since it depends on the environment. An isolated radical, without any interaction with other radicals or molecules (in interstellar space, for example) can only disappear by unimolecular decomposition. Generally, a radical is part of an environment with which it can react. Its visible stability is therefore related to its reactivity with the other molecules.

Nevertheless, the question arises as to the nature of the intrinsic structural elements which modify the stability of a radical. Two concepts, which must be defined, are often employed: *thermodynamic stability* and *kinetic stability*. *Thermodynamic stability* is quantitatively related to the enthalpy of dissociation of the R–H bond to give R$^{\bullet}$ and H$^{\bullet}$ (4.1), i.e. the R–H bond dissociation energy, BDE(R–H).

$$\text{R–H} \xrightarrow{\Delta H} \text{R}^{\bullet} + \text{H}^{\bullet} \quad (4.1)$$

The ΔH of reaction (4.1) depends primarily on the *thermodynamic stability* of R$^{\bullet}$. The main factors which determine this stability are conjugation, hyperconjugation, hybridization and the captodative effect. Certain radicals are much more stable than the ΔH of reaction (4.1) would suggest. In this case, in contrast to the thermodynamic stability, one speaks of *kinetic stability*. This is controlled to a large extent by steric crowding at the radical center.

4-1 THERMODYNAMIC STABILITY

The nature of the atom on which the unpaired electron is centered and the delocalization of this electron are the principal parameters determining the more or less marked thermodynamic stability of a radical. On the basis of its half-life ($t_{1/2}$) a radical is described as *stable* (when $t_{1/2}$ is greater than 10^{-3} s), *stabilized* or *destabilized* ($t_{1/2} < 10^{-3}$ s).

4-1-1 Stable radicals

diphenylpicrylhydrazyl radical, DPPH

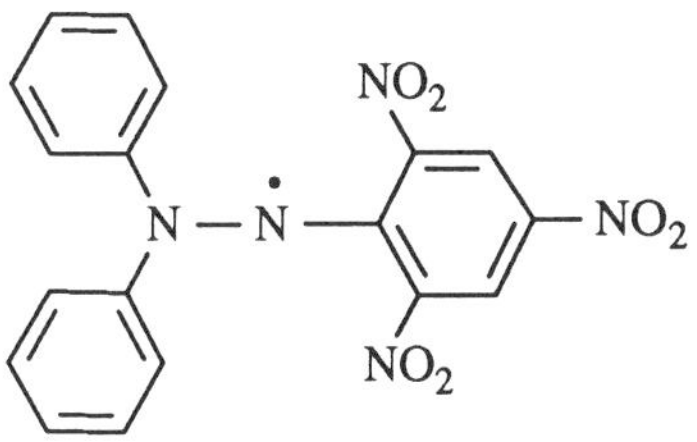

Scheme 4.1

Stable radicals are those which can be stored for several months, several years even, without special precautions (air, light, humidity, etc.). The diphenylpicrylhydrazyl radical, DPPH, is a typical example. It appears as deep violet crystals which can be handled like any other stable molecule. Its high stability is explained by the existence of several limiting resonance forms. This is a common feature of many other radicals such as galvinoxyl (Scheme 4.2) and nitroxide radicals (Scheme 4.3).

galvinoxyl radical

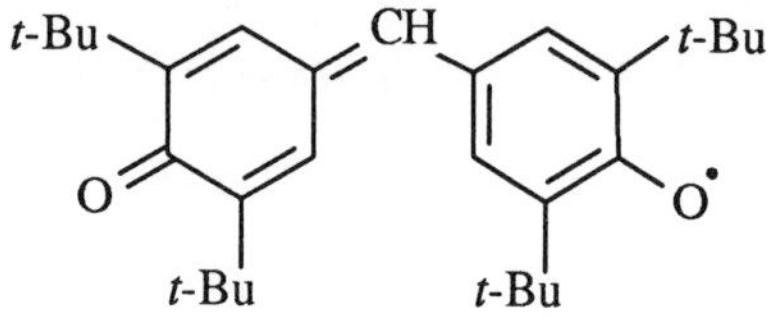

Scheme 4.2

Nitroxide radicals are best represented with a three-electron N–O bond; this explains why there is no dimerization through the oxygen or nitrogen atoms.

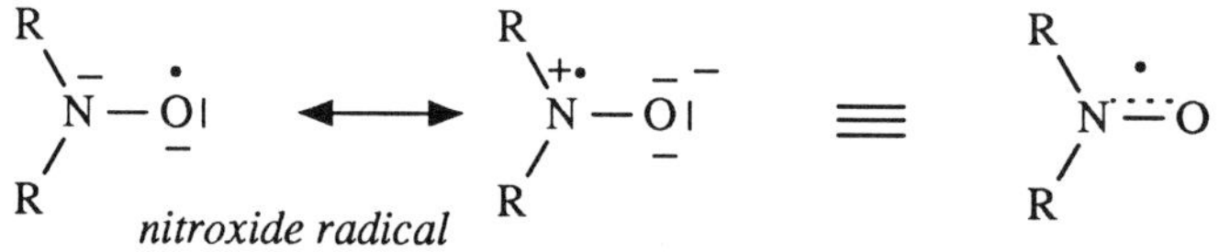

nitroxide radical

Scheme 4.3

Nitroxide radicals are only stable if there is no hydrogen on the carbon α to the nitrogen, in order to avoid conversion into nitrone (4.2).

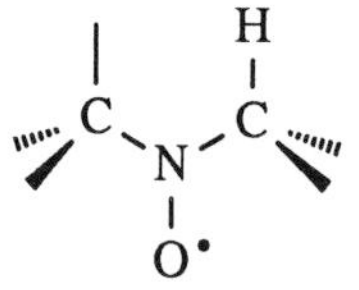

$$\text{(nitroxide with α-H)} + R^\bullet \longrightarrow \text{nitrone} + RH \qquad (4.2)$$

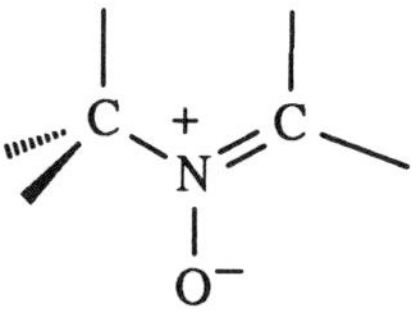

nitrone

4-1-2 Stabilized and destabilized radicals

Stabilized and destabilized radicals have half-lives generally less than 10^{-3} s and for this reason are referred to as *short-lived*. The stabilization energy of a radical, $R^\bullet$, relative to a $Me^\bullet$ radical can be defined by equation (4.3).

$$E_s(R^\bullet) = BDE(CH_3-H) - BDE(R-H) \qquad (4.3)$$

$E_s(R^\bullet)$ is the difference between the dissociation energy of a C–H bond of methane, taken as reference, and that of R–H, the molecule associated with the radical $R^\bullet$. A positive value of $E_s(R^\bullet)$ indicates that $R^\bullet$ is more stable than $Me^\bullet$, and vice versa. Since the methyl radical is planar the unpaired electron, in a 2p orbital, is not delocalized. The stabilization energy of the methyl radical is considered to be zero.

BDE(R–H), i.e. the enthalpy of the reaction (4.1), is related to the standard heats of formation of reactant and products by the following equation.

$$BDE(R-H) = \Delta H°_f(H^•) + \Delta H°_f(R^•) - \Delta H°_f(R-H) \qquad (4.4)$$

This way of defining radical stabilities is admittedly arbitrary but it allows us to compare radicals of different sizes, though strictly speaking one can only compare the thermodynamic stabilities of isomeric radicals. In terms of the standard heats of formation, the stabilization energy, E_s (Table 4.1), is expressed as the difference between two differences (equation (4.5) which comes from (4.3) and (4.4)).

$$E_s(R^•) = [\Delta H°_f(CH_3^•) - \Delta H°_f(R^•)] - [\Delta H°_f(CH_3-H) - \Delta H°_f(R-H)] \qquad (4.5)$$

R–H	BDE	$E_s(R^•)$	R–H	BDE	$E_s(R^•)$
CH_3–H	105	0	H–CH_2NH_2	95	10
CH_3CH_2–H	101	4	$BrCH_2$–H	103	2
$(CH_3)_2CH$–H	99	6	$ClCH_2$–H	103	2
$(CH_3)_3C$–H	95	10	FCH_2–H	103	2
$(CH_3)_3CCH_2$–H	100	5	Cl_3C–H	96	9
$CH_2=CHCH_2$–H	86	19	F_3C–H	108	–3
$PhCH_2$–H	85	20	cyclopropyl–H	106	–1
1,4-cyclohexadienyl–H	73	32	cyclobutyl–H	97	8
			$CH_2=CH$–H	104	1
$CH_3C(O)CH_2$–H	92	13	Ph–H	111	–6
H–$CH_2CO_2CH_3$	99	6	CH_3O–H	104	1
RC(O)–H	86	19	PhO–H	87	18
H–CH_2CHO	92	13	$HOCH_2$–H	94	11
H–CH_2NO_2	96	9	CH_3OCH_2–H	93	12
H–CH_2CO_2H	97	8	Ph_3C–H	77	28

Table 4.1 – *Bond dissociation energies (BDE) of R–H and R• stabilization energies in kcal/mol.*

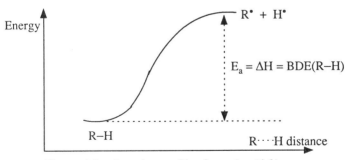

Figure 4.1 – *Reaction profile of reaction (4.1).*

In this expression the second term corrects the first term for specific internal interactions of the R skeleton. The figure obtained in this way represents the stabilization due to interaction of the unpaired electron with the skeleton. This implies that the unpaired electron in the methyl radical is not stabilized, which is reasonable since it lies in a 2p orbital orthogonal to the other orbitals of the planar radical. Nevertheless, this value must handled with caution, especially if the structure of the skeleton is significantly different in R–H and R•. It should be pointed out that $E_s(R•)$ can also be inexact because it is difficult to measure or to calculate standard heats of formation. On the other hand, this definition has the following advantages:

- its simplicity;
- many BDE(R–H) are known;
- steric effects are minimized since hydrogen is the smallest substituent which can be attached to a radical center.

The activation energy for the homolysis of an R–H bond is equal to the enthalpy of reaction and the bond dissociation energy, BDE(R–H) (Figure 4.1).

4-1-3 Factors contributing to the thermodynamic stability of radicals

In homolysis (4.1) there are several factors which stabilize the radical, R•, and therefore weaken the R–H bond.

Hyperconjugation

Hyperconjugation is responsible for the stability order in (4.6).

$$(CH_3)_3C• \quad > \quad (CH_3)_2CH• \quad > \quad CH_3CH_2• \quad > \quad CH_3• \qquad (4.6)$$

$$\text{tertiary} \qquad\qquad \text{secondary} \qquad\qquad \text{primary}$$

The interpretation in terms of orbital interactions is as follows (Figure 4.2). The 2p orbital containing the unpaired electron interacts with the π and π* orbitals of the alkyl group adjacent to the radical center. These interactions stabilize the two electrons in this orbital and slightly destabilize the unpaired electron, the overall result being energetically favorable. The presence of two (secondary radical) or three

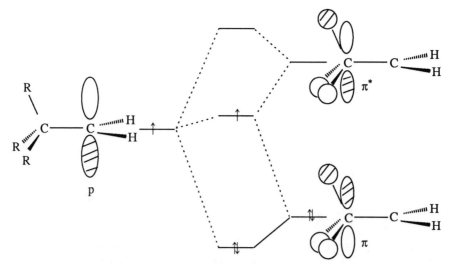

Figure 4.2 – *Orbital interactions associated with hyperconjugation in an alkyl radical.*

Thermodynamic stability

(tertiary radical) alkyl groups accentuates this effect. The slight destabilization of the unpaired electron is the resultant of two opposed effects: destabilization by the π orbital and weak stabilization by the high-energy π* orbital. The principles of orbital interactions in radicals will be developed in Chapter 6 which deals with radical reactivity.

Mesomerism

Mesomerism leads to a stabilization of a radical by resonance. It concerns radicals conjugated with unsaturated systems. This factor is the principal reason for the existence of stable radicals (para. 4-1-1). We shall give some examples of allylic (Scheme 4.4) and benzylic (Scheme 4.5) conjugation.

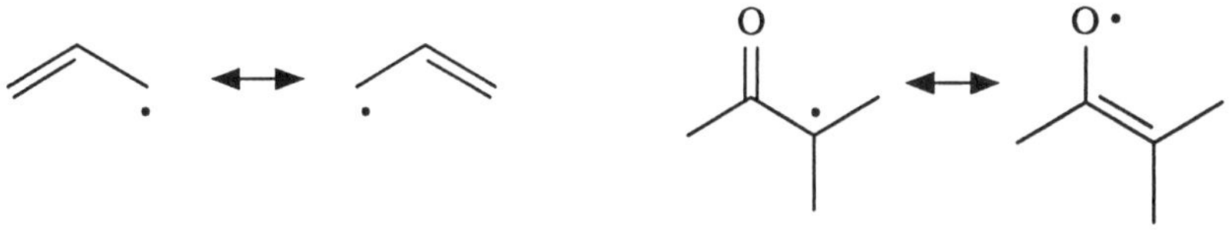

Scheme 4.4

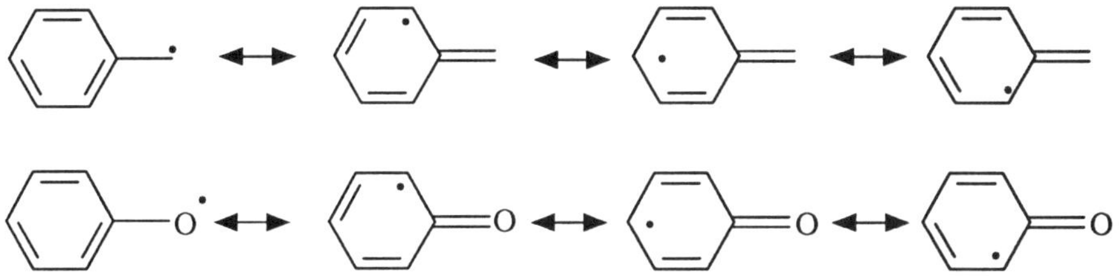

Scheme 4.5

Allylic and benzylic stabilization is also explained by an orbital interaction diagram analogous to that for hyperconjugation. Here the effects are bigger, since there are usually several π orbitals energetically close to the orbital containing the unpaired electron. Conjugation with the lone pair of a heteroatom (O, N, S, halogen, etc.) can also occur (Scheme 4.6). Generally, heteroatom stabilization is more important for X = O or N than for X = Cl or Br.

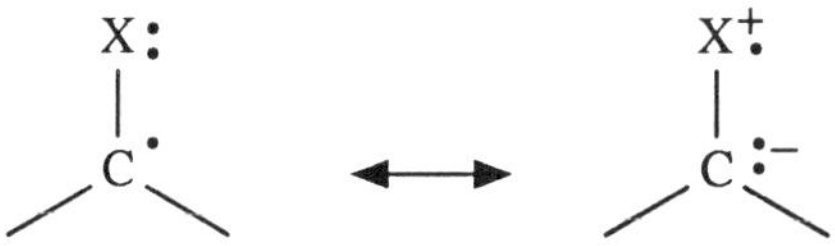

Scheme 4.6

Effect of hybridization

A σ type radical, where the orbital with the unpaired electron has a certain degree of s character, is less stable than a π type radical, where the electron is in a pure p orbital. The destabilization increases with the s character of the orbital (cf. BDE of Table 4.1) from right to left in Scheme 4.7.

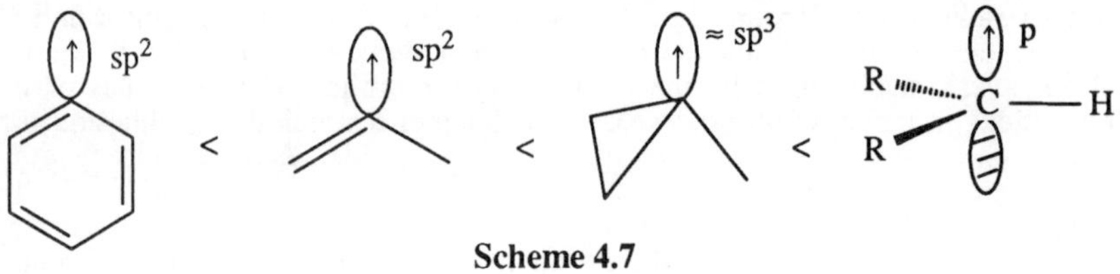

Scheme 4.7

Captodative effect

If a radical, XR•Y, includes two substituents, X and Y, capable of exerting electronic effects, the stabilization energy contributed by these substituents is not strictly additive ($\Delta E \neq 0$) (eqn. (4.7)).

$$\Delta E = [E_s(XR^\bullet H) + E_s(HR^\bullet Y)] - E_s(XR^\bullet Y) \qquad (4.7)$$

When the two substituents are of the same type (electron-donating or attracting) the resulting stabilization energy is generally less than the sum of the two effects taken separately ($\Delta E > 0$). This is referred to as an *antagonistic* effect. Conversely, if one of the substituents is an attractor and the other a donor the stabilization is greater than the sum of the two separate effects ($\Delta E < 0$). This is called a *synergistic* or *captodative* effect; this latter term refers to the nature of the two substituents. This *extra-stabilization* is such that these radicals, which can be obtained by adding a radical to a captodative olefin, are remarkably stable.

$$R^\bullet + CH_2{=}C\begin{smallmatrix}c\\ \\d\end{smallmatrix} \longrightarrow RCH_2{-}C^\bullet\begin{smallmatrix}c\\ \\d\end{smallmatrix} \qquad (4.8)$$

This effect is frequently encountered in biochemistry (aminosugars, vitamins C and E, enzymatic processes). The phenomenon is explained by a succession of orbital interactions: the acceptor stabilizes the orbital with the unpaired electron which, for this reason, interacts more strongly with the donor than in the absence of the acceptor. Conversely, in a di-acceptor or a di-donor the first interaction pushes the singly occupied orbital away from the orbitals of the second substituent and, thus, reduces the stabilization energy.

•R	BDE(R–H)	E_s	E_s^*	ΔE
•CH(CHO)$_2$	99	6	26	20
•CH(NO$_2$)$_2$	99	6	18	12
•CH(t-Bu)$_2$	98	7	18	11
•CH(OCH$_3$)$_2$	91	14	24	10
•CH(CH$_3$)OCH$_3$	91	14	16	2
•CH(NH$_2$)CHO	73	31	23	–8
•CH(NH$_2$)CO$_2$H	76	29	18	–12

Table 4.2 – *BDE(R–H), bond dissociation energies; E_s, stabilization energy calculated from BDE; E_s^* stabilization energy calculated by additivity; ΔE, cf. equation (4.7). Energies are given in kcal/mol.*

4-2 KINETIC STABILITY

The kinetic stability of a radical is generally due to steric factors. When there is a great deal of crowding around a radical center its reactivity with a substrate (another radical or a molecule) decreases considerably. The half-life of the radical is then greatly increased (> 10^{-3} s); it becomes *persistent* and can be observed by normal spectroscopic methods. The 2,4,6-tri-*tert*-butylphenyl radical (4.9) though electronically destabilized like the phenyl radical, has a half-life of the order of 0.1 s. It disappears in an energetically unfavorable intramolecular 1,4 hydrogen transfer process.

(4.9)

The other radicals shown in Scheme 4.8 are also stable and have half-lives of 6 s and 200 s, respectively.

Scheme 4.8

The stability of the triphenylmethyl radical is not due to conjugation but mainly to a steric effect; instead of being coplanar, the aromatic rings are disposed as in a threebladed propeller. Radical coupling, when it occurs, leads not to hexaphenylethane (4.10) but to a product resulting from addition to one of the phenyl groups (4.11).

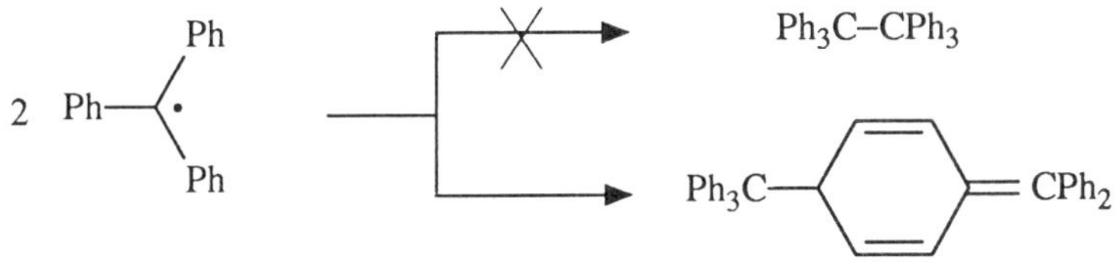

$$2 \ Ph_3C\cdot \longrightarrow Ph_3C-CPh_3 \quad (4.10)$$

$$\longrightarrow Ph_3C-\langle\rangle=CPh_2 \quad (4.11)$$

For steric reasons the Koelsch radical (Scheme 4.9) reacts extremely slowly even with oxygen.

Scheme 4.9

4-3 CONCLUSIONS

The concept of stability applied to free radicals leads to a three-category classification:

– *reactive* or *short-lived* radicals (stabilized or destabilized), with half-lives less than 10^{-3} s, and involved in most radical processes, particularly chain reactions. Their thermodynamic stabilities are defined by comparing their C–H bond dissociation energies with that of Me–H;
– *stable radicals* with half-lives greater than 10^{-3} s and whose stability is due to π delocalization factors;
– *persistent radicals* whose kinetic stability is largely the result of steric hindrance.

5 – ELEMENTARY REACTIONS AND MECHANISMS

Radical mechanisms are combinations of *elementary radical reactions*. They can be classified into two types depending on whether or not a chain process is involved.

5-1 ELEMENTARY REACTIONS

Elementary reactions can be divided into three groups:

- formation of radicals
- transformation of one radical into another
- disappearance of radicals

In ionic processes the displacement of electron pairs is represented by an arrow. Since radical reactions generally involve unpaired electrons, their displacement is usually represented by half-headed arrows.

5-1-1 Formation of radicals

Homolytic dissociation
This reaction occurs for compounds with a low-energy bond, such as benzoyl peroxide. Homolysis is provoked by photolysis or thermolysis.

$$\text{Ph-C(=O)-O-O-C(=O)-Ph} \xrightarrow{h\nu \text{ or } \Delta} 2 \ \text{Ph-C(=O)-O}^\bullet \qquad (5.1)$$

Electron transfer
Electron transfer can occur for both neutral and charged molecules, for example Fenton's reduction (5.2) and oxidation by Ag^{2+} (5.3).

$$\text{HOOH} \xrightarrow{Fe^{2+} \rightarrow Fe^{3+}} [\text{HOOH}]^{\bullet-} \longrightarrow HO^\bullet + HO^- \qquad (5.2)$$

$$\text{RCO}_2\text{H} \xrightarrow{Ag^{2+} \rightarrow Ag^+} [\text{RCO}_2\text{H}]^{\bullet+} \longrightarrow RCO_2^\bullet + H^+ \qquad (5.3)$$

The methods used for generating radicals are discussed in detail in Chapter 9.

5-1-2 Transformation of one radical into another

Transformation reactions are characteristic of the various steps of chain mechanisms which are frequently encountered in radical processes. There are two types of reaction:

- *radical-molecule* reactions, including transfer and addition reactions;
- *unimolecular* reactions, such as fragmentation or elimination, isomerization, rearrangement, cyclization and intramolecular migration of atoms or groups.

Intermolecular transfer reaction

This type of reaction is also referred to as *substitution* or *displacement* and is denoted S_H2 (bimolecular homolytic substitution). This can be:

- either the transfer of a monovalent atom (hydrogen or halogen);

$$Cl^\bullet + H-R \longrightarrow HCl + R^\bullet \quad (5.4)$$

- or a substitution reaction on a bivalent or metal atom

$$R^\bullet + R^1-O-O-R^2 \longrightarrow R-O-R^1 + R^2-O^\bullet \quad (5.5)$$

$$PhS^\bullet + Bu_3Sn-SnBu_3 \longrightarrow PhS-SnBu_3 + Bu_3Sn^\bullet \quad (5.6)$$

$$Bu_3Sn^\bullet + R-SePh \longrightarrow Bu_3Sn-SePh + R^\bullet \quad (5.7)$$

$$R^\bullet + R^1-Co(salen)(py) \longrightarrow R-Co(salen)(py) + R^{1\bullet} \quad (5.8)$$

Intramolecular transfer (migration)

Intramolecular transfers generally involve the migration of monovalent atoms (5.9). These reactions are highly regioselective and are very common in radical chemistry.

$$\overset{\diagdown\ \diagup X}{-C \underset{\smile}{\quad\quad} {}^\bullet Y} \quad \xrightarrow[X = Hal, H]{Y = C, O, N} \quad \overset{\diagdown\diagup\quad\quad\diagdown X}{C^\bullet \underset{\smile}{\quad\quad} Y} \quad (5.9)$$

Intermolecular addition

Intermolecular addition can occur on double or triple bonds (5.10) - (5.11) or on a hypervalent atom (5.12).

$$Cl^\bullet + CH_2=CH-R \longrightarrow Cl-CH_2-\overset{\bullet}{C}H-R \quad (5.10)$$

$$Cl^\bullet + CH\equiv C-R \longrightarrow Cl-CH=\overset{\bullet}{C}H-R \quad (5.11)$$

$$RO^\bullet + P(OEt)_3 \longrightarrow RO\overset{\bullet}{P}(OEt)_3 \quad (5.12)$$

Intramolecular addition (cyclization)

Intramolecular addition reactions are both regioselective and stereoselective and are frequently used for the construction of mono- and polycyclic molecules.

Elementary reactions

$$\text{CH}_2=\text{CH-CH}_2\text{-CH}_2\text{-CH}_2^\bullet \longrightarrow \text{cyclopentyl}^\bullet \quad (5.13)$$

Fragmentation reactions - elimination or rearrangement
There are two classes of fragmentation, called β-cleavage and α-cleavage. Formally, β-cleavage represents the inverse of an addition reaction.

$$(\text{CH}_3)_3\text{C-O}^\bullet \longrightarrow (\text{CH}_3)_2\text{C=O} + {}^\bullet\text{CH}_3 \quad (5.14)$$

$$R-C(=O)-O-O^\bullet \longrightarrow R^\bullet + CO_2 \quad (5.15)$$

$$R-O-\overset{\bullet}{P}(OEt)_3 \longrightarrow R^\bullet + O=P(OEt)_3 \quad (5.16)$$

Reactions that occur via α-cleavage generally involve hypervalent (5.17) or acyl radicals (5.18).

$$\overset{\bullet}{R}OPR_3 \longrightarrow R^\bullet + ROPR_2 \quad (5.17)$$

$$R-\overset{\bullet}{C}=O \longrightarrow R^\bullet + CO \quad (5.18)$$

A fragmentation reaction, which generally implies the cleavage of a C–C bond, can give rise to a molecular rearrangement.

$$\text{bicyclic}^\bullet\text{-R} \longrightarrow \text{cyclohexenyl}^\bullet\text{-R} \quad (5.19)$$

5-1-3 Reactions leading to the disappearance of radicals

Dimerization
Dimerization is the opposite of homolysis.

$$^\bullet\text{CH}_3 + {}^\bullet\text{CH}_3 \longrightarrow \text{CH}_3-\text{CH}_3 \quad (5.20)$$

Disproportionation
Disproportionation leads to a saturated product and an unsaturated one.

$$\text{CH}_3-\overset{\bullet}{\text{CH}}_2 + \text{CH}_3-\overset{\bullet}{\text{CH}}_2 \longrightarrow \text{CH}_3-\text{CH}_3 + \text{CH}_2=\text{CH}_2 \quad (5.21)$$

This is the preferred reaction for sterically hindered radicals. Stabilized radicals or radicals with little steric hindrance usually dimerize.

Electron transfer

Electron transfer produces an oxidant and a reducing agent, which in certain cases can lead to a reaction where radicals reappear.

$$A^{\bullet} + M^{n+} \xrightarrow{\text{reduction}} A^- + M^{(n+1)+} \quad (5.22)$$

$$A^{\bullet} + M^{n+} \xrightarrow{\text{oxidation}} A^+ + M^{(n-1)+} \quad (5.23)$$

$$R\overset{\bullet}{C}HCH_2R^1 + Cu^{2+} \longrightarrow RCH=CHR^1 + H^+ + Cu^+ \quad (5.24)$$

5-2 RADICAL CHAIN MECHANISMS

5-2-1 Description

These mechanisms are very common in radical chemistry. They are particularly used with synthetic goals in mind, such as polymerization or the construction of complex molecular structures. A chain mechanism involves three steps: initiation, propagation and termination. This can be represented either linearly (Figure 5.1) or as a cyclic process (Figure 5.2). The formal kinetics of such reactions are described in Chapter 7.

$$\text{initiation} \quad \text{Rad–X} \longrightarrow \text{Rad}^{\bullet} \quad (5.25)$$

propagation:
$$\text{Rad}^{\bullet} + \text{Mol} \xrightarrow{\text{addition}} \text{Rad}_a^{\bullet} \quad (5.26)$$

$$\text{Rad}_a^{\bullet} \xrightarrow{\text{isomerization}} \text{Rad}_i^{\bullet} \quad (5.27)$$

$$\text{Rad}_i^{\bullet} + \text{Rad–X} \xrightarrow{\text{transfer}} \text{Rad}_i\text{–X} + \text{Rad}^{\bullet} \quad (5.28)$$

$$\text{termination} \quad \text{two radicals} \longrightarrow \text{non-radical products} \quad (5.29)$$

Figure 5.1 – *Linear representation of a hypothetical radical chain reaction.*

The *initiation* step consists of creating the first radical of the chain, Rad•, and may be induced by one of a number of processes (Chap. 9).

The *propagation* step is a succession of *elementary reactions* in which each radical produced in one reaction is consumed in the next. This step is fundamental since it is where the required products are formed. The first radical gives rise to one or more elementary reactions (addition and/or isomerization) before being regenerated.

The rates of the propagation reactions are all equal to each other and the *overall rate* is controlled by one step. This step is called the *rate determining step*. If just one of these reactions has a rate constant which is too small, chain propagation will stop. Conversely, for the chain to be efficient, i.e. for the yield to be good, the rate constant of each reaction in the chain (5.26) - (5.28) must be high.

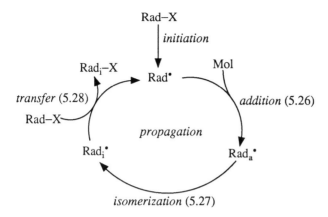

Figure 5.2 – *Cyclic representation of a hypothetical radical chain reaction.*

The *termination* step involves dimerization, disproportionation, oxidation or reduction of the radicals involved in the rate determining step.

We will describe the most common types of propagation reaction. To simplify the reaction schemes, the initiation and termination steps will not be described.

5-2-2 Exchange reactions

Exchange corresponds to the following overall reaction:

$$A-B \;+\; \Sigma-X \;\longrightarrow\; A-X \;+\; \Sigma-B \quad (5.30)$$

Propagation consists of two transfer steps. An example of this is the functionalization of hydrocarbons, R–H, by halogenation.

$$X^\bullet \;+\; R-H \;\longrightarrow\; H-X \;+\; R^\bullet \quad (5.31)$$

$$R^\bullet \;+\; X_2 \;\longrightarrow\; X^\bullet \;+\; R-X \quad (5.32)$$

$$R-H \;+\; X_2 \;\longrightarrow\; R-X \;+\; H-X \quad (5.33)$$

Another example is the reduction of halides by Bu_3SnH, which can be considered as a defunctionalization of the halide, R–X.

$$Bu_3Sn^\bullet \;+\; R-X \;\longrightarrow\; Bu_3Sn-X \;+\; R^\bullet \quad (5.34)$$

$$R^\bullet \;+\; Bu_3Sn-H \;\longrightarrow\; R-H \;+\; Bu_3Sn^\bullet \quad (5.35)$$

$$R-X \;+\; Bu_3Sn-H \;\longrightarrow\; R-H \;+\; Bu_3Sn-X \quad (5.36)$$

5-2-3 Aromatic substitution

Homolytic aromatic substitution corresponds to the following overall reaction:

$$X\text{-}C_6H_4\text{-}H + \Sigma\text{-}Y \longrightarrow X\text{-}C_6H_4\text{-}\Sigma + Y\text{-}H \quad (5.37)$$

Some of these reactions follow a chain mechanism, in particular the alkylation of benzene by a diacyl peroxide. Propagation involves two steps: an addition reaction (5.38) and an oxidation (5.39). It should be noted, however, that most aromatic substitutions do not proceed via chain mechanisms.

$$X\text{-}C_6H_4\text{-}H + R^\bullet \longrightarrow X\text{-}C_6H_4(H)(R)^\bullet \quad (5.38)$$

$$X\text{-}C_6H_4(H)(R)^\bullet + (RCO_2)_2 \longrightarrow X\text{-}C_6H_4\text{-}R + RCO_2H + R^\bullet + CO_2 \quad (5.39)$$

5-2-4 Addition reactions

Radical addition to an unsaturated bond creates C–C and C–heteroatom bonds (5.40) and gives an addition product or one-to-one adduct.

$$R\text{-}X + \text{C=C} \longrightarrow R\text{-}C\text{-}C\text{-}X \quad (5.40)$$

This reaction occurs in two steps, addition (5.41) and transfer (5.42).

$$R^\bullet + \text{CH}_2\text{=CHY} \xrightarrow{k_{add}} R\text{-}CH_2\text{-}CHY^\bullet \quad (5.41)$$

$$R\text{-}CH_2\text{-}CHY^\bullet + R\text{-}X \xrightarrow[X = H \text{ or halogen}]{k_{trans}} R\text{-}CH_2\text{-}CHXY + R^\bullet \quad (5.42)$$

If $k_{add}[\text{=-Y}]$ is greater than $k_{trans}[R\text{-}X]$ then the intermediate radical will undergo a further addition leading to telomers (small n) or polymers (large n) instead of the one-to-one adduct.

$$R\text{-}CH_2\text{-}CHY^\bullet + n\,[\text{CH}_2\text{=CHY}] \xrightarrow{k_{add}} R\text{-}[CH_2\text{-}CHY]_n\text{-}CH_2\text{-}CHY^\bullet \quad (5.43)$$

For example:

Radical chain mechanisms

$$Ph-CH=CH_2 + CCl_4 \xrightarrow[\text{yield 96\%}]{\text{(50 moles excess)}} \text{telomers} \quad (5.44)$$

$$Ph-CH=CH_2 + BrCCl_3 \longrightarrow Ph-CHBr-CH_2-CCl_3 \quad (5.45)$$

5-2-5 Isomerizing addition reaction

In some cases the intermediate radical obtained in an addition step can isomerize or rearrange before the transfer step. Thus, the addition of R–X to β-pinene gives rise to a compound in which the 4-membered ring has opened.

$$R^\bullet + \text{[β-pinene]} \longrightarrow \text{[bicyclic radical with R]} \quad (5.46)$$

$$\text{[bicyclic radical with R]} \longrightarrow \text{[ring-opened radical with R]} \quad (5.47)$$

$$\text{[ring-opened radical with R]} + R-X \longrightarrow X\text{[ring-opened with R]} + R^\bullet \quad (5.48)$$

5-2-6 Allylation

Allylic substitution corresponds to the following overall reaction:

$$R-Y + C_1{=}C_2{-}C_3{-}X \longrightarrow R{-}C_1{-}C_2{=}C_3 + X-Y \quad (5.49)$$

The propagation reaction occurs in three steps: addition (5.50), β-elimination (5.51) and transfer (5.52).

$$R^\bullet + C_1{=}C_2{-}C_3{-}X \longrightarrow R{-}C_1{-}\overset{\bullet}{C_2}{-}C_3{-}X \quad (5.50)$$

$$R{-}C_1{-}\overset{\bullet}{C_2}{-}C_3{-}X \longrightarrow R{-}C_1{-}C_2{=}C_3 + X^\bullet \quad (5.51)$$

$$X^\bullet + R-Y \longrightarrow R^\bullet + X-Y \quad (5.52)$$

Allylation is efficient only when the C–X bond is weak, i.e. when X = Br, SR, SnR_3, etc. The β-elimination and transfer can be replaced by an oxidation and a reduction, as shown by the following example:

$$PhCO_2^\bullet + CH_2{=}CHCH_2R \longrightarrow PhCO_2CH_2\overset{\bullet}{C}HCH_2R \quad (5.53)$$

$$PhCO_2CH_2\overset{\bullet}{C}HCH_2R + Cu^{2+} \longrightarrow PhCO_2CH_2CH{=}CHR + H^+ + Cu^+ \quad (5.54)$$

$$(PhCO_2)_2 + Cu^+ \longrightarrow PhCO_2^\bullet + PhCO_2^- + Cu^{2+} \quad (5.55)$$

5-2-7 Cyclization

Isomerization generally transforms an unsaturated compound into a cyclic compound.

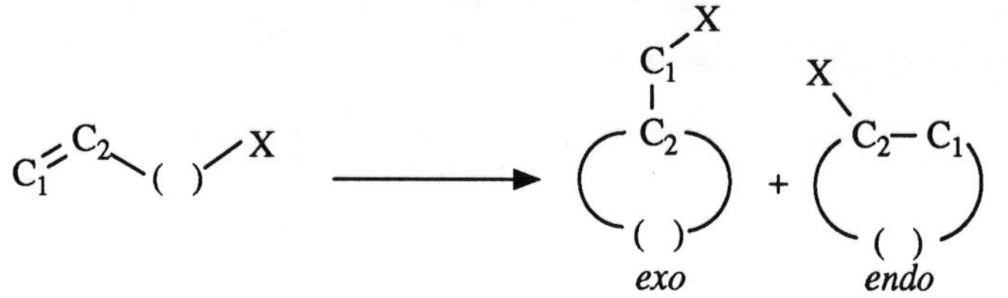

$$C_1\!\!=\!\!C_2\text{---}(\ )\text{---}X \longrightarrow \underset{exo}{\begin{pmatrix} C_1\text{---}X \\ C_2 \\ (\ ) \end{pmatrix}} + \underset{endo}{\begin{pmatrix} X \\ C_2\text{---}C_1 \\ (\ ) \end{pmatrix}} \quad (5.56)$$

The mechanism (Scheme 5.1) consists of a chain with two steps, a cyclization to give a 5- or 6-membered ring (5.58) and a transfer (5.57). This reaction is much used, because of its regio- and stereoselectivity, in the synthesis of mono- and polycyclic products.

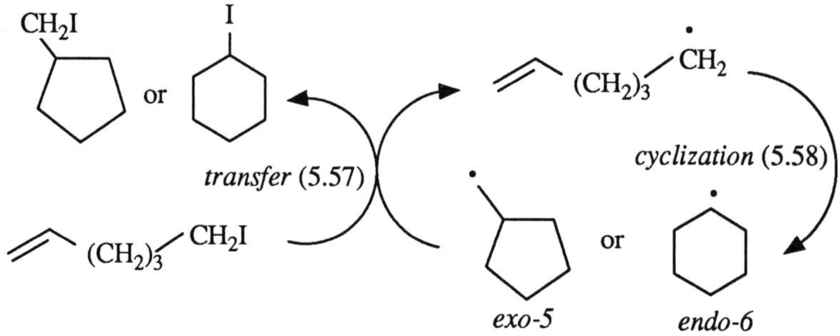

Scheme 5.1

5-2-8 Degradation reactions

The Hunsdiecker reaction (5.59) - (5.60), the degradation of acid hypobromite to bromide, and the degradation of peracid to alcohol (5.60) - (5.61), are radical chain mechanisms with a step involving CO_2 elimination and a transfer step.

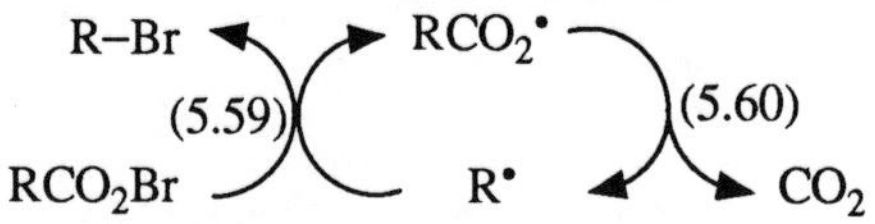

$$RCO_2^\bullet \longrightarrow R^\bullet + CO_2 \quad (5.60)$$

$$R^\bullet + RCO_3H \longrightarrow R\text{--}OH + RCO_2^\bullet \quad (5.61)$$

5-2-9 Nucleophilic radical substitution, $S_{RN}1$

The mechanism of nucleophilic radical substitution involves reactions between

Non-chain radical mechanisms

radicals and fragmentation, addition and electron transfer.

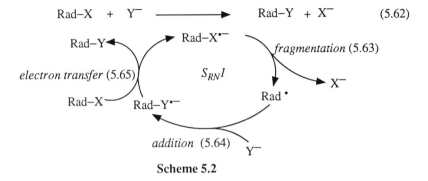

Scheme 5.2

5-3 NON-CHAIN RADICAL MECHANISMS

In somes cases the radicals formed in an elementary reaction are too unreactive to propagate the chain, but either dimerize or give a non-radical product by addition of an oxidant.

5-3-1 Dimerization of stabilized radicals

The addition of a radical to a captodative double bond gives a stable radical that dimerizes.

$$R^\bullet \; + \; CH_2{=}\!\!\!\begin{array}{c} c \\ \\ d \end{array} \longrightarrow \longrightarrow \left[RCH_2{-}\!\!\!\begin{array}{c} c \\ C \\ \diagdown \\ d \end{array} \right]_2 \qquad (5.66)$$

5-3-2 Oxidation of a radical intermediate

We will cite two frequently encountered examples.

Aromatic substitution
Rearomatization (5.68) requires an oxidant (O_2, metal salt).

$$X{-}C_6H_5{-}H \; + \; R^\bullet \longrightarrow X{-}C_6H_5({\cdot}){\langle}^H_R \qquad (5.67)$$

$$X{-}C_6H_5({\cdot}){\langle}^H_R \xrightarrow[M^+]{M^{2+}} X{-}C_6H_5{-}R \; + \; H^+ \qquad (5.68)$$

Addition of enolizable compounds

This reaction occurs in the presence of $Mn(OAc)_3$.

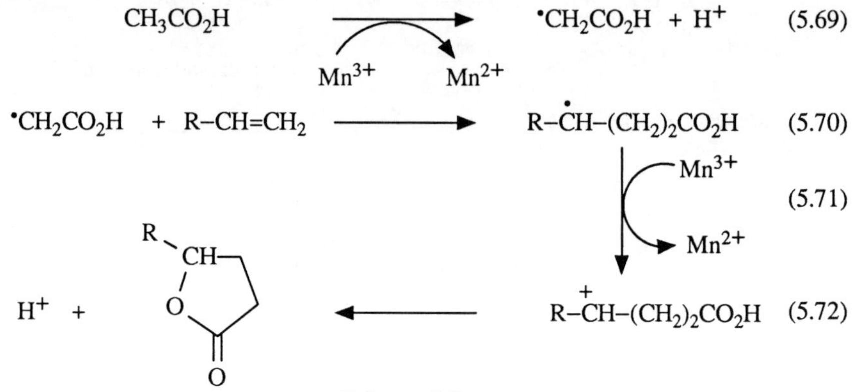

$$CH_3CO_2H \xrightarrow[Mn^{3+} \quad Mn^{2+}]{} {}^\bullet CH_2CO_2H + H^+ \quad (5.69)$$

$$^\bullet CH_2CO_2H + R-CH=CH_2 \longrightarrow R-\overset{\bullet}{C}H-(CH_2)_2CO_2H \quad (5.70)$$

$$\downarrow {}^{Mn^{3+}}_{Mn^{2+}} \quad (5.71)$$

$$H^+ + \underset{\underset{O}{\|}}{\overset{R}{\underset{O}{\diagdown}} \overset{CH}{\diagup}} \longleftarrow R-\overset{+}{C}H-(CH_2)_2CO_2H \quad (5.72)$$

Scheme 5.3

6 – REACTIVITY OF FREE RADICALS

In this chapter we describe:

- the various factors which control reactivity:
    - energy effects (enthalpy and entropy control);
    - structural effects (polar, steric, conformational and orbital control);
- the state correlation diagram model.

6-1 INTRODUCTION TO REACTIVITY

The reactivity of a free radical in a given reaction is expressed by the rate constant of the reaction. By studying the temperature-dependence of the rate the activation energy, E_a, and the pre-exponential factor, A, can be determined from the empirical Arrhenius equation (6.1).

$$k = A \exp{-E_a/RT} \tag{6.1}$$
$$k = C \exp{\Delta S^*/R} \exp{-\Delta H^*/RT} \tag{6.2}$$
$$A = C \exp{\Delta S^*/R} \tag{6.3}$$
$$E_a \approx \Delta H^* \tag{6.4}$$

The factor A, which is higher for the transfer of an atom (log A ≈ 11) than for a group of atoms (log A = 8 ~ 9), is approximately constant for reactions of the same type. The activation energy, E_a, varies from 0 to about 10 kcal/mol for the most common reactions, i.e. it is generally smaller than for ionic reactions.

In the framework of transition state theory the rate constant is related to the thermodynamic activation parameters, ΔH^* and ΔS^*, by equation (6.2). By comparing equations (6.1) and (6.2) the pre-exponential factor is associated with the activation entropy, ΔS^*, and E_a with the activation enthalpy, ΔH^*. Between the initial structure and the transition structure steric, conformational, stereoelectronic, polar, orbital, etc. effects arise.

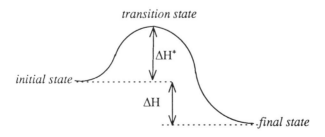

Scheme 6.1

These effects concern ΔH^* and/or ΔS^*, hence the difficulty of analyzing and interpreting rate constants and the related parameters. We shall examine these effects and their roles in terms of classical factors, orbital interaction and correlation diagrams.

The reaction profile of a reaction or elementary step is given in Scheme 6.1. ΔH represents the heat exchanged with the external medium, and ΔH^* the minimum energy required to reach the pass or transition state of the reaction.

The activation entropy expresses the probability of getting over the pass. One can draw an analogy with the ridge of a mountain (Scheme 6.2) where the passes can be:

– either very open and requiring little precision for their passage; they are therefore entropically favorable;
– or very narrow and requiring great precision for their passage; they are entropically unfavorable.

In Scheme 6.2 the reaction path is approximately perpendicular to the ridge-line.

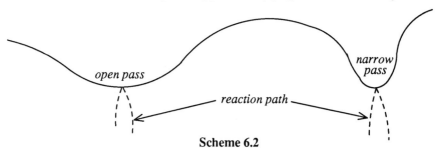

Scheme 6.2

6-2 FACTORS CONTROLLING REACTIVITY

The factors which determine the reactivity fall into two groups: (i) *energy factors* which are directly related to energy values such as the bond dissociation energy and the enthalpy of reaction; (ii) *structural factors* which depend on polar, steric and conformational effects. These two groups of factors are interdependent, since structural changes imply energy variations and vice-versa.

6-2-1 Enthalpy control

Of the various factors, enthalpy control is undoubtedly the most important. It relates the enthalpy of reaction, ΔH, to the enthalpy of activation, ΔH^*, or, if the activation entropy varies little, to the reaction rate.

Consequently, the more the reaction enthalpy, ΔH, is favorable, the more likely it is that the enthalpy of activation, ΔH^*, will be small. In other words, the less stable the attacking radical and the more stable the radical formed, the smaller the enthalpy of activation, ΔH^*. Thus, Table 6.1 shows that the rate constant for hydrogen abstraction by the Ph• radical increases as the breaking bond becomes weaker and, therefore, the exothermicity increases.

In the same way, Table 6.2 indicates that the rate constant for hydrogen transfer

Factors controlling reactivity

Ph• + R–H	→	Ph–H + R•	
R–H	primary	secondary	tertiary
k x 10^{-5} (M^{-1}s^{-1})	0.35	3.3	16.0
BDE (kcal/mol)	101	99	95
ΔH (kcal/mol)	–10	–12	–16

Table 6.1 – *Absolute rate constants for hydrogen transfer by the Ph• radical at 45°C.*

by tri-*n*-butyltin hydride, Bu$_3$SnH, to various radicals increases with the energy of the bond formed and, therefore, with the exothermicity of the reaction.

R• + Bu$_3$Sn–H	→	R–H + Bu$_3$Sn•	
R•	k(25°C) x 10^{-6} (M^{-1}s^{-1})	BDE(R–H) (kcal/mol)	ΔH (kcal/mol)
Me•	10.0	105	–31
Et•	2.3	101	–27
i-Pr•	1.5	99	–25

Table 6.2 – *Absolute rate constants for the reaction of various radicals with Bu$_3$Sn–H.*

Table 6.3 shows that unstable radicals, such as Me• and Ph•, are unselective. On the other hand, the Br• radical is less reactive and much more selective. Since the reaction is endothermic the transition state is close to the reaction products (Hammond postulate, see below) and the factor which determines the selectivity is the difference in the stabilities of the radicals formed.

Σ• + R–H	→	Σ–H + R•	
Σ• \ R–H	•CH$_3$ 110°C 105^a	Ph• 60°C 111^a	Br• 40°C 88^a
aliphatic: primary	1	1	1
secondary	4.3	9.3	8 x 10
tertiary	45	44	24 x 10^2
allylic: primary	9	15	
secondary	21	30	3 x 10^5
benzylic: primary	10	9.1	8 x 10^3
secondary	130	88	3 x 10^5

Table 6.3 – *Selectivity: k_{rel} for the homolytic transfer of various types of hydrogen. a) Bond dissociation energy (BDE) for Σ–H in kcal/mol.*

In the case of hydrogen transfer (6.5), where H–R represents a series of analogous substrates, the Evans and Polanyi equation can be written as in (6.6). Here the gy is related to the bond dissociation energy, BDE(R–H). The coefracteristic of the entering radical, $\Sigma^{\bullet}$ (α(I) = 0.97, α(NF$_2$) = 0.90, α(Br) = α(CF$_3$) = 0.49). The value of α is always less than unity and decreaion enthalpy becomes more favorable. A value close to unity indicates a transition structure close to the products. This applies to endothermic reactions for which the Evans and Polanyi equation works the best. For exothermic reactions other factors, such as the polar effect, are involved and reduce the utility of this type of relationship.

$$\Sigma^{\bullet} + H–R \xrightarrow{E_a, \Delta H} \Sigma–H + {}^{\bullet}R \qquad (6.5)$$

$$E_a = \alpha[BDE(R–H) + \beta] \qquad (6.6)$$

A corollary of the Evans and Polanyi equation is the *Hammond postulate*, which says that the more a reaction is exothermic, the closer the structure of the transition state is to that of the reactants. Conversely, the transition state of endothermic reactions resembles that of the final products.

6-2-2 Entropy control

Entropy-controlled reactions are much less common than those controlled by enthalpy. One reaction when entropy is dominant is 1,5 and 1,6 intramolecular hydrogen transfer (n = 3 and 4) from carbon to carbon, carbon to oxygen or carbon to nitrogen.

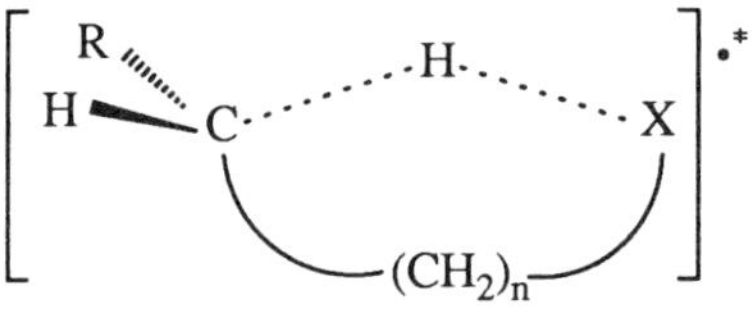

$$X = CH_2, NH_2^+, O$$

Scheme 6.3

In intermolecular hydrogen transfers a linear C...H...X geometry is preferred. In intramolecular reactions (Scheme 6.3) an almost linear geometry can be achieved from 1,5 onwards (n ≥ 3), the C...H...X angle being 150 ~ 160°. The predominance of 1,5 over 1,6 transfer comes mainly from entropy control. Beyond 1,6 the reaction fails because the activation entropy becomes highly unfavorable. On the other hand, transfers below 1,5 are not generally observed, because the activation enthalpy increases as the deviation from C...H...X linearity gets worse.

6-2-3 Polar effect

The first publications which mention a polar effect on radical reactions date from 1948. The transition state of a radical reaction can be described by several canonical forms, and the more there are of these forms of similar energy, the lower the energy of the system.

Consider the radical reaction in Scheme 6.7. The system goes through a transition state which can be described by two canonical forms, I^i and I^f. Between atoms A,

Factors controlling reactivity

B and C there are three electrons concerned in the reaction:

– the form I^i has an electron localized on A and two electrons in the B–C bond;
– the form I^f has an electron localized on C and two electrons in the A–B bond.

$$A^\bullet + B\!-\!C \longrightarrow A\!-\!B + C^\bullet \quad (6.7)$$

$$\left[\underset{I^i}{[A^\bullet \ B\!-\!C]} \longleftrightarrow \underset{I^f}{[A\!-\!B \ {}^\bullet C]} \right]^* \quad (6.8)$$

These two canonical forms show no net charge; the first resembles the initial state, the second the final state.

If atom A has a high electron affinity and the B–C bond a low ionization potential, one can imagine that the reaction will consist of a one-electron transfer from the B–C bond to atom A. The transition state will then be represented by the canonical forms in (6.9). In this type of reaction the entering radical, $A^\bullet$, is *electrophilic*.

$$\left[\underset{I^i}{[A^\bullet \ B\!-\!C]} \longleftrightarrow \underset{I^+}{[A\!:\!\!\bar{\ } \ B \stackrel{+}{\cdot} C]} \right]^* \quad (6.9)$$

The opposite situation can arise if the affinity of the B–C bond is high and the ionization potential of $A^\bullet$ low. The transition state will then be represented by the canonical forms in (6.10). In this reaction the entering radical, $A^\bullet$, is *nucleophilic*.

$$\left[\underset{I^i}{[A^\bullet \ B\!-\!C]} \longleftrightarrow \underset{I^-}{[A^+ \ B \stackrel{\bar{\ }}{\cdot} C]} \right]^* \quad (6.10)$$

In conclusion, it is important to know if the reaction is exothermic or endothermic. The structure of the transition state of an endothermic reaction is close to the final products (I^f). The reactivity of the system will depend mainly on the stability of the radical formed, $^\bullet C$, and therefore on the enthalpy of reaction. In an exothermic reaction the transition state resembles the reactants. If partial charge development is possible the reactivity of the system is enhanced. This enhancement increases with the weight of the charged form, which depends on the nature of the radical and of the substrate, and on the substituents. This effect, known as the *polar effect*, can appear in all types of radical reactions, such as hydrogen transfer, addition to double bonds, etc.

A rough, qualitative definition of the electrophilic or nucleophilic character of a radical is the following: a nucleophilic radical leads more easily to a cation by electron loss than to an anion by electron uptake, and the opposite for an electrophilic radical (6.11). The nucleophilicity of $R^\bullet$ is related to its ionization potential (IP): the lower the IP the more nucleophilic the radical. The electrophilicity of $R^\bullet$ is related to its electron affinity (EA): the higher the EA the more electrophilic the radical.

$$\xrightarrow{+e^-}_{\text{trophilic}} \quad R^\bullet \quad \xrightarrow{-e^-}_{\text{nucleophilic}} \quad R^+ \qquad (6.11)$$

12) are very electrophilic.

$$RO^\bullet, \; RCO_2^\bullet, \; >\overset{\bullet+}{N}-H, \; F^\bullet, \; Cl^\bullet \qquad (6.12)$$

Carbon radicals, X–C•<, are:

— nucleophilic if X is an electron-donor (RO–, R_2N–, R_3Si– or alkyl). In the last case the nucleophilicity increases with the degree of substitution, in the same way as the stability of the carbocations (6.13);

$$Me^\bullet \; < \; Et^\bullet \; < \; i\text{-}Pr^\bullet \; < \; t\text{-}Bu^\bullet \qquad (6.13)$$
$$\xrightarrow{\textit{radical nucleophilicity and cation stability increase}}$$

— electrophilic if X is an electron-attractor (–NO_2, –CN, –COR, –CO_2R, –CF_3).

It is important to note that the "*philicity*" of a radical is a kinetic property, not thermodynamic, i.e. it depends on whether the substrate is a donor or an attractor. In certain extreme cases the same radical can be nucleophilic or electrophilic depending on the nature of the substrate (para. 6-4-2). A more complete interpretation of radical philicity in terms of orbital interactions is given in paragraph 6-3-4.

X• + H–Σ		→	X–H + Σ•	
X•	H–Σ		ΔH	E_a
Cl•	H–C(CH$_3$)$_3$		– 8	0.2
Cl•	H–CCl$_3$		– 7	6.5
•CH$_3$	H–C(CH$_3$)$_3$		–10	8.1
•CH$_3$	H–CCl$_3$		– 9	5.8

Table 6.4 – *Reaction enthalpies and activation energies (kcal/mol).*

Hydrogen transfer
A few of the many examples of hydrogen atom transfer are listed in Table 6.4. The four reactions have about the same reaction enthalpy; the origin of the differences in the activation energies is therefore not enthalpic. The activation energy for hydrogen abstraction from isobutane by the chlorine radical is practically zero, while hydrogen abstraction from chloroform requires 6.5 kcal/mol. The generally accepted interpretation is that the polarized form (6.14) contributes strongly to the electronic structure of the transition state and lowers its energy. In this sort of reaction the Cl• radical is electrophilic.

$$[Cl\colon^- \; H \; \overset{+}{\cdot} \; C(CH_3)_3] \qquad (6.14)$$

Factors controlling reactivity

In hydrogen transfer from chloroform, since it is electrophilic the equivalent structure (6.15) is unimportant and, consequently, has little effect upon the transition state energy. On the other hand, in hydrogen transfer from chloroform to the methyl radical, the radical is nucleophilic and the situation is the opposite of the previous case.

$$[Cl:^- \; H \; ^+\!\cdot CCl_3] \qquad (6.15)$$

The polar form (6.16) contributes to the stabilization of the transition state, which explains why the methyl radical abstracts hydrogen from chloroform more easily than from isobutane.

$$[CH_3^+ \; H \!-\!\cdot \; CCl_3] \qquad (6.16)$$

The polar effect induces interesting regioselectivities in hydrogen transfer reactions. Thus, in the case of a carboxylate ester a nucleophilic radical, such as Me•, abstracts preferentially the hydrogen on the carbon α to C=O, while an electrophilic radical, such as MeO•, abstracts hydrogen α to oxygen, despite the fact that the BDEs of the bonds in question are about the same (Scheme 6.4). Similar arguments explain why the electrophilic Cl• radical preferentially abstracts H_d of propionic acid while the nucleophilic Me• radical attacks H_c (Scheme 6.4).

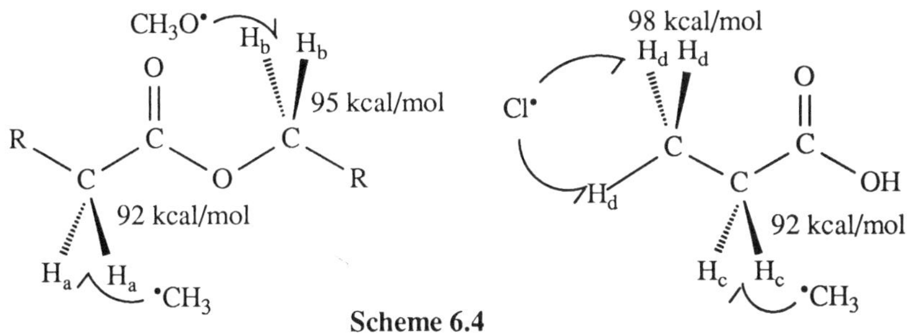

Scheme 6.4

Addition

The nucleophilic cyclohexyl radical is 6000 times as reactive with an electron-poor double bond such as that in methyl acrylate than with 1-hexene.

The polar effect can be highly important in determining the efficiency of chain reactions to obtain one-to-one addition products and in the alternance of the moieties in copolymerization Thus, in Scheme 6.5 the radical α to an electrophilic carbonyl group reacts preferentially with vinyl acetate, the most nucleophilic substrate, to give a nucleophilic radical α to the acetate group. The latter reacts preferentially with the more electrophilic substrate, maleic anhydride, to give an electrophilic radical α to the C=O group, and so on.

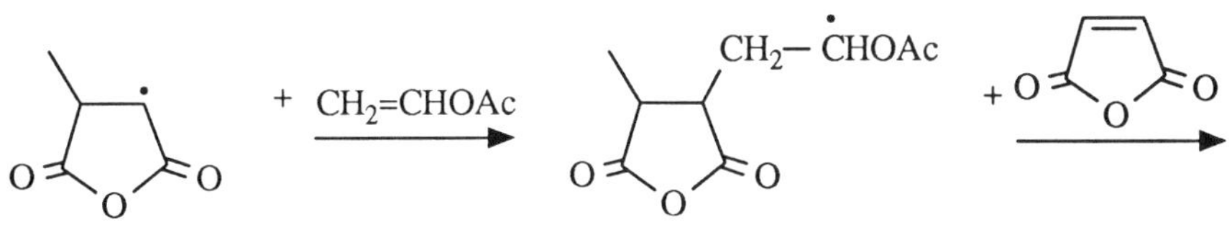

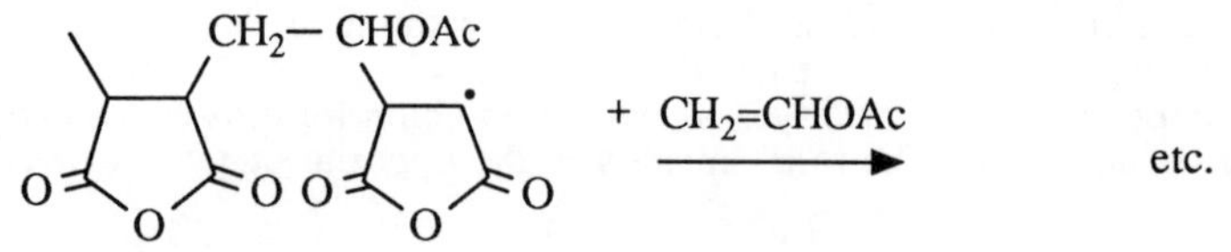

[Scheme 6.5 diagram]

Scheme 6.5

6-2-4 Steric effect

Steric effects are relatively unimportant in *hydrogen transfer reactions* because attack is front-side and linear. However, it can be significant in certain cases. For example, persistent radicals should be very reactive, but this reactivity is *masked* by a large steric effect which prevents a substrate approaching the radical center. The 2,4,6-tri-*tert*-butylphenyl radical (Scheme 6.6) is *persistent*. The *ortho tert*-butyl groups sterically hinder any attack. Its lifetime is about 0.1 s and it can therefore be observed by EPR, whereas the very reactive phenyl radical cannot. Nonetheless, the radical centers have very similar structures and the energies of the bonds formed by hydrogen transfer are the same ($\approx$ 112 kcal/mol).

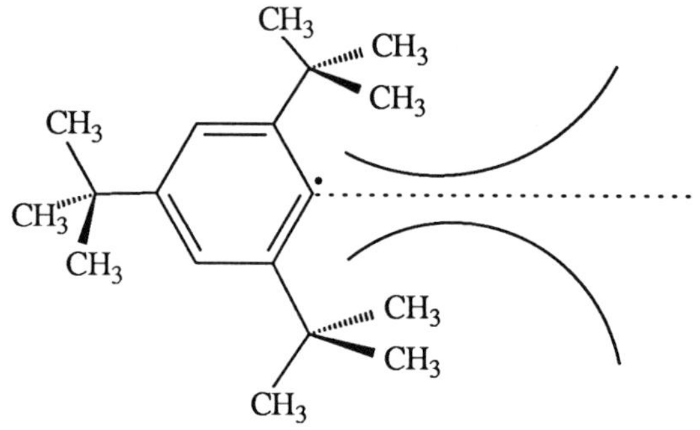

[Scheme 6.6 diagram of 2,4,6-tri-tert-butylphenyl radical]

Scheme 6.6

In radical *addition reactions* (6.17) steric hindrance of group Y controls the regioselectivity rather than the relative stability of the two radicals formed.

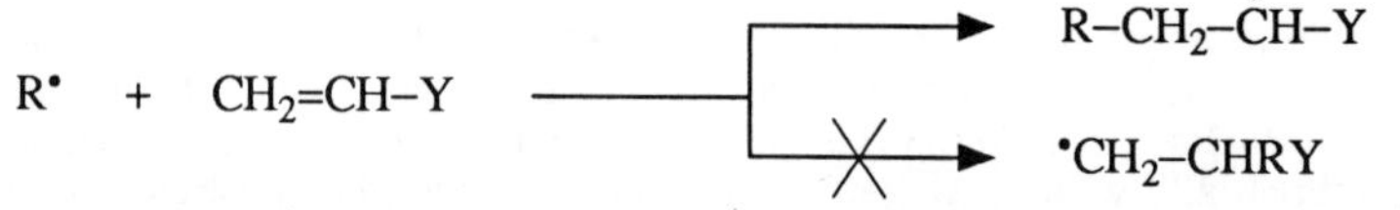

$$R^\bullet + CH_2=CH-Y \longrightarrow \begin{array}{l} R-CH_2-\overset{\bullet}{C}H-Y \\ {}^\bullet CH_2-CHRY \end{array} \qquad (6.17)$$

Steric hindrance can also govern the regioselectivity in *homolytic substitution reactions*, S_H2. Thus, alkyl radicals react readily at the peroxide oxygen of peracids. This oxygen is clearly unencumbered. On the other hand, with peresters the same type of reaction takes place on the non-peroxide oxygen of the O–O bond. The generally

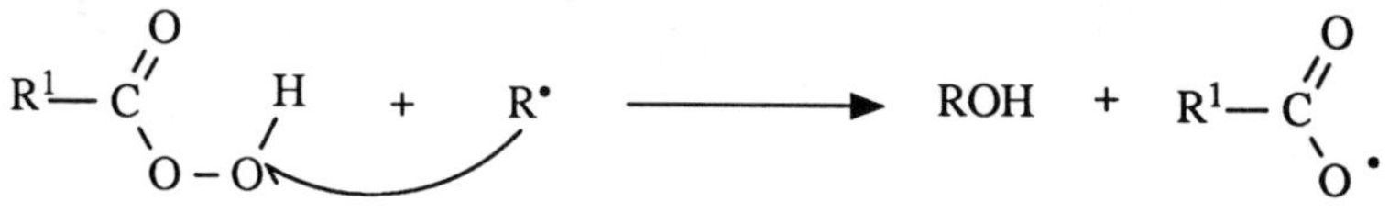

[Equation 6.18 diagram: $R^1-C(=O)-O-OH + R^\bullet \longrightarrow ROH + R^1-C(=O)-O^\bullet$] (6.18)

$$R^{\bullet} \quad R^1-C(=O)-O-O-C(Me)(Me)(Me) \longrightarrow R^1CO_2R \;+\; t\text{-BuO}^{\bullet} \quad (6.19)$$

accepted explanation of this regioselectivity is that the peroxide oxygen of peresters, in the neopentyl position with respect to the *tert*-butyl group, is inaccessible for steric reasons.

To summarize, the reactivity of a radical is controlled by a set of factors which can be complementary or contradictory. These factors are hard to quantify and, when they conflict, the final outcome is sometimes difficult to foresee. A more quantitative approach can be attempted by applying quantum theory and orbital interactions.

6-3 ORBITAL ANALYSIS OF RADICAL REACTIVITY

6-3-1 Theoretical representation of a radical

In quantum chemistry molecular systems are classified as closed shell systems and open shell systems depending on whether there are unpaired electrons or not. In *closed shell* molecules all the electrons are paired (Scheme 6.7 A). Each orbital, characterized by its energy and its spatial expansion, is either doubly occupied or unoccupied. The highest occupied orbital is called the HOMO (Highest Occupied Molecular Orbital) and the lowest empty orbital the LUMO (Lowest Unoccupied Molecular Orbital). In molecules with *open shells* there is, in addition to the paired electrons, at least one unpaired electron. This is the case of free radicals. One can have a restricted or an unrestricted representation.

In the *restricted representation*, apart from the SOMO (Singly Occupied Molecular Orbital) which contains only one electron, each orbital is either doubly occupied or unoccupied (Scheme 6.7 B). The *unrestricted representation* makes use of spin orbitals. In quantum mechanics a spin orbital is a function, the product of a space function and a spin function, characterized by its energy and its spatial expansion. Each spin orbital is either empty or occupied by a single electron (Scheme 6.7 C). The α and β spin orbitals are orthogonal since the spin functions of the α and β electrons are. To each orbital of the restricted representation corresponds an α and a β spin orbital characterized by very similar space functions. The numbers of α spin orbitals and β spin orbitals are the same. By convention, the α part of a free radical has one more spin orbital occupied than the β part. Particular attention must be paid to the equivalent of the SOMO in the restricted representation: the αSOMO and the βSOMO. These are two spin orbitals with very similar spatial parts but the αSOMO is occupied by an electron whereas the βSOMO is empty. In the orbital filling system, generally the αSOMO is the highest occupied α spin orbital and the βSOMO the lowest empty β spin orbital. Moreover, the energy of the βSOMO (empty) is much higher than that of the αSOMO (occupied).

6-3-2 Orbital interactions between a radical, R•, and a substrate, S

In the orbital approach the total interaction energy for a radical, R•, and a substrate, S, is the sum of the energies for the interaction of the orbitals taken two-by-two

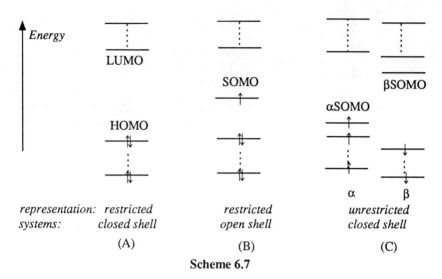

Scheme 6.7

(second-order approximation). Depending on the electronic occupation of these orbitals we can distinguish several types of interaction.

In the *restricted representation* (Scheme 6.8) a radical fragment and a *closed shell* fragment call into play, depending on the number of electrons involved, four types of orbital interaction:

– one-electron SOMO(R•) ↔ UMO(S) interactions **1** between the SOMO of the radical and the vacant orbitals (Unoccupied Molecular Orbital) of the substrate;
– two-electron OMO(R•) ↔ UMO(S) and UMO(R•) ↔ OMO(S) interactions **2** between doubly occupied orbitals (Occupied Molecular Orbital) and vacant orbitals;
– three-electron SOMO(R•) ↔ OMO(S) interactions **3** between the SOMO and the doubly occupied orbitals of the substrate;
– four-electron OMO(R•) ↔ OMO(S) interactions **4** between doubly occupied orbitals.

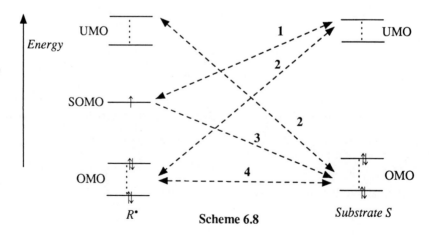

Scheme 6.8

In the *unrestricted representation* there are only two types of interaction between spin orbitals (Scheme 6.9):
- one-electron OMO(R•) ↔ UMO(S) and UMO(R•) ↔ OMO(S) interactions **1α** and **1β** between empty and filled orbitals;
- two-electron OMO(R•) ↔ OMO(S) interactions **2α** and **2β** between occupied orbitals.

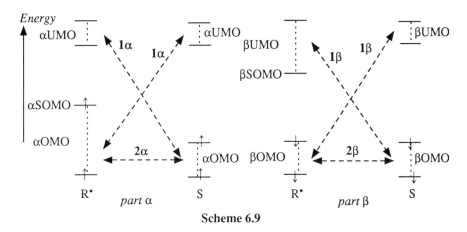

Scheme 6.9

All interactions between vacant and filled orbitals stabilize (one- or two-electron interactions **1**, **2**, **1α** and **1β**). The associated energetic effects, ΔE^n (6.20), are proportional to the square of the overlap and inversely proportional to the energy difference between the interacting orbitals. H_{ii} and H_{jj} are the energies of the orbitals before interaction, S_{ij} and H_{ij} the overlap and the matrix element for interaction between the two orbitals.

$$\Delta E^n = n \frac{k\, S_{ij}^2}{H_{jj} - H_{ii}} \qquad k = (H_{ij} - H_{jj}/S_{ij})^2 \qquad n = 1 \text{ or } 2 \text{ electrons} \qquad (6.20)$$

The energetic effect associated with two full orbitals destabilizes (four- or two-electron interactions **4**, **2α** and **2β**). This repulsive effect increases with the square of the overlap and the mean value of the energies of the initial orbitals (6.21).

$$\Delta E^n = n\, S_{ij}^2\, [(H_{ii} + H_{jj})/2 - (H_{ij}/S_{ij})] \,/\, (1 - S_{ij}^2) \qquad n = 2 \text{ or } 4 \text{ electrons} \qquad (6.21)$$

The effect associated with three electrons can be positive or negative. A complete study shows that, for three-electron SOMO(R•) ↔ OMO(S) interactions, the smaller the energy difference the more they stabilize. These three-electron interactions can destabilize when the orbitals are energetically distant.

The SOMO of a restricted framework is divided in the unrestricted framework into an occupied αSOMO spin orbital and an empty βSOMO spin orbital. The αSOMO, like any other occupied spin orbital, gives rise to stabilizing **1** and destabilizing **2** interactions, whereas the βSOMO, like any empty spin orbital, gives only stabilizing interactions **1**. The overall effect on the energy associated with αSOMO and βSOMO depends on the relative weights of the various interactions.

6-3-3 Frontier orbital approximation

In a qualitative approach one takes into account only the most important stabilizing effects which are associated with the interaction between the highest occupied orbital (HOMO) of the reactant and the lowest unoccupied orbital (LUMO) of the substrate (*nucleophilic reaction*), or between the LUMO of the reactant and the HOMO of the substrate (*electrophilic reaction*) (Scheme 6.10 A).

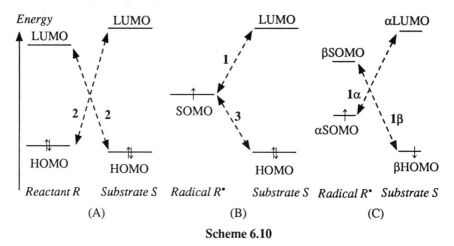

Scheme 6.10

If one assumes that the overlaps are of the same order, the HOMO ↔ LUMO interaction is the most stabilizing of the *full orbital-empty orbital* interactions, since these are the two MOs which are the closest in energy. The frontier orbital approximation does not consider the destabilizing effects of the full orbitals. This amounts to neglecting repulsive effects or considering that they are constant within a given series. In the framework of the frontier orbital approximation Scheme 6.10 A describes the interactions between two closed shell systems which must be taken into account.

The frontier interactions between two sub-systems, of which one is a radical, are given in Schemes 6.10 B and C for the restricted and unrestricted representations, respectively. In the *restricted representation* the frontier orbitals are the SOMO of the radical system and the HOMO and LUMO of the non-radical system. There are now only two interactions to consider: the one-electron SOMO ↔ LUMO stabilizing interaction **1** and the three-electron SOMO ↔ HOMO interaction **3**. This three-electron interaction can either stabilize or destabilize. In the *unrestricted representation* the occupied αSOMO and the empty βSOMO are the frontier orbitals of the radical. They interact with the substrate αLUMO and the βHOMO, respectively, the frontier spin orbitals of the non-radical sub-system. Schemes 6.10 A and C are closely analogous; a radical is nucleophilic if the αSOMO ↔ αLUMO interaction **1α** is stronger than the βSOMO ↔ βHOMO interaction **1β**, and electrophilic in the opposite case. Interaction **3** (Scheme 6.10 B) should be compared with the sum of the interactions **1β + 2α** (Scheme 6.10 C) where **2α** is the destabilizing αSOMO ↔ αHOMO interaction.

6-3-4 Nucleophilic or electrophilic character of a radical

The nucleophilic, electrophilic or neutral character of a radical in a radical reaction depends on the relative magnitudes of the interactions between the orbitals of the radical, R•, and the substrate, S. From this point of view three types of interaction are distinguished:

 – *nucleophilic interactions* between the occupied MOs of R• and the empty ones of S;
 – *electrophilic interactions* between the empty MOs of R• and the filled ones of S;
 – *repulsive interactions* between the occupied MOs of R• and S.

The contribution of the SOMO to these interactions must be analyzed in detail:

 – in the *restricted representation* the SOMO, which is a singly occupied orbital, clearly contributes to the nucleophilicity, particularly by the SOMO ↔ LUMO frontier interaction. The SOMO ↔ OMO interactions (two orbitals, three electrons), among which is included the SOMO ↔ HOMO frontier interaction, have a more complex role, since they contribute both to the electrophilicity (interaction with the "empty part of the SOMO") and to the repulsion (interaction with the "filled part of the SOMO"). In the restricted representation it is not possible to separate these two contributions;
 – in the *unrestricted representation* the contributions of the SOMO are easier to analyse. The three types of interaction, αSOMO ↔ αUMO, βSOMO ↔ βOMO and αSOMO ↔ αOMO, which include the frontier interactions, αSOMO ↔ αLUMO, βSOMO ↔ βHOMO and αSOMO ↔ αHOMO, contribute to the nucleophilicity, electrophilicity and repulsion, respectively.

The energy effect associated with electrophilic and nucleophilic interactions increases when the energy difference between the orbitals decreases. The repulsive interactions decrease with the mean energy of the orbitals. Since the orbital energies are decreased or increased by donor and attractor substituents, respectively, it follows that:

 – nucleophilic interactions are favored by donor substituents on the radical, R•, or acceptors on the substrate, S. Conversely, they are disfavored by acceptors on R• and donors on S. *With respect to a given substrate, the nucleophilicity of a radical increases with the SOMO energy (i.e. when its ionization potential (IP) decreases)*;
 – electrophilic interactions behave in the opposite way. *The electrophilicity of the SOMO increases when its energy decreases (its IP increases)*.

6-4 APPLICATIONS OF ORBITAL ANALYSIS

The energy effect associated with orbital interactions concerns enthalpy; entropy effects are not taken into account. In a study of reactivity orbital interactions are calculated at the beginning of the reaction coordinate from the initial structures of the reactants, the radical and the substrate. To compare reaction systems successfully the effects calculated in the initial state must appear in the transition state, that is, the reaction pathways must not cross (Figure 6.1: the non-crossing principle).

This principle is not always obeyed; however, if one studies reactions which are comparable and, moreover, exothermic (transition state close to the reactants) there is

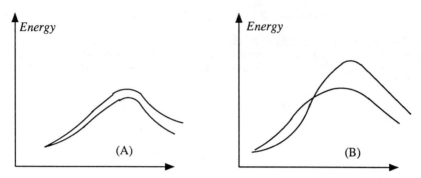

Figure 6.1 – *Non-crossing principle.*

some hope of finding conditions where this principle is obeyed, i.e. the potential energy curves can be deduced from each other by proportionality (Figure 6.1 A). If this is not the case (Figure 6.1 B) the potential surface has to be calculated by the *ab initio* method.

6-4-1 Addition

For this type of reaction the frontier orbital approximation is justified since the energies of the substrate π (HOMO) and π^* (LUMO) MOs are very well separated from all the other MOs. The frontier orbital interactions to be considered in the restricted and unrestricted representations are those indicated in Schemes 6.10 B and C.

The radical is *nucleophilic* in the case of an addition where it has donor substituents and the substrate attractor groups. The SOMO ↔ LUMO interaction (restricted representation) or αSOMO ↔ αLUMO (unrestricted representation) is then the most important. The stabilization associated with these interactions increases as the energy difference between the orbitals decreases. We have then two cases:

– the reaction depends on the SOMO energy. Thus, the reactivity with an electron-poor olefin varies in the order: t-Bu$^\bullet$ > n-Bu$^\bullet$ > Et$^\bullet$ > Me$^\bullet$ (Table 6.5). It should be noted that this is the order of radical stability. *This example therefore contradicts the rule which states that the more stable the radical, the less reactive it is;*

R$^\bullet$ + CH$_2$=CH–Z $\xrightarrow{k}$ R–CH$_2$–ĊH–Z			
R$^\bullet$	Z	k (M^{-1}s^{-1})	T (°C)
Me$^\bullet$	P(O)(OEt)$_2$	2.5 x 10^3	–40
Et$^\bullet$	P(O)(OEt)$_2$	2.6 x 10^3	–40
n-Bu$^\bullet$	P(O)(OEt)$_2$	5.0 x 10^3	–40
t-Bu$^\bullet$	P(O)(OEt)$_2$	5.9 x 10^3	–40
n-Hex$^\bullet$	CN	5.9 x 10^5	0
t-Bu$^\bullet$	CN	1.0 x 10^6	23

Table 6.5 – *Rate constants for addition of R$^\bullet$ to CH$_2$=CH–Z.*

Applications of orbitals analysis 63

– the reaction depends on the LUMO. Thus, the activation energy for the reaction of the *t*-Bu• radical with mono- and 1,1-disubstituted olefins decreases linearly with the increase in the LUMO energy level.

Normally, the more reactive a radical, the less selective it is. This is not always the case; with electron-poor olefins *t*-Bu• is more reactive than *n*-Hex• but it is also more selective. Orbital interactions (Scheme 6.11) explain perfectly this apparent anomaly. The stabilization energy (ΔE^1) and, therefore, the reactivity, is inversely proportional to the energy difference between the orbitals (SOMO ↔ LUMO or αSOMO ↔ αLUMO). The fact that the SOMO energy of *t*-Bu• is higher than that of *n*-Hex• and that the LUMO of $CH_2=CH-CN$ is lower than that of $CH_2=CH-CO_2Me$ implies that $\Delta\epsilon_1/\Delta\epsilon_2 > \Delta\epsilon_3/\Delta\epsilon_4$ and, therefore, that $\Delta E^1_2 - \Delta E^1_1 > \Delta E^1_4 - \Delta E^1_3$; consequently $k_2/k_1 > k_4/k_3$. The *t*-Bu• radical is much more reactive with both olefins than *n*-Hex• ($\Delta E^1_2 > \Delta E^1_4$ and $\Delta E^1_1 > \Delta E^1_3$) but is, nevertheless, more selective ($\Delta E^1_2 - \Delta E^1_1 > \Delta E^1_4 - \Delta E^1_3$).

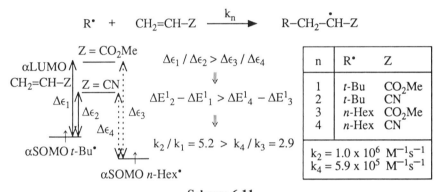

Scheme 6.11

The SOMO of *t*-Bu•, which is the closer to the LUMO, therefore makes this radical more reactive but also more selective, since it distinguishes more clearly between the LUMOs of the two olefins.

In the case of an addition where the radical is *electrophilic* the SOMO ↔ HOMO frontier interaction (restricted representation) or βSOMO ↔ βHOMO (unrestricted representation) is dominant. Analysis of the effects of the SOMO ↔ HOMO interaction on the energy is complex, and it is easier to examine the βSOMO ↔ βHOMO interaction, which always stabilizes. Electron-attracting substituents on the radical and electron-donors on the olefin reduce the βSOMO ↔ βHOMO energy difference; they therefore favor this interaction. Thus, •CCl_2CO_2Me adds to 1-decene 2.5 times

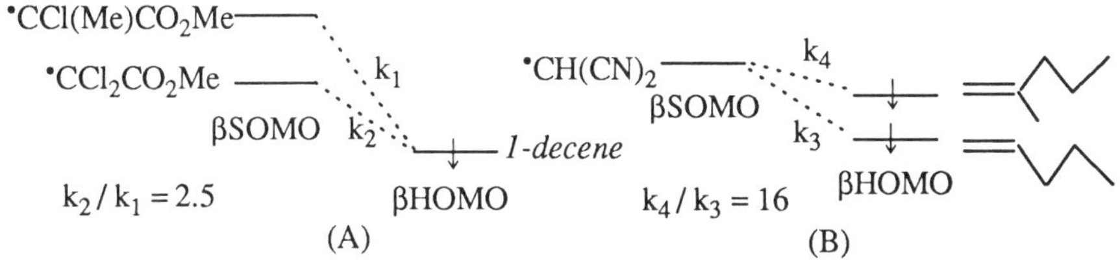

Scheme 6.12

faster than ˙CClMeCO$_2$Me (Scheme 6.12 A), and ˙CH(CN)$_2$ adds to 2-methyl-1-pentene 16 times faster than to 1-pentene (Scheme 6.12 B). In the same way, it has been calculated that the SOMO ↔ HOMO interaction is dominant in the addition to various ethylenic compounds of electrophiles, such as phenyl radicals with electron-attracting *para* substituents and alkoxyl or trifluoromethyl radicals.

Depending on whether the radical is nucleophilic or electrophilic, one observes opposite tendencies in the selectivity of addition to two olefins, one electron-rich and the other electron-poor. Thus, Table 6.6 shows that the selectivity of the electrophilic radical, ˙CH(CO$_2$Et)$_2$, increases with the electron-richness of the olefin. This indicates that the βSOMO ↔ βHOMO interaction is the most important, since the selectivity correlates with the βHOMO energy.

R˙ + CH$_2$=C(Ph)Z		⟶ R–CH$_2$–Ċ(Ph)Z
Z	R = CH(CO$_2$Et)$_2$	R = cy-C$_6$H$_{11}$
CO$_2$Et	1	42.0
Ph	3.5	3.6
Me	3.7	1.0

Table 6.6 – *Relative rate constants for the addition of R˙ to olefins.*

The opposite selectivity is observed for the nucleophilic cy-C$_6$H$_{11}$˙ radical. Here the αSOMO ↔ αLUMO interaction determines the result and the selectivity is correlated with the αLUMO energy. In this case the stability of the radical formed has little bearing upon the selectivity. When Z = Ph or Me, the reactivity of the malonyl radical, ˙CH(CO$_2$Et)$_2$, is about the same whereas the stabilities of the radicals formed are different, and the cyclohexyl radical is more reactive for Z = CO$_2$Et than for Z = Ph, while the stabilities of the radicals formed would suggest the opposite. In the typical example of the alternate copolymerization of maleic anhydride and vinyl acetate (Scheme 6.5) the maleic anhydride radical, which is an electrophile, only adds to vinyl acetate, the more electron-rich olefin, whereas the vinyl acetate radical, which is nucleophilic, only adds to maleic anhydride, the least electron-rich olefin.

One of the explanations of the high reactivity of captodative olefins is that the small HOMO ↔ LUMO energy difference (or αHOMO ↔ αLUMO and βHOMO ↔ βLUMO) makes the SOMO ↔ LUMO and SOMO ↔ HOMO interactions (or αSOMO ↔ αLUMO and βSOMO ↔ βHOMO) very favorable. In alkynes this energy difference is large, which explains why they are less reactive with electrophilic and nucleophilic radicals.

To summarize, radicals behave as electrophiles or nucleophiles with unsaturated substrates, depending on the relative magnitude of the energy effects associated with interactions between the SOMO and the frontier orbitals of the substrate.

6-4-2 Homolytic aromatic substitution

Homolytic aromatic substitution goes by a two-step mechanism: an addition to the aromatic nucleus (6.22), which is often the rate determining step, and an oxidation step which leads to rearomatization of the intermediate cyclohexadienyl intermediate (6.23).

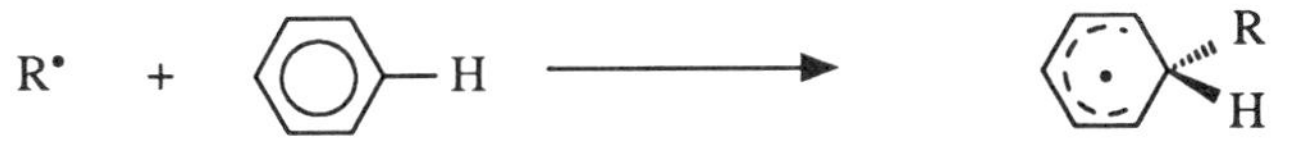

Applications of orbitals analysis

$$R^{\bullet} + \langle\text{C}_6\text{H}_5\rangle-H \longrightarrow \langle\text{C}_6\text{H}_5(\bullet)\rangle\begin{smallmatrix}R\\H\end{smallmatrix} \qquad (6.22)$$

$$Ox + \langle\text{C}_6\text{H}_5(\bullet)\rangle\begin{smallmatrix}R\\H\end{smallmatrix} \longrightarrow \langle\text{C}_6\text{H}_5\rangle-R + H^+ + Ox^{\bullet -} \qquad (6.23)$$

Orbital interactions, calculated for the addition of p-X-phenyl radicals (X = Me, H, NO$_2$) to 4-methylpyridine and protonated 4-methylpyridine, correctly interpret the results. It is necessary to take into account interactions of the SOMO with the πHOMO and πLUMO, and also those with the orbitals immediately below (πHOMO−1) and above (πLUMO+1) (Scheme 6.13).

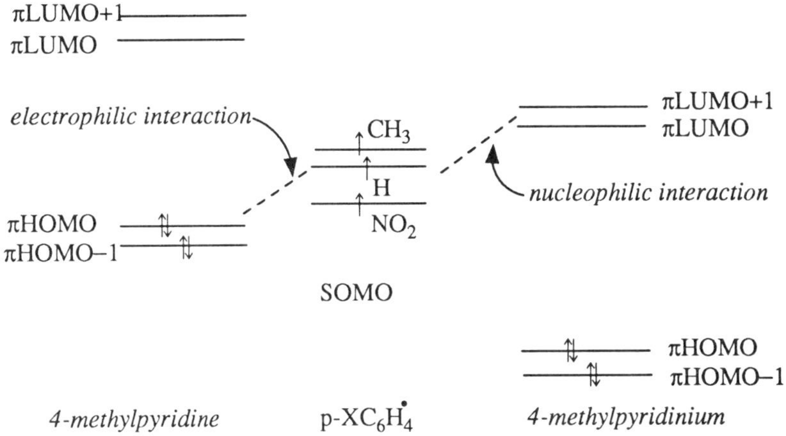

Scheme 6.13

For addition to neutral 4-methylpyridine the SOMO ↔ πHOMO and SOMO ↔ πHOMO−1 interactions are the most important; the radicals then behave as electrophiles and their reactivity increases as the SOMO energy falls. The polarization of the πHOMO, πLUMO, πHOMO−1 and πLUMO+1 at the 2 and 3 positions of the pyridine nucleus is in good agreement with the experimentally observed lack of regioselectivity (2/3 = 48/52 for the phenyl radical). If 4-methylpyridine is protonated all the energy levels of the orbitals are lowered; the energy effects associated with the SOMO ↔ πLUMO and SOMO ↔ πLUMO+1 interactions predominate. The regioselectivity increases (2/3 = 84/16 for the phenyl radical), which shows that the SOMO ↔ πLUMO interaction is the most important. Consequently, the radicals behave as nucleophiles and their reactivity increases with the SOMO energy. The reactivity of the p-X-phenyl radicals with protonated pyridines varies in the reverse order to that observed for neutral pyridines.

The regioselectivity depends on the localization of the substrate orbitals (cf. numerator of equation 6.20). Because of protonation the LUMO of the pyridine is enhanced at the 2 position (Figure 6.2), which clearly reflects the preferential attack at this position. The same type of reasoning explains why the reactions of radicals with aromatic heterocycles depend on the donor or attractor character of the substituents on the radical or the substrate. *The same radical can, therefore, act as an electrophile or as a nucleophile depending on the substrate.*

```
        Me                                                              Me
         |                                                               |
      ╱═╲  − 0.49       Me                   Me                       ╱═╲  − 0.51
     ║   ║                |                    |                     ║ + ║
      ╲═╱  + 0.51      ╱═╲  − 0.58          ╱═╲  − 0.56               ╲═╱  + 0.49
         N            ║   ║  + 0.26        ║ + ║  + 0.11                 NH
                      ╲═╱  + 0.31          ╲═╱  + 0.45
    πLUMO+1              N  − 0.56            NH + 0.49              πLUMO+1
    ε = 6.9eV                                                        ε = −0.8eV
                       πLUMO                  πLUMO
                      ε = 6.6eV              ε = −0.9eV
```

```
                        Me                    Me
                         |                     |                         Me
        Me            ╱═╲  + 0.51           ╱═╲  − 0.48                   |
         |           ║   ║                 ║ + ║                       ╱═╲  − 0.50
      ╱═╲  − 0.55     ╲═╱  + 0.49           ╲═╱  + 0.52               ║   ║  − 0.44
     ║   ║  − 0.33       N                    NH                       ╲═╱  + 0.03
      ╲═╱  + 0.23                                                         NH + 0.50
         N + 0.57      πHOMO                  πHOMO
                      ε = −8.2eV             ε = −14.3eV              πHOMO−1
    πHOMO−1                                                           ε = −15.2eV
    ε = −8.6eV
                    4-methylpyridine                               4-methylpyridinium
```

Figure 6.2 – *Structures and energies of the frontier molecular orbitals and the next higher and low orbitals of 4-methylpyridine and 4-methylpyridinium calculated at the STO-3G level.*

The *reactivity* of radicals with protonated pyridine increases with their nucleophilicity (t-Bu$^\bullet$ > cy-C$_6$H$_{11}^\bullet$ > Et$^\bullet$), i.e. *in the reverse order of their stabilities*. Furthermore, the regioselectivity increases with the reactivity (Table 6.7). These two observations are in contradiction with the usual reactivity-stability or reactivity-regioselectivity relationships. Since the SOMO ↔ LUMO interaction is the most important, the reactivity varies inversely with the stability, and the regioselectivity (attack at two different sites) rises when the SOMO ↔ LUMO separation falls. This example is analogous to that of nucleophilic addition to a double bond (dominant SOMO ↔ LUMO interaction) where the selectivity follows the reactivity (see above) The philicity of a radical is, therefore, a kinetic rather than a thermodynamic property.

radical	C_2	C_4
Et$^\bullet$	56	44
cy-C$_6$H$_{11}^\bullet$	63	37
t-Bu$^\bullet$	68	32

Table 6.7 – *Attack (%) on the C_2 and C_4 carbons of pyridinium cation.*

6-4-3 Hydrogen transfer

The first index for the reactivity of free radicals, f(R), sums the squares of the

LCAO atomic coefficients of the LUMO and the HOMO at the reaction center of the substrate. Values of f(R) for hydrogen abstraction from hydrocarbons confirm the order of relative reactivities observed: $H_{prim} < H_{sec} < H_{tert}$. This reactivity index does not take into account the attacking radical.

$$f(R) = C^2_{LUMO} + C^2_{HOMO} \qquad (6.24)$$

In 1961 wasintroduced the delocalizability index, D(R), where the attacking radical is represented by ϵ_{SOMO}, the SOMO energy (6.25). ϵ_{occ} and ϵ_{unocc} are the energies of the occupied and unoccupied MOs of the substrate, C_{occ} and C_{unocc} the LCAO coefficients of the MOs for the atom attacked. D(R) is similar to the superdelocalizability index, S(R), used for π systems. S(R) and D(R) are generally expressed in units of β, which is the resonance integral in the Hückel method. It should be noted that these indices are not limited to the frontier orbitals of the substrate but include the entire set of MOs. They only take account of stabilizing interactions (*empty orbital* ↔ *full orbital* interactions); repulsive interactions are completely neglected. The values of D(R), obtained from EHMO or MINDO/3 calculations for hydrogen transfer reactions, correlate well with the activation energies. The correlation is better if D(R) is multiplied by C^2_{SOMO}, C_{SOMO} being the atomic coefficient of the SOMO at the reaction center.

$$S(R)/(-\beta) \text{ or } D(R)/(-\beta) = \sum_{occ} \frac{C^2_{occ}}{\epsilon_{SOMO} - \epsilon_{occ}} + \sum_{unocc} \frac{C^2_{unocc}}{\epsilon_{unocc} - \epsilon_{SOMO}} \qquad (6.25)$$

D(R) indices, since they depend on the energy level of the SOMO (ϵ_{SOMO}), are suitable for interpreting the regioselectivity of hydrogen transfer in terms of the electro- or nucleophilicity of the radical. Consider the change in regioselectivity observed when hydrogen is abstracted from esters by Me• or MeO• radicals, or from acids by Me• and Cl• (Scheme 6.4). The reaction of Me• is controlled by the SOMO ↔ LUMO interaction, therefore by the form of the LUMO. That of MeO• and Cl• is controlled by the SOMO ↔ HOMO interaction, therefore by the form of the HOMO. Generally speaking, nucleophilic radicals react preferentially with hydrogen α to an attractor group such as C=O, and electrophilic radicals with hydrogen α to a donor group (ether, amine, etc.) or at the position furthest from an attractor.

In hydrogen abstraction from hydrocarbons radicals tend to be electrophilic, since the SOMO ↔ HOMO interaction is stronger than the SOMO ↔ LUMO interaction. This is due to the fact that the LUMO energy is generally high in hydrocarbons, unless it is reduced by an electron-attracting substituent, as in haloalkanes.

6-4-4 Homolytic substitution on a peroxide bond

Alkyl radicals react with the O–O bond of a peracid by transfer of an OH group (6.26).

$$R^\bullet + \underset{\underset{O-O}{|}}{\overset{\overset{O}{\|}}{H-C-R}} \longrightarrow R-OH + RCO_2^\bullet \qquad (6.26)$$

Since peracids have a low-energy σ^*_{O-O} orbital localized on the peroxide bond, the reaction is controlled by the SOMO ↔ LUMO interaction. In such a reaction alkyl radicals are therefore nucleophilic and their reactivity increases with the SOMO energy. Thus, the reactivity of alkyl radicals increases in the order: $R^{\bullet}_{prim} < R^{\bullet}_{sec} < R^{\bullet}_{tert}$, when the IPs of the radicals decrease. However, benzyl radicals, which have IPs close to those of tertiary alkyl radicals, are distinctly less reactive. Therefore, the SOMO ↔ LUMO energy difference is not the only factor controlling the reaction. One must also take into account the delocalization of the SOMO, which enters into the value of the SOMO ↔ LUMO overlap. Since the benzyl radical is delocalized, the spin density on the SOMO orbital at the reaction center is lower than that on a localized alkyl radical. If the SOMO energy is low, as in the case of an alkoxyl radical, reaction (6.26) does not work; the remaining possibility is radical transfer of the peroxide hydrogen of the peracid (6.27).

$$\Sigma O^{\bullet} + \underset{O-O}{\overset{H}{|}}\!\!\!\!\diagdown\!\!\overset{O}{\underset{}{\overset{\|}{C}}}\!\!-R \longrightarrow \Sigma O{-}H + RCO_3^{\bullet} \quad (6.27)$$

6-4-5 β-Fragmentation

Alkoxyl radicals can evolve, simultaneously or not, following various reaction processes which depend essentially on their geometry. When the experimental conditions are such that intermolecular hydrogen transfer can be avoided, the major reactions are:

- intramolecular hydrogen transfer;
- oxidation, for primary and secondary alkoxyl radicals;
- β-fragmentation.

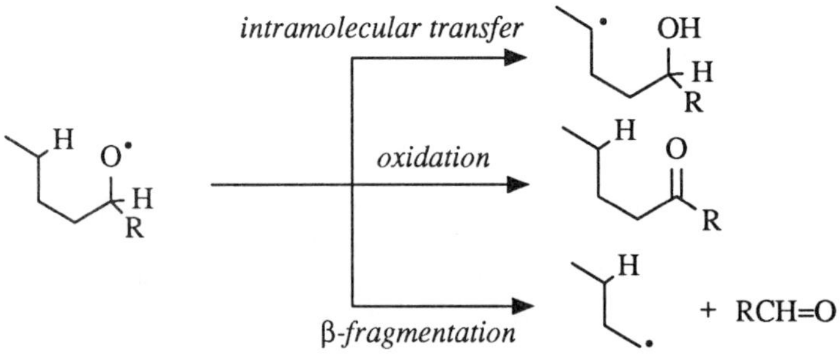

Scheme 6.14

β-Fragmentation appears to be facilitated if an alkyl group is *anti* to the orbital containing the unpaired electron.

Alkoxyl radicals with a plane of symmetry have two distinct electronic states, $^2A'$ and $^2A"$, depending on whether the SOMO has a' or a" symmetry (Chap. 3). The a' and a" SOMOs are mainly localized on the $2p_x$ and $2p_y$ AOs of the oxygen situated in or orthogonal to the plane of symmetry, respectively (Scheme 6.15). For the methoxyl radical, where $R^1 = R^2 = R^3 = H$, the $^2A'$ state is slightly more stable than the $^2A"$ state. The principal difference between the two states lies in the H–C–O angle in the plane of symmetry, which is more open in structure ^{2}A" (112.8°) than in ^{2}A' (106.1°). The final fragmentation product is a ketone or aldehyde obtained by loss of an alkyl radical. In the reaction the ^{2}A' state correlates with the final product in its ground state, whereas the ^{2}A" correlates with the excited state, n→π*; only the ^{2}A' configuration is favorable to β-fragmentation. For this reason β-fragmentation is often not observed when the alkoxyl radical has a plane of symmetry and when R^1 = H or Me.

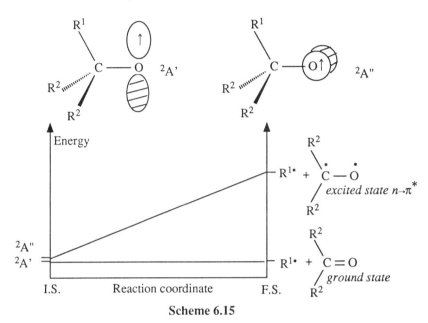

Scheme 6.15

6-4-6 Stereoelectronic control

Stereoelectronic interactions play an important part in bond breaking or formation. For example, in the case of a cyclohexyl radical, abstraction of the pseudoaxial hydrogen, H_{ax}, to give compound **2** is much preferred (ca. 8 times) to that of H_{eq}, despite a steric and thermodynamic disadvantage (6.28). This preference comes from stereoelectronic control resulting from a situation of maximum orbital overlap between the C–H bond broken and the singly occupied orbital.

In the same way, the glycopyranosyl radical gives principally the axial addition product with acrylonitrile (6.29). Antiperiplanar overlap between the doublet of the ring-oxygen and the forming C–C bond favors axial attack despite steric hindrance.

70   Chapter 6: reactivity of free radicals

![Reaction scheme 6.29 showing acetylated sugar radical reacting with CH2=CH-CN to form product with CH2-CH2CN group] (6.29)

6-5 STATE CORRELATION DIAGRAMS

The energy profile of a reaction reflects the evolution of the electronic structure from the reactants to the products. To obtain a qualitative description of this profile one can follow the evolution of the active electrons. In general, the active electrons are those of the bonds made and broken in the reaction. In the case of a simple radical reaction there are three active electrons. The basic principle of any State Correlation Diagram (SCD) is to interrelate certain known properties of the reactants with those of the products. Here there is correlation between electronic structures which have conserved the way in which the electrons are coupled.

6-5-1 Radical reaction

In an elementary radical reaction the ground state of the reactants (R↑ + A↑↓B) correlates with a "pseudo excited state" of the products (R↑↑A+ ↓B), and the ground state of the products (R↑↓A + ↓B) correlates with a "pseudo excited state" of the reactants (R↑ + A↓↓B). The reaction profile can therefore be decomposed into two diabatic curves (curves which cross). By mixing these two diabatic functions one obtains two new curves, the lower of which represents the profile of the reaction in question. The coupling of two diabatic functions depends, amongst other things, on their energy difference. Coupling is maximal where the diabatics cross, this point corresponding to the position of the transition state (Scheme 6.16).

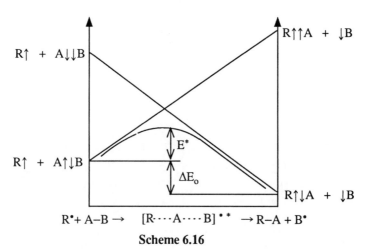

Scheme 6.16

State correlation diagrams

The lower the crossing point, the lower the transition state energy. The energy difference, ΔE_o, corresponds to the enthalpy of reaction. The difference between the ground state and the pseudo excited state is about 3/4 of the triplet-singlet difference, ΔE_{T-S}, which is closely related to the bond strength ($\Delta E_{T-S} = \gamma\ 2\ BDE$ with $\gamma < 1$). The height of the crossing point falls therefore with the strength of the bond broken or formed. Thus, this explains why the energy of an athermic reaction decreases with the X–R bond energy and why the rate constant for halogen abstraction by an alkyl or stannyl radical follows the order indicated in Table 6.8: I > Br > Cl.

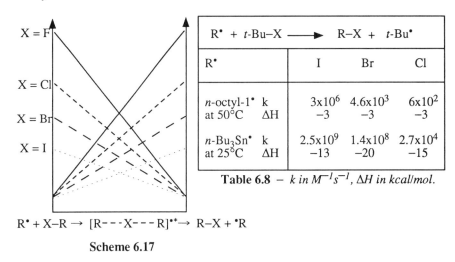

R• + t-Bu–X	→	R–X + t-Bu•	
R•	I	Br	Cl
n-octyl-1• k at 50°C ΔH	3×10^6 −3	4.6×10^3 −3	6×10^2 −3
n-Bu$_3$Sn• k at 25°C ΔH	2.5×10^9 −13	1.4×10^8 −20	2.7×10^4 −15

Table 6.8 – k in $M^{-1}s^{-1}$, ΔH in kcal/mol.

R• + X–R → [R- - -X- - -R]•* → R–X + •R

Scheme 6.17

6-5-2 Interpretation of the enthalpy effect

The reactivity-stability relationship, when it is confirmed, which is not always the case, can also be explained by the SCD. Consider the reaction in Scheme 6.18 where different types of alkyl radicals abstract hydrogen from a hydrocarbon, Σ–H. Since

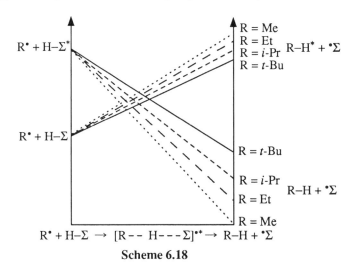

R• + H–Σ → [R- - H- - -Σ]•* → R–H + •Σ

Scheme 6.18

always the same the difference between its ground state and its pseudo fixed. Since the strength of the bond formed increases in the order: ...–H < Et–H < Me–H, the ground states of the products, P, lie in the order indicated in the Scheme, so as to respect the variation of the enthalpy of reaction. It can be shown by various equations that the pseudo excited states are energetically closer to each other than the ground states. The lines for the diabatics give, under these conditions, the following order for the activation barriers: Me• < Et• < i-Pr• < t-Bu•, and show, moreover, that the closer the transition state is to the reactants, the lower the barrier. These results are in good agreement with experiment. Scheme 6.18 is, of course, valid for the reverse reaction: abstraction of a tertiary, secondary or primary hydrogen by a given radical.

6-5-3 Interpretation of the polar effect

The reduction of the activation energy of a reaction by a polar effect is well explained by the introduction of a third diabatic curve into the SCD. This curve represents the correlation between the charge transfer states. Scheme 6.19 illustrates this for hydrogen abstraction from t-Bu–H by Cl• and Me• radicals.

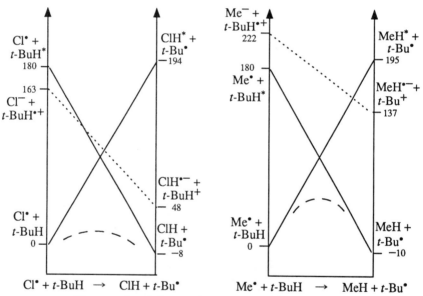

The energies of the pseudo-states are given in kcal/mol.

Scheme 6.19

These two reactions have identical ΔH; nevertheless, the activation energy of the first is much lower than that of the second. The diabatics of the ground state and the pseudo excited state of the reactants are identical for the two reactions and therefore lead to identical activation energies. The difference arises from the contribution of a third diabatic constructed from the charge transfer states (dashed line in Scheme 6.19). This diabatic is very low for abstraction by Cl•; consequently, it strongly stabilizes the transition state. This is not the case for transfer to the Me• radical, where the energy of the charge transfer diabatic is much higher.

7 – RADICAL KINETICS

Most radical reactions occur by a chain mechanism. Knowledge of the absolute rate constants of the elementary reactions is essential for explaining and predicting radical processes. Their values are obtained by physical methods (EPR, flash photolysis, etc.) but also by comparing rates with those of certain radicals called for this reason *radical clocks*. We shall give the basis of this methodology.

7-1 KINETICS OF CHAIN REACTIONS

To write the analytical kinetic equations we shall take an exchange reaction as an example (7.1):

$$AB + XY \longrightarrow AX + YB \quad (7.1)$$

This goes through a chain radical mechanism, with two transfer reactions (7.4) and (7.5). In the presence of a chemical initiator, P_2, we have:

initiation	$P_2 \xrightarrow{k_2} 2 P^\bullet$			(7.2)
	$P^\bullet + AB \xrightarrow{k_3} PB + A^\bullet$			(7.3)
propagation	$A^\bullet + XY \xrightarrow{k_4} AX + Y^\bullet$			(7.4)
	$Y^\bullet + AB \xrightarrow{k_5} YB + A^\bullet$			(7.5)
termination	$2 Y^\bullet \xrightarrow{k_6} Y_2$			(7.6)
	$2 A^\bullet \xrightarrow{k_7} A_2$			(7.7)
	$A^\bullet + Y^\bullet \xrightarrow{k_8} AY$			(7.8)

Scheme 7.1

The radicals can also disappear by disproportionation in the termination reactions.

7-1-1 Chain length, L

Each radical formed during initiation ($P^\bullet$, reaction (7.2)) induces a propagation chain which in each *cycle* produces a molecule of AX and of YB. This chain stops when one of the radicals ($A^\bullet$ or $Y^\bullet$) disappears in a termination reaction. The number of *cycles* of the propagation chain per initiator radical formed is called the *chain length*, L. This can be defined as the ratio of the rate of product formation to that of the appearance of initiator radicals.

$$L = (d[AX]/dt) / (-2\, d[P_2]/dt) \quad (7.9)$$

This chain length is determined experimentally by measuring the rates of appearance of the products and of disappearance of the initiator. Since the latter is equal to the rate of formation of the termination products one obtains an estimate of the chain length from the ratio of the amount of transfer products (AX or YB) to that of termination products at the end of the reaction. An efficient chain mechanism implies L > 100; L can reach very high values.

7-1-2 Steady state approximation

To set up the kinetic equations we adopt the *steady state approximation which stipulates that the rate of appearance of each radical is equal to the rate of its disappearance*. The corollary of this approximation is that the concentration of each radical species is constant. Hence, in our example:

$$d[P^\bullet]/dt = d[A^\bullet]/dt = d[Y^\bullet]/dt = 0 \qquad (7.10)$$

It is clear that at time $t = 0$ and $t = \infty$ the radical concentration is zero whereas during the reaction it has a certain value; consequently the concentration is time-dependent. Nevertheless, it is considered that when the reaction is between about 10% and 80% complete a practically stable regime is established and that the steady state approximation is valid.

7-1-3 Kinetic equations

The overall rates of appearance of AX and YB and disappearance of AB and XY are:

$$d[AX]/dt = -d[XY]/dt = k_4[A^\bullet][XY] \qquad (7.11)$$

$$d[YB]/dt = k_5[Y^\bullet][AB] \qquad (7.12)$$

$$-d[AB]/dt = k_5[Y^\bullet][AB] + k_3[P^\bullet][AB] \qquad (7.13)$$

Depending on the values of the concentrations, [XY] and [AB], one of the three termination reactions occurs. We shall now set up the overall rate expressions, which are different in the three cases, by writing for each radical:

$$\boxed{d[\text{Rad}^\bullet]/dt = \text{rate of appearance} - \text{rate of disappearance} = 0} \qquad (7.14)$$

Equation (7.14) is the formal expression of the steady state approximation.

Termination by reaction (7.6): $2\,Y^\bullet \rightarrow Y_2$
From the equations,

$$d[Y^\bullet]/dt = k_4[A^\bullet][XY] - k_5[Y^\bullet][AB] - 2k_6[Y^\bullet]^2 = 0 \qquad (7.15)$$

$$d[A^\bullet]/dt = k_5[Y^\bullet][AB] + k_3[P^\bullet][AB] - k_4[A^\bullet][XY] = 0 \qquad (7.16)$$

$$d[P^\bullet]/dt = 2\,k_2[P_2] - k_3[P^\bullet][AB] = 0 \qquad (7.17)$$

Kinetics of chain reactions 75

we obtain: $\quad 2 k_2[P_2] = 2 k_6[Y^\bullet]^2 \Rightarrow [Y^\bullet] = (k_2/k_6)^{1/2}[P_2]^{1/2}$ (7.18)

and: $k_4[A^\bullet][XY] = k_5[Y^\bullet][AB] + 2 k_2[P_2] = k_5(k_2/k_6)^{1/2}[P_2]^{1/2}[AB] + 2 k_2[P_2]$ (7.19)

hence: $d[AX]/dt = -d[XY]/dt = -d[AB]/dt = k_5(k_2/k_6)^{1/2}[P_2]^{1/2}[AB] + 2 k_2[P_2]$ (7.20)

and: $\quad d[YB]/dt = k_5(k_2/k_6)^{1/2}[P_2]^{1/2}[AB]$ (7.21)

If the chain length is great enough, $2 k_2[P_2]$ can be neglected. Consequently:

$$k_4[A^\bullet][XY]/dt \approx k_5[Y^\bullet][AB] = k_5(k_2/k_6)^{1/2}[P_2]^{1/2}[AB] \quad (7.22)$$

and: $\boxed{d[AX]/dt = -d[XY]/dt = -d[AB]/dt \approx d[YB]/dt = k_5(k_2/k_6)^{1/2}[P_2]^{1/2}[AB]}$ (7.23)

The overall rates of appearance and disappearance of the products are, therefore, equal to the rate of chain propagation. They depend on the initiator and AB concentrations, as well as on the rate constants of the initiation (k_2) and termination (k_6) steps and on that (k_5) of step (7.5). On the other hand, they are independent of the concentration of YX and of the rate constant (k_4) of step (7.4). Therefore, step (7.4) is not involved in the kinetic equation, contrary to step (7.5) which is called the *rate determining step*.

Termination by reaction $2 A^\bullet \rightarrow A_2$ (7.7)
From the equations:

$$d[Y^\bullet]/dt = k_4[A^\bullet][XY] - k_5[Y^\bullet][AB] = 0 \quad (7.24)$$

$$d[A^\bullet]/dt = k_5[Y^\bullet][AB] + k_3[P^\bullet][AB] - k_4[A^\bullet][XY] - 2 k_7[A^\bullet]^2 = 0 \quad (7.25)$$

$$d[P^\bullet]/dt = 2 k_2[P_2] - k_3[P^\bullet][AB] = 0 \quad (7.17)$$

we obtain: $\quad 2 k_2[P_2] = 2 k_7[A^\bullet]^2 \Rightarrow [A^\bullet] = (k_2/k_7)^{1/2}[P_2]^{1/2}$ (7.26)

and: $\quad k_4[A^\bullet][XY] = k_5[Y^\bullet][AB] = k_4(k_2/k_7)^{1/2}[P_2]^{1/2}[XY]$ (7.27)

hence: $\quad d[YB]/dt = d[AX]/dt = -d[XY]/dt = k_4(k_2/k_7)^{1/2}[P_2]^{1/2}[XY]$ (7.28)

and: $\boxed{-d[AB]/dt = k_4(k_2/k_7)^{1/2}[P_2]^{1/2}[XY] + 2 k_2 [P_2] \approx k_4(k_2/k_7)^{1/2}[P_2]^{1/2}[XY]}$ (7.29)

$2 k_2[P_2]$ can be neglected if the chain is long enough.

Therefore the overall rate depends only on the initiation, termination and step (7.4) reactions. Consequently, (7.4) is the *rate determining step*.

Termination by reaction $A^\bullet + Y^\bullet \rightarrow AY$ (7.8)
From the equations:

$$d[Y^\bullet]/dt = k_4[A^\bullet][XY] - k_5[Y^\bullet][AB] - k_8[A^\bullet][Y^\bullet] = 0 \quad (7.30)$$

$$d[A^\bullet]/dt = k_5[Y^\bullet][AB] + k_3[P^\bullet][AB] - k_4[A^\bullet][XY] - k_8[A^\bullet][Y^\bullet] = 0 \quad (7.31)$$

$$d[P^\bullet]/dt = 2 k_2[P_2] - k_3[P^\bullet][AB] = 0 \quad (7.17)$$

we obtain: $\quad 2 k_2[P_2] = 2 k_8[A^\bullet][Y^\bullet]$ (7.32) and: $\quad k_4[A^\bullet][XY] = k_5[Y^\bullet][AB] + k_2[P_2]$ (7.33)

When the chain is long enough $k_2[P_2]$ can be neglected. Hence:

$$k_4[A^\bullet][XY] \approx k_5[Y^\bullet][AB] \quad (7.34)$$

and, given that $k_2[P_2] = k_8[A^\bullet][Y^\bullet]$, we obtain:

$$[A^\bullet] \approx (k_2k_5/k_8k_4)^{1/2}[P_2]^{1/2}[AB]^{1/2}[XY]^{1/2} \quad (7.35)$$

hence: $\quad k_4[A^\bullet][XY]/dt \approx k_5[Y^\bullet][AB] \approx k_4(k_2k_5/k_8k_4)^{1/2}[P_2]^{1/2}[AB]^{1/2}[XY]^{1/2}$ (7.36)

and:

$$\boxed{\begin{array}{c} d[AX]/dt = -d[XY]/dt \approx -d[AB]/dt \approx d[YB]/dt = \\ k_4(k_2k_5/k_8k_4)^{1/2}[P_2]^{1/2}[AB]^{1/2}[XY]^{1/2} \end{array}} \quad (7.37)$$

In this kinetic equation all the elementary steps except (7.3) are involved.

Conclusions
From this formal example the following principles for chain mechanisms have been deduced:

- the rates of appearance and of disappearance of radicals in the initiation (7.2) and termination steps (7.6) and (7.8) are equal;
- the rates of each step of the propagation chain as well as the overall rates of appearance and disappearance of the products and the reactants are equal.

The reactants can appear in the kinetic equations with a fractional order. In this example the initiator, P_2, appears as the square root of its concentration, as do AB and XY when the reaction ends with (7.8). If either the termination reaction or the form of the kinetic equation is known it is possible to identify the overall rate determining step of the chain mechanism. By altering the value of the ratio [AB]/[XY] one can modify the experimental conditions so that one or another step of the chain determines the overall rate. Generally, the reactant with the lowest concentration imposes the rate determining step.

7-2 EXAMPLES OF KINETICS

7-2-1 Addition reaction

Radical addition reactions are very much used for forming C–C bonds and obtaining one-to-one addition products.

$$AB \quad + \quad C{=}C \quad \longrightarrow \quad ACCB \quad (7.38)$$

The chain is often initiated by a peroxide, P_2 (Scheme 7.2):

Examples of kinetics

$$
\begin{array}{lllll}
\textit{initiation} & P_2 & \xrightarrow{k_{39}} & 2\ P^\bullet & (7.39) \\
 & P^\bullet + AB & \xrightarrow{k_{40}} & PB + A^\bullet & (7.40) \\
\textit{propagation} & A^\bullet + C{=}C & \xrightarrow{k_{41}} & ACC^\bullet & (7.41) \\
 & ACC^\bullet + AB & \xrightarrow{k_{42}} & ACCB + A^\bullet & (7.42) \\
\textit{termination} & 2\ A^\bullet & \xrightarrow{k_{43}} & A_2 & (7.43) \\
 & 2\ ACC^\bullet & \xrightarrow{k_{44}} & (ACC)_2 & (7.44) \\
 & A^\bullet + ACC^\bullet & \xrightarrow{k_{45}} & ACCA & (7.45)
\end{array}
$$

Scheme 7.2

In order to avoid telomers

$$ACC^\bullet \xrightarrow{+\ C=C} ACCCC^\bullet \xrightarrow{+\ C=C} ACCCCCC^\bullet \xrightarrow{+\ C=C} \quad (7.46)$$

the reaction is performed with a deficiency of olefin, which makes step (7.41) rate determining. Termination then occurs by (7.43) since radicals $A^\bullet$ accumulate. Under these conditions the kinetic equation is:

$$d[ACCB]/dt = k_{41}(k_{39}/k_{43})^{1/2}[P_2]^{1/2}[C{=}C] \qquad (7.47)$$

The competition between reactions (7.42) and (7.46) has been much studied and has led to the determination of the relative or absolute rate constants for addition (k_{41}) of various radicals to olefins.

Obviously, in the absence of AB the initiator radical, $P^\bullet$, plays the part of $A^\bullet$ and there is extensive polymerization, which causes the radical $P(CC)_nCC^\bullet$ to disappear by dimerization or disproportionation. In the presence of AB the olefin concentration decreases during this reaction so that, after a few olefin molecules have been attached, transfer to $A(CC)_nCC^\bullet$ can become preferred to addition to a new olefin molecule. In this way one obtains a telomer, $A(CC)_{n+1}B$. It is easy to appreciate the economic interest of this type of reaction if it can be rationally controlled, since it offers a route to small functionalized molecules (by A and B) from AB and an olefin.

7-2-2 Thermal decomposition of peracids

In this case the peracid, RCO_3H, is at the same time the initiator, the substrate and the radical source. It has been established that there is a chain mechanism involving reactions (7.49) to (7.53). The formal expression for the rate of disappearance of the peracid is:

$$d[RCO_3H]/dt = k_{52}(k_{49}/k_{53})^{1/2}[RCO_3H]^{3/2} \qquad (7.48)$$

which corresponds to the order found experimentally. It confirms that (7.52) is the rate determining step.

$$
\begin{array}{lllll}
\textit{initiation} & RCO_3H & \xrightarrow{k_{49}} & RCO_2^\bullet + HO^\bullet & (7.49) \\
 & HO^\bullet + RCO_3H & \xrightarrow{k_{50}} & H_2O + RCO_3^\bullet & (7.50)
\end{array}
$$

propagation
$$RCO_2^{\bullet} \xrightarrow{k_{51}} R^{\bullet} + CO_2 \quad (7.51)$$
$$R^{\bullet} + RCO_3H \xrightarrow{k_{52}} ROH + RCO_2^{\bullet} \quad (7.52)$$

termination
$$2\ R^{\bullet} \xrightarrow{k_{53}} R_2 \text{ or } RH + R(-H) \quad (7.53)$$

Scheme 7.3

7-2-3 Photochemical chlorination by *t*-BuO–Cl

Radical chlorination of a hydrocarbon by *tert*-butyl hypochlorite proceeds by the chain mechanism indicated in Scheme 7.4.

initiation
$$t\text{-BuO–Cl} \xrightarrow{k_{54}} t\text{-BuO}^{\bullet} + Cl^{\bullet} \quad (7.54)$$
$$Cl^{\bullet} + R\text{–H} \xrightarrow{k_{55}} H\text{–Cl} + R^{\bullet} \quad (7.55)$$

propagation
$$t\text{-BuO}^{\bullet} + R\text{–H} \xrightarrow{k_{56}} t\text{-BuO–H} + R^{\bullet} \quad (7.56)$$
$$R^{\bullet} + t\text{-BuO–Cl} \xrightarrow{k_{57}} R\text{–Cl} + t\text{-BuO}^{\bullet} \quad (7.57)$$

termination
$$2\ R^{\bullet} \xrightarrow{k_{58}} R_2 \text{ or } RH + R(-H) \quad (7.58)$$
$$2\ t\text{-BuO}^{\bullet} \xrightarrow{k_{60}} (t\text{-BuO})_2 \quad (7.59)$$
$$t\text{-BuO}^{\bullet} + R^{\bullet} \xrightarrow{k_{61}} t\text{-BuOR} \quad (7.60)$$

Scheme 7.4

The initiation rate is given by equation (7.61) where I is the incident light intensity.

$$-d[t\text{-BuO–Cl}]/dt = k_{54}[t\text{-BuO–Cl}] = k'_{54}\ I\ [t\text{-BuO–Cl}] \quad (7.61)$$

The kinetic law depends on RH. For toluene (R = $PhCH_2$) one finds experimentally equation (7.62):

$$-d[RH]/dt = k\ I^{0.55}\ [t\text{-BuO–Cl}]^{0.65}[RH]^{0.92} \quad (7.62)$$

where k is a term which combines all the rate constants for the reactions involved, including the efficiency of the initiation reaction (7.54). The corresponding kinetic equation is:

$$-d[RH]/dt \approx k_{56}(k'_{54}/k_{59})^{0.5}\ I^{0.5}\ [t\text{-BuO–Cl}]^{0.5}\ [RH] \quad (7.63)$$

This result shows that the partial orders are not strictly integer or half-integer and that the termination step is (7.59).

In the case of chloroform (RH = $CHCl_3$) one finds experimentally equation (7.64)

$$-d[t\text{-BuO–Cl}]/dt = k\ I^{0.5}\ [t\text{-BuO–Cl}]^{1.3} \quad (7.64)$$

which corresponds to the formal kinetic equation (7.65), indicating that the termination step is (7.58).

$$-d[t\text{-BuO–Cl}]/dt \approx k_{57}(k'_{54}/k_{58})^{0.5} I^{0.5} [t\text{-BuO–Cl}]^{1.5} \qquad (7.65)$$

7-3 COMPETITIVE SYSTEMS: RATE CONSTANT DETERMINATION

In order to understand elementary radical reactions and to exploit them subsequently it is important to determine rate constants. The absolute rate constants of certain reference reactions have been measured by physicochemical experiments using spectroscopic methods (EPR, flash photolysis, etc.). The absolute rate constants of other reactions can be calculated from these reference constants by using competitive reaction systems.

7-3-1 Competitive reaction of a radical with two substrates

Let us consider the chlorination of a mixture of two hydrocarbons, R^1H and R^2H, by t-BuOCl. This involves two propagation chains:

$$\text{chain } R^1\text{–H} \quad \begin{vmatrix} t\text{-BuO}^\bullet + R^1\text{–H} \xrightarrow{k_{66}} t\text{-BuO–H} + R^{1\bullet} & (7.66) \\ R^{1\bullet} + t\text{-BuO–Cl} \xrightarrow{k_{67}} R^1\text{–Cl} + t\text{-BuO}^\bullet & (7.67) \end{vmatrix}$$

$$\text{chain } R^2\text{–H} \quad \begin{vmatrix} t\text{-BuO}^\bullet + R^2\text{–H} \xrightarrow{k_{68}} t\text{-BuO–H} + R^{2\bullet} & (7.68) \\ R^{2\bullet} + t\text{-BuO–Cl} \xrightarrow{k_{69}} R^2\text{–Cl} + t\text{-BuO}^\bullet & (7.69) \end{vmatrix}$$

Application of the steady state approximation to $R^{1\bullet}$ and $R^{2\bullet}$ therefore gives equations (7.70) and (7.71),

$$k_{66}[R^1H][t\text{-BuO}^\bullet] = k_{67}[R^{1\bullet}][t\text{-BuO–Cl}] \qquad (7.70)$$

$$k_{68}[R^2H][t\text{-BuO}^\bullet] = k_{69}[R^{2\bullet}][t\text{-BuO–Cl}] \qquad (7.71)$$

hence the rates of formation of R^1Cl and R^2Cl are (7.72) and (7.73), from which follows (7.74).

$$d[R^1Cl]/dt = k_{67}[R^{1\bullet}][t\text{-BuO–Cl}] = k_{66}[R^1H][t\text{-BuO}^\bullet] \qquad (7.72)$$

$$d[R^2Cl]/dt = k_{69}[R^{2\bullet}][t\text{-BuO–Cl}] = k_{68}[R^2H][t\text{-BuO}^\bullet] \qquad (7.73)$$

hence: $\qquad d[R^1Cl]/d[R^2Cl] = k_{66}/k_{68}[R^1H]/[R^2H] \qquad (7.74)$

Generally one uses a large excess of hydrocarbon with respect to hypochlorite. The concentrations of R^1Cl and R^2Cl are measured at the beginning of the reaction, that is, for less than 20% reaction. Under these conditions the concentrations of R^1H and R^2H are practically constant and the result of integration reduces to equation (7.75).

$$k_{66}/k_{68} \approx [R^1Cl][R^2H] / [R^2Cl][R^1H] \qquad (7.75)$$

In this way the relative value of k_{66} with respect to k_{68} and, if k_{68} is known, the absolute value of k_{66} can be obtained.

7-3-2 Use of radical clocks

The exact rate constants for the rearrangement of certain radicals are known and can be used as references in a competitive system; hence their name, *radical clocks*. The cyclization of the 5-hexenyl radical has been studied in detail and is widely used to determine the absolute rate constants for hydrogen and halogen transfer (Scheme 7.5).

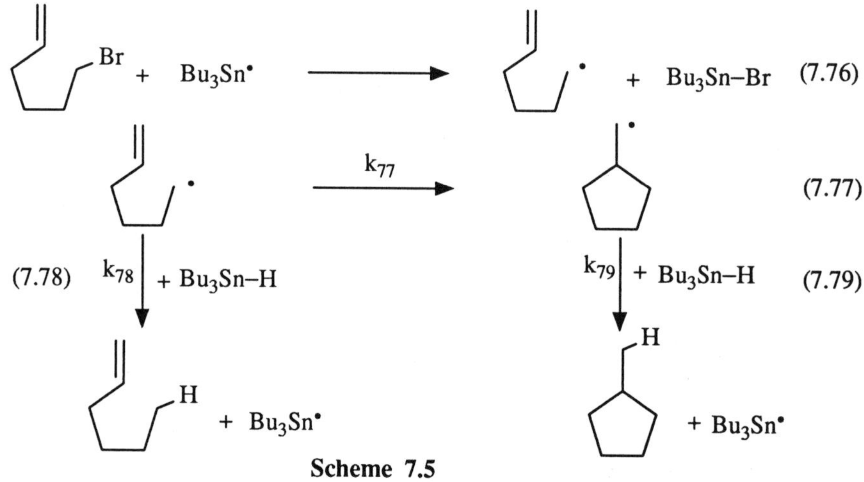

Scheme 7.5

By making the steady state approximation for the cyclized radical and writing the rate of appearance of each of the products, hexene and methylcyclopentane (Scheme 7.5), one obtains the expression:

$$k_{77}/k_{78} \approx [Bu_3SnH][methylcyclopentane]/ [hexene] \qquad (7.80)$$

Since k_{77} is known exactly (2.5 x 10^5 s^{-1} at 25°C) k_{78} can be obtained. This expression is only valid, as in the previous case, provided that the radicals undergo only the reactions indicated, that the products do not subsequently evolve and that the product concentrations are determined at the beginning of the reaction. Moreover, the effect of the temperature must not be neglected; for example, k_{77} = 1.7 x 10^3 s^{-1} at −40°C and 5.3 x 10^5 s^{-1} at 80°C.

The combination of various hydrogen atom donors (Table 7.1 and Chap. 11) and radical clocks (Chaps 12, 13 and Table 13.3) provides a range of reference rate constants for determining the absolute rate constants of a given reaction.

H or X donor	BDE	radical				
		primary	tertiary	benzyl	phenyl	t-BuO$^\bullet$
PhS–H	80	9.2×10^7	1.5×10^8	3×10^5	1.9×10^9	
THF	92	6×10^3	2×10^3		4.8×10^6	8.3×10^6
Bu$_3$Sn–H	74	2.4×10^6	1.8×10^6	3.6×10^4	5.9×10^8	
Bu$_3$Ge–H	82	1×10^5			2.6×10^8	
Et$_3$Si–H	90	7×10^3	3×10^3			5.7×10^6
(cy-C$_6$H$_{11}$)$_2$P–H	74	1×10^6		2.5×10^3		
t-Bu–I	54	3×10^6				
t-Bu–Br	70	4.6×10^3			1×10^6	
t-Bu–Cl	81	6×10^2				
Br–CCl$_3$	56	2×10^8	2.6×10^8		1.4×10^9	
Cl–CCl$_3$	71	1×10^5	4.9×10^4		6×10^6	

Table 7.1 - *Hydrogen and halogen atom transfer reactions. Rate constants at room temperature are in $M^{-1}s^{-1}$ and the BDE(R–H or R–X) in kcal/mol.*

8 – RADICALS CENTERED ON AN ATOM OTHER THAN CARBON

Many radical species which have the unpaired electron centered on an atom other than carbon (halogen, heteroatom, metal atom, etc.) play an important role in organic chemistry. The presentation of this chapter will follow the various groups of the Periodic Table: 14 (Si, Ge, Sn), 15 (N, P), 16 (O, S) and 17 (halogens). The structures of these radicals are described in Chapter 3.

8-1 Si, Ge AND Sn-CENTERED RADICALS

Radicals $R_3M^{\bullet}$ ($R_3Si^{\bullet}$, $R_3Ge^{\bullet}$ and $R_3Sn^{\bullet}$) are significantly different from the carbon-centered radicals $R_3C^{\bullet}$. This is a result of the fundamental characteristics of these elements:

- the size;
- the less pronounced aptitude to form double bonds, M=X;
- an electropositivity greater than that of carbon;
- bond dissociation energies, BDE(M–H), lower but BDE(M–Hal) and BDE(M–O) greater than those observed with carbon (Table 8.1).

X	H	Cl	Br	I	Me	OH	OEt	NH_2
Me_3C-X	95	84	70	54	86	95	82	85
Me_3Si-X	90	112	96	77	90	128	111	100
Me_3Ge-X	82	116	104	63	76		107	
Me_3Sn-X	74	94	83	69	65	110	84	

Table 8.1 – *Bond dissociation energies $BDE(Me_3M-X)$ in kcal/mol.*

8-1-1 Production

These radicals are produced:

- by hydrogen transfer from a hydride to a radical, $R^{\bullet}$. This reaction becomes easier, because the values of BDE(M–H) decrease, on going from silicon to germanium to tin (Table 8.1);

$$R_3M-H + R^{\bullet} \longrightarrow R_3M^{\bullet} + R-H \qquad (8.1)$$

- by photolysis of hydrides (8.2) or dimers (8.3).

$$R_3M-H \xrightarrow{h\nu} R_3M^\bullet \qquad (8.2)$$
$$R_3M-MR_3 \xrightarrow{h\nu} 2\ R_3M^\bullet \qquad (8.3)$$

8-1-2 Reactions

Hydrogen transfer

Analysis of the bond energies (Table 8.1) suggests that hydrogen transfer from an organic compound to $R_3M^\bullet$ (the reverse of reaction (8.1)) is endothermic, therefore difficult. However, such a reaction is observed with $Me_3Si^\bullet$ and toluene, because the benzylic C–H bond is weak.

Halogen transfer

These reactions have been much studied, in particular those of radicals $R_3Sn^\bullet$. They can be used to reduce halides R–X to R–H, via the carbon radical $R^\bullet$, by the following chain mechanism:

$$\begin{array}{lll}
\text{initiation} & R_3M-H \longrightarrow R_3M^\bullet & (8.4) \\
\text{chain} \left| \begin{array}{l} R_3M^\bullet + R-X \longrightarrow R_3M-X + R^\bullet \\ R^\bullet + R_3M-H \longrightarrow R_3M^\bullet + R-H \end{array} \right. & & \begin{array}{l}(8.5)\\(8.6)\end{array}
\end{array}$$

The rate of reaction (8.5) depends on the degree of substitution of the carbon radical formed ($R^\bullet_{tert} > R^\bullet_{sec} > R^\bullet_{prim}$) and follows the order: RI > RBr > RCl > RF. Thus, $Bu_3Sn^\bullet$ reacts 2000 times faster with hexyl bromide than with pentyl chloride. This makes it possible to perform selective reactions of the type:

$$\text{cyclohexane-CClBr} \xrightarrow{Bu_3SnH/0°C} \text{cyclohexane-CClH} \quad 97\% \qquad (8.7)$$

Polyhalo derivatives react more rapidly than dihalo derivatives and these more rapidly than monohalo compounds, giving selective reduction;

$$PhCCl_3 \xrightarrow{Bu_3SnH/25°C} PhCHCl_2\ 88\% \qquad (8.8)$$

Direct halogen transfers from aromatic halides are possible:

$$R_3M^\bullet + Ar-X \longrightarrow R_3M-X + Ar^\bullet \qquad (8.9)$$

The determination of Hammett's ρ for halogen transfer from substituted benzyl halides shows that $R_3M^\bullet$ radicals behave as nucleophiles, which is consistent with the electropositivity of M.

Reaction (8.6), which is very fast due to its exothermicity, is able to trap radical $R^\bullet$ before an eventual rearrangement (Scheme 8.1). One can also obtain information about the structure of radical intermediates and their isomerization rates. Use of R_3Si-H, R_3Ge-H and R_3Sn-H gives a range of rates for reaction (8.6) which can be adapted to various situations.

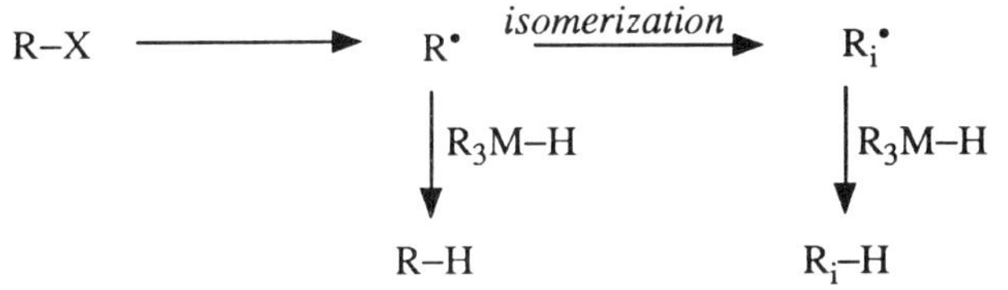

Si, Ge and Sn-centered radicals

$$R-X \longrightarrow R^{\bullet} \xrightarrow{\text{isomerization}} R_i^{\bullet} \quad (8.10)$$

$$\downarrow R_3M-H \qquad\qquad \downarrow R_3M-H$$

$$R-H \qquad\qquad\qquad R_i-H$$

Scheme 8.1

When the overall reaction (8.11) is very exothermic, as in the case of Ge and Sn hydrides with bromides and iodides, an initiator is not necessary. It is likely that initiation occurs by a *molecule-molecule* reaction (Chap. 9). Both steps of the chain, (8.5) and (8.6), are exothermic (except with fluorides), which makes the process very efficient.

$$R_3M-H + R-X \longrightarrow R_3M-X + R-H \quad (8.11)$$

Addition to carbon-carbon double and triple bonds

Addition of the hydrides, R_3Si-H, R_3Ge-H and R_3Sn-H, to C=C double bonds proceeds by a radical chain mechanism:

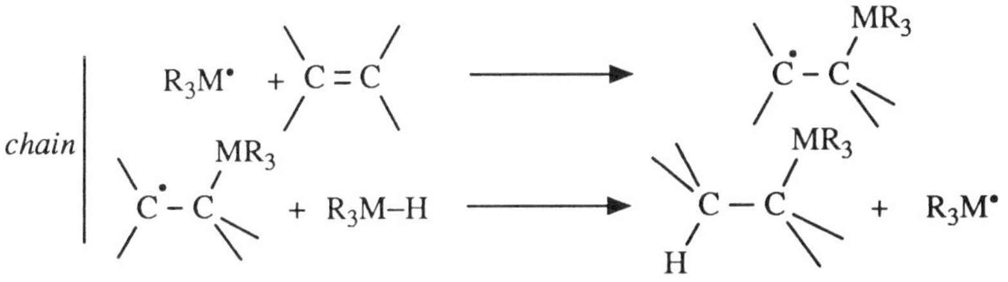

$$\text{chain} \begin{cases} R_3M^{\bullet} + \text{C=C} \longrightarrow \text{C}-\text{C}(MR_3) & (8.12) \\ \text{C}^{\bullet}-\text{C}(MR_3) + R_3M-H \longrightarrow \text{C}(H)-\text{C}(MR_3) + R_3M^{\bullet} & (8.13) \end{cases}$$

In agreement with the dissociation energies given in Table 8.1, reaction (8.12) is irreversible with $R_3Si^{\bullet}$ but reversible with $R_3Sn^{\bullet}$ even at room temperature. However, the exothermicity of reaction (8.13) makes it easy to obtain the addition product. Many organometallic compounds have been synthesized by this procedure.

Triple bonds are as reactive as double bonds in this addition process. Intramolecular additions (cyclizations) also occur but the regioselectivity can be different from that observed with carbon radicals. Thus, with silicon the principal reaction is the formation of a 6-membered ring.

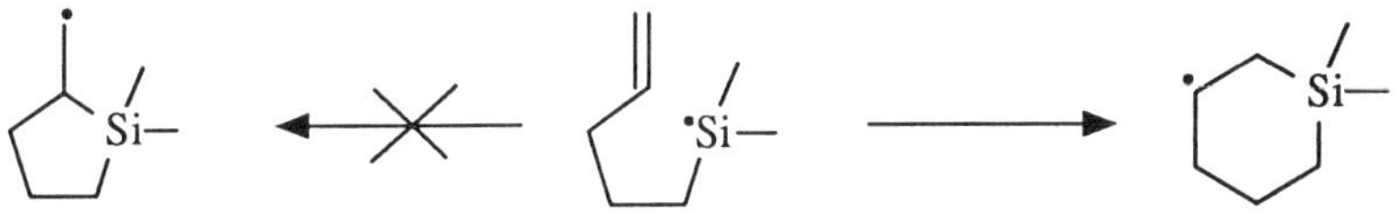

$$\text{[5-membered ring with Si]} \xleftarrow{\times} \text{[open-chain Si radical]} \longrightarrow \text{[6-membered ring with Si]} \quad (8.14)$$

Addition to C=heteroatom double bonds

Radicals $R_3M^{\bullet}$ add to the heteroatom. This regioselectivity is explained by the fact that the M–heteroatom bond is stronger than the M–C bond (Table 8.1). Such reactions are observed with various functions or groups.

Carbonyl group,

$$\text{chain} \begin{vmatrix} R_3M^\bullet + O{=}C{\diagup\atop\diagdown} \longrightarrow R_3M{-}O{-}\overset{\bullet}{C}{\diagup\atop\diagdown} & (8.15) \\ R_3M{-}O{-}\overset{\bullet}{C}{\diagup\atop\diagdown} + R_3M{-}H \longrightarrow R_3M{-}O{-}C{\overset{H}{\diagup}\atop\diagdown} + R_3M^\bullet & (8.16) \end{vmatrix}$$

Imino group,

$$t\text{-}Bu_2C{=}NH + R_3Si^\bullet \longrightarrow t\text{-}Bu_2\overset{\bullet}{C}{-}NH{-}SiR_3 \quad (8.17)$$

Pyridine,

(8.18)

Thiocarbonyl group. This reaction is very easy and in the case of xanthates is a source of carbon radicals.

$$R{-}OH \longrightarrow R{-}O{-}\overset{S}{\overset{\|}{C}}{-}X \xrightarrow{Bu_3Sn^\bullet} R{-}O{-}\overset{S{-}SnBu_3}{\overset{|}{\underset{\bullet}{C}}}{-}X \xrightarrow[\beta\text{-}elimination]{} R^\bullet \xrightarrow{Bu_3SnH} R{-}H \quad (8.19)$$

Nitro group. Since the M–O bond is stronger than the M–N bond nitro compounds can be reduced:

$$RNO_2 \xrightarrow{Bu_3Sn^\bullet} R{-}\underset{\underset{O^\bullet}{|}}{N}{-}O{-}SnBu_3 \xrightarrow{\beta\text{-}elimination} R^\bullet \xrightarrow{Bu_3SnH} R{-}H \quad (8.20)$$

Radical-radical reactions

The reaction between two radicals $R_3M^\bullet$ leads to dimerization and rarely to disproportionation, whereas both are commonly found with carbon radicals. This difference can be attributed to the fact that it is difficult to form M=C double bonds.

$$2\ R_3M^\bullet \begin{array}{c} \longrightarrow R_3M{-}MR_3 \quad (8.21) \\ \xcancel{\longrightarrow} R_3M{-}H + R{-}CH{=}MR_2 \quad (8.22) \end{array}$$

8-2 NITROGEN-CENTERED RADICALS

These radicals are important in organic and bio-organic chemistry. The radical chemistry of nitrogen compounds is controlled by the weakness of N–Cl and N–NO bonds as compared with N–H bonds. Moreover, the BDE values for $R_2N–H$, $R_2NH^+–H$ and $Me–NH_2$ and $CH_3–NH_3^+$ (91, 96, 85 and 112 kcal/mol, respectively) explain in part why nitrogen-centered radicals are more reactive when protonated.

8-2-1 Production

Nitrogen radicals, neutral or protonated, are generally obtained in three ways.

Transfer reaction
This is general for N–Cl and N–Br bonds. Hydrogen transfer, however, is rarely used since N–H bonds are strong (91 kcal/mol).

$$\ce{>N-X} \text{ or } \ce{>NH^+-X} + R^{\bullet} \longrightarrow \ce{>N^{\bullet}} \text{ or } \ce{>NH^{+\bullet}} + RX \quad (8.23)$$

Unimolecular homolysis
Nitrogen radicals are also obtained by homolysis of N–Cl or N–NO bonds by UV irradiation (reactions (8.24) and (8.25)):

$$\ce{>N-Cl} \text{ or } \ce{>NH^+-Cl} \xrightarrow{h\nu} \ce{>N^{\bullet}} \text{ or } \ce{>NH^{+\bullet}} + Cl^{\bullet} \quad (8.24)$$

$$\ce{>N-NO} \xrightarrow{h\nu} \ce{>N^{\bullet}} + {}^{\bullet}NO \quad (8.25)$$

and by thermal or photochemical decomposition of tetrazenes (8.26).

$$\begin{array}{c} \diagup \\ -N \\ \diagdown \\ N=N \\ \diagdown \\ N- \\ \diagup \end{array} \xrightarrow{\Delta \text{ or } h\nu} N_2 + 2 \ce{>N^{\bullet}} \quad (8.26)$$

Conversely, thermal homolysis of the N–N bond (BDE(MeNH–NHMe) = 65 kcal/mol) is too difficult except for certain fluoro compounds.

$$F_2N-NF_2 \xrightleftharpoons{\Delta H = 21 \text{ kcal/mol}} 2 \; {}^{\bullet}NF_2 \quad (8.27)$$

Oxido-reduction
The reduction of N-halogen compounds by metal salts is a very common source of nitrogen radicals.

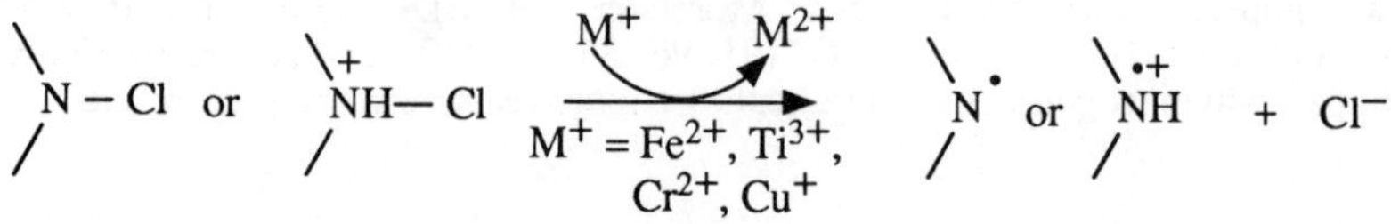

$$\diagup\!\!\!\!\!\diagdown N-Cl \text{ or } \diagup\!\!\!\!\!\diagdown \overset{+}{N}H-Cl \xrightarrow[M^+ = Fe^{2+}, Ti^{3+}, Cr^{2+}, Cu^+]{M^+ \quad M^{2+}} \diagup\!\!\!\!\!\diagdown N^\bullet \text{ or } \diagup\!\!\!\!\!\diagdown \overset{\bullet+}{N}H + Cl^- \quad (8.28)$$

Nitroxide (or aminoxyl) and iminoxyl radicals are obtained, from secondary amines or hydro- xylamines, by oxidation with hydrogen peroxide (in the presence of tungsten, molyb- denum or vanadium compounds) or by peracids.

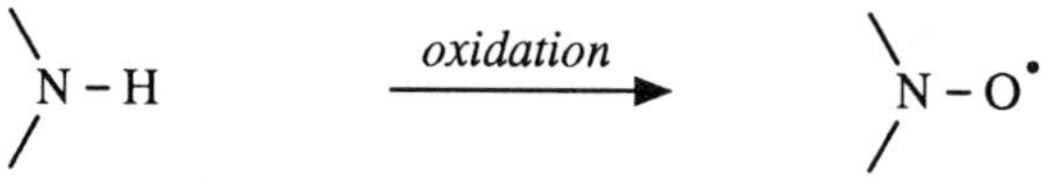

$$\diagup\!\!\!\!\!\diagdown N-H \xrightarrow{\text{oxidation}} \diagup\!\!\!\!\!\diagdown N-O^\bullet \quad (8.29)$$

8-2-2 Reactions

Aminyl radicals

Non-protonated dialkylaminyl radicals are unreactive. This low reactivity, which is surprising in view of the strength of the N–H and N–C bonds, can be explained by the electron repulsion of the nitrogen lone pair. They behave as electrophiles in the hydrogen transfer reactions of substituted toluenes and as nucleophiles in addition to α-methylstyrenes, but in both cases the reactions are sluggish. Nevertheless, cyclization reactions (8.30) have been studied though little used, since protonated aminyl radicals are more reactive.

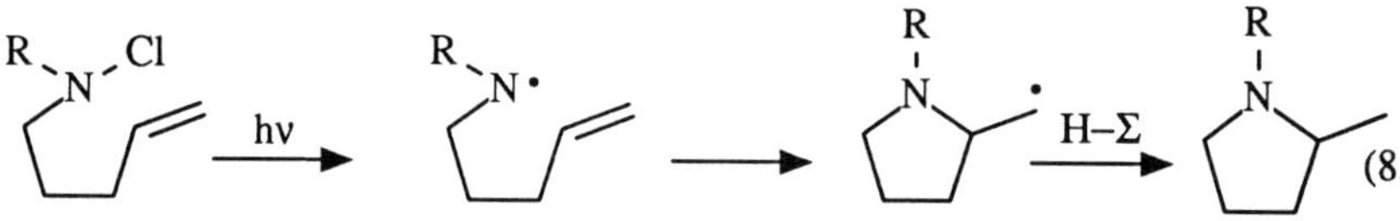

$$\text{(scheme 8.30: R-N(Cl)-CH}_2\text{CH}_2\text{CH=CH}_2 \xrightarrow{h\nu} \text{R-N}^\bullet \longrightarrow \text{pyrrolidinyl radical} \xrightarrow{H-\Sigma} \text{N-R pyrrolidine)} \quad (8.30)$$

On the other hand, dimerization (8.31) and disproportionation (8.32) are common.

$$2\,(CH_3CH_2)_2N^\bullet \longrightarrow (CH_3CH_2)_2N-N(CH_2CH_3)_2 \quad (8.31)$$

$$\longrightarrow CH_3CH_2-N=CHCH_3 + (CH_3CH_2)_2NH \quad (8.32)$$

Modification of the electron density at the nitrogen atom by protonation, by complexation with metals or by an electron-attracting substituent, leads to a remarkable increase in the reactivity, and makes the radical electrophilic.

Protonated aminyl radicals

Hydrogen transfer reactions of protonated aminyl radicals owe their efficiency to the strength of the N^+–H bond (BDE(Me_2NH^+–H) = 96 kcal/mol); because of their marked electrophilicity they are highly regioselective. This explains, amongst other things, why the ω–1 hydrogens in long aliphatic chains bearing an electron-attracting group react with the aminium radical (ω–1 effect).

$$R_2\overset{+\bullet}{N}-H + CH_3\underset{H}{\overset{|}{C}H}(CH_2)_nZ \xrightarrow{Z = Cl, CO_2CH_3} R_2\overset{+}{N}H_2 + CH_3\overset{\bullet}{C}H(CH_2)_nZ \quad (8.33)$$

$$CH_3\overset{\bullet}{C}H(CH_2)_nZ + R_2\overset{+}{N}HCl \longrightarrow CH_3\underset{Cl}{\overset{|}{C}H}(CH_2)_nZ + R_2\overset{+\bullet}{N}-H \quad (8.34)$$

When R is a long aliphatic chain an intramolecular 1,5 hydrogen transfer leads to a δ-chloroamine (Hofmann-Löffler-Freytag reaction) and then, by means of a base, to a pyrrolidine derivative:

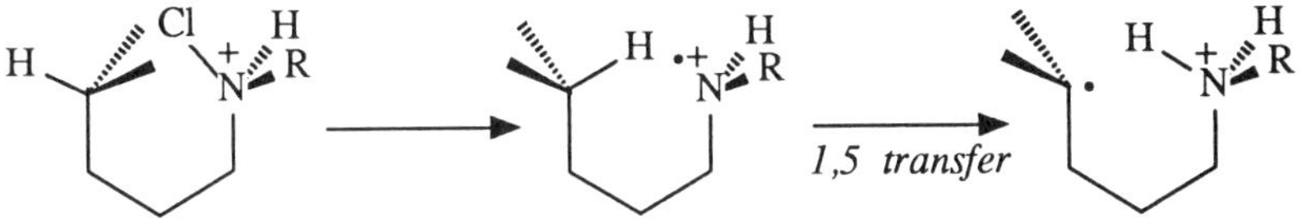

(8.35)

N-chloroamine   1,5 transfer

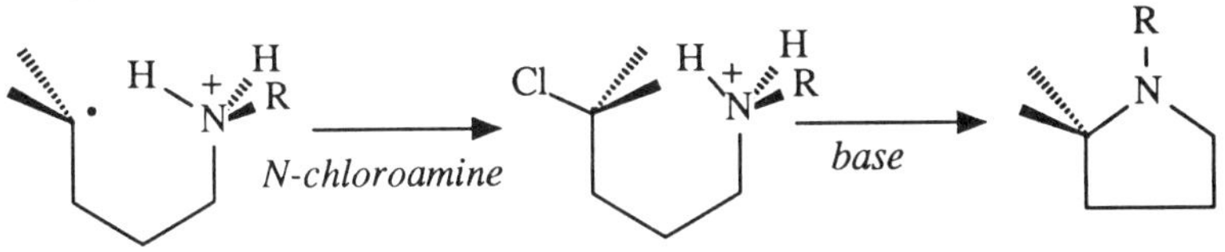

(8.36)

                N-chloroamine   base

β-Chloroamines are obtained by inter- or intramolecular addition to double bonds (8.37) and (8.38). Aromatic substrates can be aminated by aromatic substitution (Chaps 14 and 22).

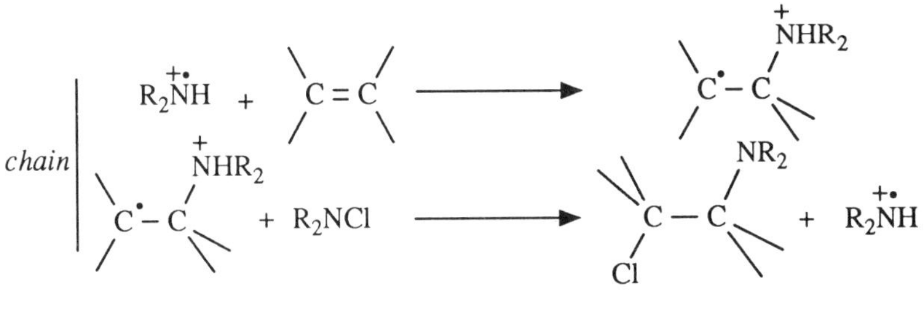

(8.37)

chain

(8.38)

Complexed aminyl radicals

Dialkylaminyl radicals complexed by metals are also electrophilic and perform the same reactions as protonated aminyl radicals. Moreover, unlike these, their

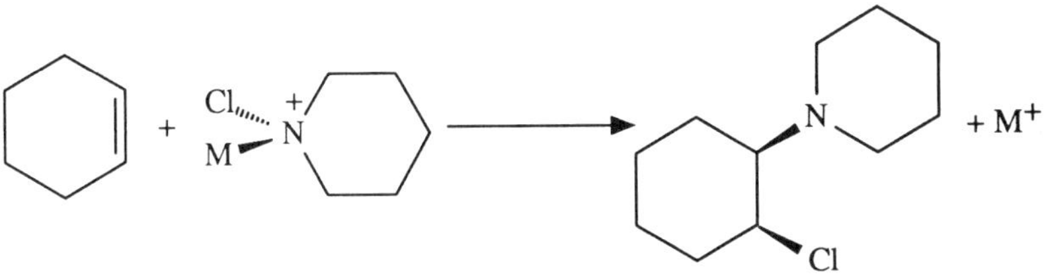

$$+ M^+ \quad (8.39)$$

addition reactions are stereoselective. Thus, with cyclohexene the major product is the *cis* isomer.

Amidyl radicals

The electrophilicity and the reactivity of amidyl radicals are intermediate between those of aminyl and protonated aminyl radicals. They are delocalized π-allyl radicals which only react at nitrogen. They add to olefins slowly and show a preference for allylic hydrogen transfers. Intramolecular addition leads to lactams.

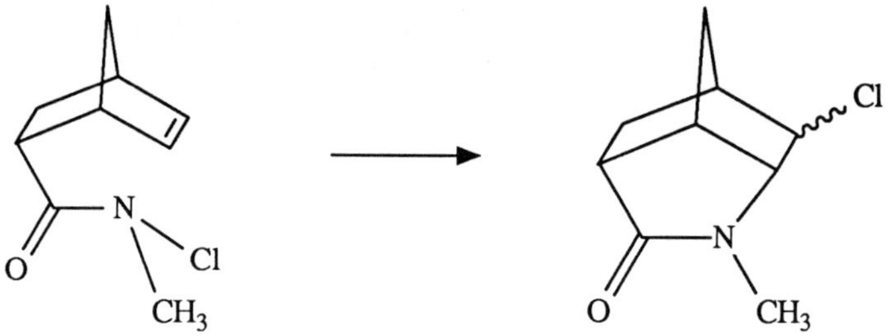

(8.40)

Amidyl radicals bearing an aliphatic chain undergo a 1,5 intramolecular hydrogen transfer. δ-Chlorocarboxamides can be used to prepare both lactams and lactones.

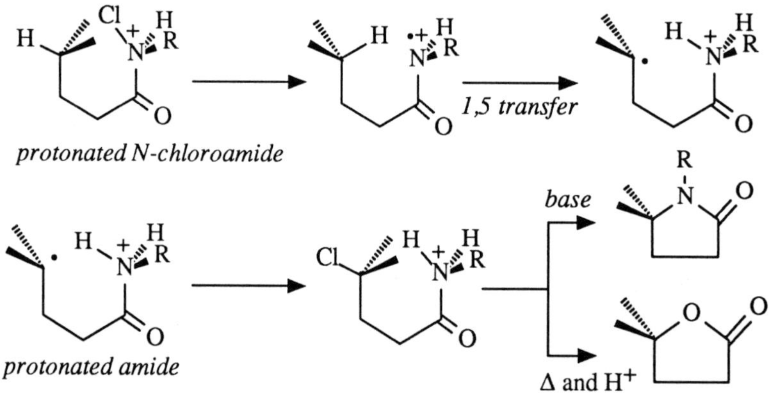

(8.41)

(8.42)

(8.43)

Nitroxide radicals

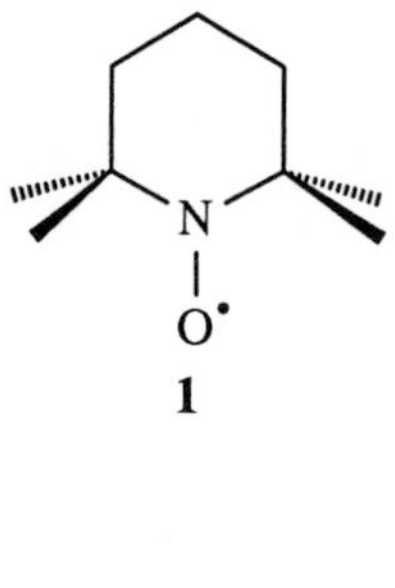

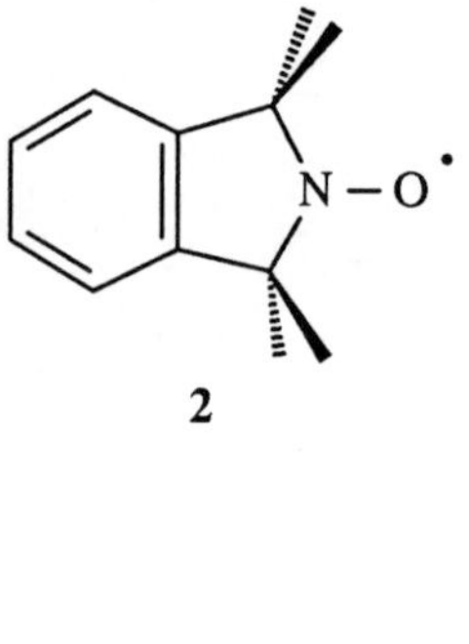

Nitroxide radicals with no α hydrogen, such as the 2,2,6,6-tetramethyl-1-piperidinyloxyl free radical (TEMPO), **1**, and the 1,1,3,3-tetramethyl-2-isoindolinyloxyl free radical, **2**, which are very stable, are used as traps for short-lived radicals since reaction (8.44) is very rapid (10^8 to 10^9 $M^{-1}s^{-1}$). Radical $R^\bullet$ can be identified from the alkoxylamine obtained, which is stable and can be isolated. They are also used as *spin markers* to study the structure of complex biological molecules.

$$R^\bullet \ + \ \underset{}{\diagup\!\!\!N\text{-}O^\bullet\diagdown} \ \longrightarrow \ \underset{alkoxylamine}{\diagup\!\!\!N\text{-}OR\diagdown} \qquad (8.44)$$

8-3 PHOSPHORUS-CENTERED RADICALS

The multiple valence states of phosphorus explain why there are potentially so many types of phosphorus-centered radicals. Unlike nitrogen compounds, phosphorus compounds are characterized by weak P–H bonds; for this reason phosphorus-centered radicals are particularly stable.

8-3-1 Production

The various types of phosphorus radicals are most often produced from P–H or P–Hal bonds.

Transfer

$$Z_nP\text{–}X \ + \ R^\bullet \ \xrightarrow[n = 2 \text{ or } 4]{X = H, Hal} \ Z_nP^\bullet \ + \ R\text{–}X \qquad (8.45)$$

These reactions are exothermic. For example, the ΔH of reaction (8.46) is -5, -1 and -1.5 kcal/mol for $X = H$, Cl and Br, respectively.

$$PX_3 \ + \ Et^\bullet \ \xrightarrow{\Delta H < 0} \ ^\bullet PX_2 \ + \ Et\text{–}X \qquad (8.46)$$

Photolysis

$$Z_nP\text{–}X \ \xrightarrow[X = H, Cl, n = 2 \text{ or } 4]{h\nu} \ Z_nP^\bullet \ + \ X^\bullet \qquad (8.47)$$

Addition

Phosphoranyl radicals are also obtained by addition of a radical to a trivalent phosphorus compound.

$$R^\bullet \ + \ PZ_3 \ \longrightarrow \ R\dot{P}Z_3 \qquad (8.48)$$

This process can be very rapid. Thus, for reaction (8.49), k_1 is 9×10^8 $M^{-1}s^{-1}$, much greater than k_{-1}.

$$t\text{-BuO}^{\bullet} + P(OEt)_3 \underset{k_{-1}}{\overset{k_1}{\rightleftharpoons}} t\text{-BuOP}^{\bullet}(OEt)_3 \qquad (8.49)$$

Reduction
The reduction of a phosphonium salt leads to a phosphoranyl radical.

$$R_4P^+ \xrightarrow{+e^-} R_4P^{\bullet} \qquad (8.50)$$

8-3-2 Reactions

The chemistry of these radicals is controlled by the weakness of the P–H bond (BDE(P–H) ≈ 74 kcal/mol). This implies that:

– phosphines are good hydrogen donors able to function as inhibitors in chain reactions;
– phosphorus radicals only abstract weakly bonded hydrogens (allylic or benzylic C–H);
– it is easy to chlorinate phosphines or phosphine oxides by a radical process using CCl_4.

$$Ph_2P(=O)\text{-H} + {}^{\bullet}CCl_3 \longrightarrow Ph_2P^{\bullet}(=O) + HCCl_3 \qquad (8.51)$$

$$Ph_2P^{\bullet}(=O) + CCl_4 \longrightarrow Ph_2P(=O)\text{-Cl} + {}^{\bullet}CCl_3 \qquad (8.52)$$

Addition reactions are used to prepare various types of phosphorus compounds. The addition step is only slightly exothermic, which suggests that the reaction is reversible. However, the second step is strongly exothermic.

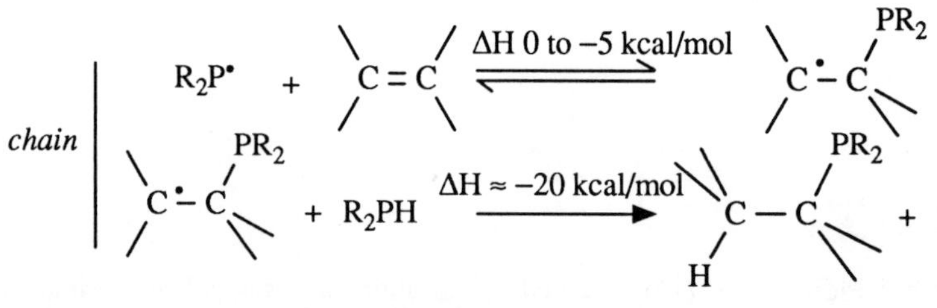

$$R_2P^{\bullet} + \text{C=C} \underset{}{\overset{\Delta H\ 0\ \text{to}\ -5\ \text{kcal/mol}}{\rightleftharpoons}} \text{C}^{\bullet}\text{-C(PR}_2) \qquad (8.53)$$

chain

$$\text{C}^{\bullet}\text{-C(PR}_2) + R_2PH \xrightarrow{\Delta H \approx -20\ \text{kcal/mol}} \text{C(H)-C(PR}_2) + R_2P^{\bullet} \qquad (8.54)$$

Phosphoranyl radicals typically undergo α- and β-fragmentation. The competition between these two fragmentations is not related to their exothermicity. In the case of the radical $Bu_3P^{\bullet}Ot\text{-Bu}$ both fragmentations are exothermic, by 22 and 46 kcal/mol, respectively. Nonetheless, the first has a lower activation energy than the second, the product consisting of 80% $Bu_2POt\text{-Bu}$ and 20% $Bu_3P=O$.

$$\text{structure with O-R, P, R groups} \xrightarrow{\alpha\text{-fragmentation}} R\!-\!P(R)\!-\!OR + R^\bullet \quad (8.55)$$

$$\xrightarrow{\beta\text{-fragmentation}} R\!-\!P(=O)(R)\!-\!R + R^\bullet \quad (8.56)$$

8-4 OXYGEN-CENTERED RADICALS

The most common oxygen-centered radicals belong to three families: oxyl, acyloxyl and peroxyl. Their reactivities are very different but all are electrophilic. Radicals such as $O_2^{\bullet-}$, $HOO^\bullet$ and $HO^\bullet$ are specially important in biological systems (Chap. 16).

8-4-1 Production

There are several methods for obtaining oxygen radicals.

Homolysis
Weak O–O (peroxides), O–Cl (hypochlorite) and O–NO (nitrite) bonds can be broken thermally or photochemically.

$$RO\text{–}X \xrightarrow[X = OH, OR, Cl, NO]{\Delta \text{ or } h\nu} RO^\bullet + X^\bullet \quad (8.57)$$

$$R\text{–}C(=O)\text{–}O\text{–}O\text{–}Z \xrightarrow[Z = R\text{–}C=O, R, H]{\Delta \text{ or } h\nu} R\text{–}C(=O)\text{–}O^\bullet + ZO^\bullet \quad (8.58)$$

Transfer
Hydrogen transfer from RO–H or RCO_2–H bonds is not a general method for preparing oxygen radicals since the relevant bonds are too strong (Table 8.2).

XO$^\bullet$	RCO$_2^\bullet$	PhCO$_2^\bullet$	RO$^\bullet$	RCO$_3^\bullet$	ROO$^\bullet$	PhO$^\bullet$
BDE(XO–H)	106	110	103	93	88	86

Table 8.2 – *Oxygen radicals: BDE(XO–H) in kcal/mol.*

On the other hand, the BDE(O–H) of phenols and hydroperoxides are small, because the radicals are conjugatively stabilized by the aromatic nucleus or the electron pair of the α-oxygen. Aryloxyl, ArO$^\bullet$, and peroxyl, R–O–O$^\bullet$, radicals are therefore easily obtained by hydrogen abstraction if R$^\bullet$ is electrophilic; the reaction is usually exothermic. Phenols are therefore efficient hydrogen-donors which inhibit chain reactions, hence the importance of tocopherols (vitamin E) in the protection of biological systems against radical reactions.

$$R^\bullet + ArO-H \longrightarrow R-H + ArO^\bullet \quad (8.59)$$

$$R^\bullet + R^1OO-H \longrightarrow R-H + R^1OO^\bullet \quad (8.60)$$

Oxygen radicals are also easily obtained by halogen transfer from hypohalites or the hydroxy group of peracids if $R^\bullet$ is nucleophilic.

$$R^\bullet + R^1O-Hal \longrightarrow R-Hal + R^1O^\bullet \quad (8.61)$$

$$R^\bullet + R^1-\underset{O-O}{\overset{O}{C}}\diagdown H \longrightarrow ROH + R^1-\underset{O^\bullet}{\overset{O}{C}} \quad (8.62)$$

Oxidation

Oxygen-containing molecules can easily be oxidized by loss of one of the electrons of the oxygen lone pair. If the radical obtained has a hydrogen bonded to oxygen an oxygen radical is obtained by deprotonation.

$$Z-O-Y \xrightarrow{-e^-} Z-\overset{\bullet+}{O}-Y \quad (8.63)$$

$$Z-O-H \xrightarrow{-e^-} Z-\overset{\bullet+}{O}-H \rightleftharpoons ZO^\bullet + H^+ \quad (8.64)$$

Oxygen radicals can therefore be obtained by oxidation:

– by metal salts;

$$RO-H + M^{n+} \xrightarrow{Ce^{4+}, Pb^{4+}, Fe^{3+}, MnO_2} RO^\bullet + H^+ + M^{(n-1)+} \quad (8.65)$$

$$ROO-H + Ce^{4+} \longrightarrow ROO^\bullet + H^+ + Ce^{3+} \quad (8.66)$$

$$RCO_2H + M^{n+} \xrightarrow{Ag^{2+}, Pb^{4+}} RCO_2^\bullet + H^+ + M^{(n-1)+} \quad (8.67)$$

– by electrochemical oxidation of phenols, alcoholates or carboxylates (Kolbe reaction).

$$ArO-H \xrightarrow{-e^-} ArO^\bullet + H^+ \quad (8.68)$$

$$ArO^- \xrightarrow{-e^-} ArO^\bullet \quad (8.69)$$

$$R-CO_2^- \xrightarrow{-e^-} R-CO_2^\bullet \quad (8.70)$$

Reduction

The antibonding orbital σ^*_u of the O–O bond of peroxides is of low energy. It therefore readily accepts an electron from metal reducing agents (Cu^+, Fe^{2+}, Cr^{2+}, Co^{2+}).

$$ZO-OY \xrightarrow{+e^-} ZO\overset{\bullet-}{-}OY \rightleftharpoons ZO^\bullet + {}^-OY \quad (8.71)$$

In this way a peroxide is transformed into a radical and an oxygen anion. The H_2O_2–Fe^{2+} system, which is a source of $HO^\bullet$, is known as Fenton's reagent.

$$XO–OH + M^{n+} \xrightarrow{X = H, R} XO^\bullet + HO^- + M^{(n+1)+} \quad (8.72)$$

$$XO–OX + M^{n+} \xrightarrow{X = R, PhCO} XO^\bullet + XO^- + M^{(n+1)+} \quad (8.73)$$

$$RCO_3R^1 + M^{n+} \longrightarrow \begin{cases} RCO_2^\bullet + R^1O^- + M^{(n+1)+} \quad (8.74) \\ R^1O^\bullet + RCO_2^- + M^{(n+1)+} \quad (8.75) \end{cases}$$

Reaction with oxygen

Peroxyl compounds arise in most cases from the autoxidation of organic compounds by the oxygen of the air. In this chemical process peroxyl radicals are formed by the reaction of carbon radicals with oxygen. This reaction, which is extremely rapid ($k \approx 10^9$ $M^{-1}s^{-1}$) is probably diffusion controlled but it can be reversible; the equilibrium depends on the oxygen pressure and the stability of the radical $R^\bullet$. This equilibrium can be observed for $Ph_3C^\bullet$, for example, whereas for most aliphatic radicals the reverse reaction, involving loss of oxygen, is practically nonexistent at normal temperatures. Peroxyl radicals, because of their low reactivity, dimerize to tetroxides which decompose to oxyl radicals.

$$R^\bullet + O_2 \rightleftharpoons ROO^\bullet \quad (8.76)$$

$$2\ ROO^\bullet \longrightarrow ROO–OOR \longrightarrow 2\ RO^\bullet + O_2 \quad (8.77)$$

8-4-2 Reactions

The reactivity of oxygen radicals depends mainly on the strength of the bond formed (O–H and O–metal bonds) and on their electrophilic character. Another characteristic of these radicals is that they undergo β-fragmentation.

Hydrogen transfer reactions

All oxygen radicals are electrophilic but the rate of hydrogen transfer depends mainly on the enthalpy balance of the reaction.

$$ZO^\bullet + R–H \longrightarrow ZO–H + R^\bullet \quad (8.78)$$

Since the BDEs of RO–H and $ArCO_2$–H (Table 8.2) are distinctly higher than those of most C–H bonds, hydrogen transfer reactions to $RO^\bullet$ and $ArCO_2^\bullet$ are exothermic and therefore common. When R is aliphatic the very short lifetime of $RCO_2^\bullet$, due to its rapid decarboxylation, means that this reaction cannot be observed. On the other hand, since the BDEs of ArO–H, ROO–H and RCO_3–H are low, the corresponding oxygen radicals, $ArO^\bullet$, $ROO^\bullet$ and $RCO_3^\bullet$, abstract only weakly bonded hydrogens (allylic, benzylic, of thiophenols, etc.). For example, the rate constants for hydrogen transfer from toluene to radicals t-$BuO^\bullet$ and t-$BuOO^\bullet$ are 10^5 and 1.2×10^{-2} $M^{-1}s^{-1}$, respectively.

In the autoxidation of aldehydes the weakly bonded aldehydic hydrogen is transferred to the acylperoxyl radical.

$$RCO_3^\bullet + RCH=O \longrightarrow RCO_3H + R\overset{\bullet}{C}=O \quad (8.79)$$

Addition reactions

The various oxygen radicals add to double bonds at very different rates:

$$HO^\bullet > ArCO_3^\bullet > ArCO_2^\bullet \geq RO^\bullet > t\text{-BuO}^\bullet > ROO^\bullet \quad (8.80)$$

With radicals of the type $RO^\bullet$ β-fragmentation and hydrogen abstraction, especially from an allylic position, are usually faster than addition. Addition is only observed in the very favorable case of intramolecular reactions. With acylperoxyl and peroxyl radicals the radical obtained by addition leads to an epoxide by a fast intramolecular homolytic substitution, S_Hi (8.81).

$$\underset{/}{\overset{\backslash}{C}}=\underset{\backslash}{\overset{/}{C}} \xrightarrow{+ XOO^\bullet} \overset{XO-O}{\underset{C-\overset{\bullet}{C}}{\diagup\diagdown}} \xrightarrow[S_Hi]{X=R-C=O, R} \overset{O}{\underset{C-C}{\diagup\diagdown}} + XO^\bullet \quad (8.81)$$

When the double bond has allylic hydrogens there is competition between addition and allylic hydrogen transfer.

$$XO^\bullet + CH_2=CH-CH_2R \begin{array}{c} \xrightarrow{k_{all}} XOH + CH_2=CH-\overset{\bullet}{C}HR \quad (8.82) \\ \xrightarrow{k_{add}} XO-CH_2-\overset{\bullet}{C}H-CH_2R \quad (8.83) \end{array}$$

The chemoselectivity k_{all}/k_{add} favors allylic transfer for $t\text{-BuO}^\bullet$ and addition for $PhCO_2^\bullet$ (Table 8.3). Since it is known that k_{add} and k_{all} are both greater for $PhCO_2^\bullet$ than for $t\text{-BuO}^\bullet$ this result shows that the chemoselectivity depends on the addition reaction. Addition of oxygen radicals to aromatic compounds is relatively uncommon, except for the radical $PhCO_2^\bullet$ which leads to aryl benzoates.

$XO^\bullet$	$RCO_2^\bullet$	$PhCO_2^\bullet$	$t\text{-BuO}^\bullet$	$ROO^\bullet$	$PhCO_3^\bullet$
$k_{frag}\ s^{-1}$	10^{10}	10^6	10^5		
k_{all}/k_{add}		0.2	30	1	0.5

Table 8.3 – *Oxygen radicals : rate constants for β-fragmentation and rate constant ratio, k_{all}/k_{add}.*

Reactions with metals

Since O–metal bonds are strong, oxygen radicals react rapidly with metallic centers. Thus, the S_H2 reaction of $t\text{-BuO}^\bullet$ on tributylborane is very exothermic, since BDE(B–C) and BDE(B–O) are 83 and 125 kcal/mol, respectively.

$$Me_3CO^\bullet + BBu_3 \xrightarrow{S_H2} Me_3CO-BBu_2 + Bu^\bullet \quad (8.84)$$

Reaction at phosphorus

Since P–O bonds are strong they are formed very easily by oxygen radicals. Thus, one goes from triethyl phosphite to a phosphoranyl radical intermediate, which can be observed by EPR before it decomposes by β-fragmentation.

$$Me_3CO^\bullet + P(OEt)_3 \longrightarrow Me_3C-O-\overset{\bullet}{P}(OEt)_3 \longrightarrow Me_3C^\bullet + O=P(OEt)_3 \quad (8.85)$$

Fragmentation reactions

The β-fragmentation of alkoxyl radicals (8.86) is a very general reaction,

$$\underset{\underset{Me}{Me}}{\overset{Me}{\diagdown}}C-O^\bullet \longrightarrow \underset{Me}{\overset{Me}{\diagdown}}C=O + {}^\bullet CH_3 \quad (8.86)$$

which can suggest interesting synthetic procedures for ring enlargement (8.87).

$$(8.87)$$

The decarboxylation of acyloxyl radicals (8.88) is a β-fragmentation.

$$R-C\underset{O^\bullet}{\overset{\overset{O}{\|}}{\diagdown}} \longrightarrow R^\bullet + CO_2 \quad (8.88)$$

The rate of this reaction depends on the stability of the radical R$^\bullet$ formed (Table 8.3). This reaction is therefore a good source of alkyl radicals but is much less efficient for aryl radicals. The β-fragmentation of a peroxyl radical is the reverse of trapping by oxygen (8.76). Solvation of oxygen radicals by polar and polarizable solvents favors β-fragmentation.

8-5 SULFUR-CENTERED RADICALS

These much-studied radicals are particularly involved in biological processes. Thiyl radicals, RS$^\bullet$, where the sulfur atom is bonded to only one substituent, are the most common and the most important of this family. Mono- and dioxygenated thiyl radicals (sulfinyl RS$^\bullet$=O and sulfonyl RSO$_2^\bullet$) are hypervalent radicals.

8-5-1 Production

Thiyl radicals

Thiyl radicals are obtained from thiols:

– by hydrogen transfer (8.89). This reaction, which is easy because BDE(RS–H) is small (88 kcal/mol), whereas it is difficult for alcohols, ROH, explains why thiols inhibit radical reactions by hydrogen transfer;

$$RS-H + R^{1\bullet} \longrightarrow RS^{\bullet} + R^1-H \qquad (8.89)$$

– by photolysis (8.90);

$$RS-H \xrightarrow{h\nu} RS^{\bullet} \qquad (8.90)$$

– by oxidation by a metal cation (8.91);

$$RS-H + Ce^{4+} \longrightarrow RS^{\bullet} + H^+ + Ce^{3+} \qquad (8.91)$$

The thermolysis of disulfides (8.92) is little used because it requires higher temperatures than for peroxides, since BDE(RS–SR) ≈ 65 kcal/mol whereas BDE(RO–OR) ≈ 38 kcal/mol. For example, the rate constant of reaction (8.93) is 2×10^{-8} s^{-1} whereas that for O–O homolysis in benzoyl peroxide at the same temperature is 5×10^{-4} s^{-1}.

$$RS-SR \longrightarrow 2\ RS^{\bullet} \qquad (8.92)$$

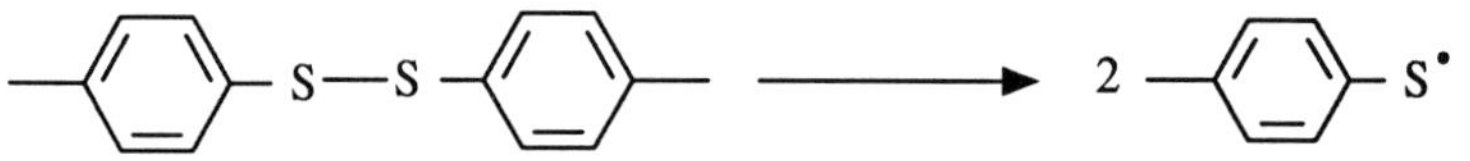

$$\text{Ar}-S-S-\text{Ar} \longrightarrow 2\ \text{Ar}-S^{\bullet} \qquad (8.93)$$

Sulfinyl radicals

Various radical species are produced by the thermolysis of sulfinyls, sulfones, thiosulfinates or arylsulfenyl nitrates.

$$\underset{O}{R-S-SO_2-R} \longrightarrow \underset{sulfinyl}{R-\overset{\bullet}{S}=O} + \underset{sulfonyl}{R\overset{\bullet}{S}O_2} \qquad (8.94)$$

$$\underset{O}{R-S-S-R} \longrightarrow R-\overset{\bullet}{S}=O + \underset{thiyl}{R-S^{\bullet}} \qquad (8.95)$$

$$ArS-O-NO_2 \longrightarrow Ar-\overset{\bullet}{S}=O + {}^{\bullet}NO_2 \qquad (8.96)$$

Sulfonyl radicals

The most important methods for obtaining sulfonyl radicals are:

– reversible addition of a radical to SO_2 (8.97);

$$R^{\bullet} + SO_2 \rightleftharpoons R-SO_2^{\bullet} \qquad (8.97)$$

– chlorine abstraction from an alkyl or arylsulfonyl chloride (8.98).

$$RSO_2-Cl + R^{1\bullet} \longrightarrow RSO_2^{\bullet} + R^1-Cl \quad (8.98)$$

8-5-2 Reactions

Thiyl radicals

In contrast to alkoxyl radicals they do not undergo β-fragmentation reactions (8.99).

$$\underset{\underset{Me}{Me}}{\overset{Me}{\diagdown}}C-S^{\bullet} \quad \xrightarrow{\quad \times \quad} \quad \underset{Me}{\overset{Me}{\diagdown}}C=S + {}^{\bullet}CH_3 \quad (8.99)$$

They abstract hydrogen with difficulty but add to multiple bonds very well, leading to a wide variety of sulfur compounds ((8.100) and (8.101)).

chain
$$RS^{\bullet} + \diagdown C = C \diagup \underset{k_{-1}}{\overset{k_1}{\rightleftharpoons}} \diagdown \overset{\bullet}{C} - C \diagup^{SR} \quad (8.100)$$

$$\diagdown \overset{\bullet}{C} - C \diagup^{SR} + RSH \xrightarrow{k_2} \diagdown \underset{H}{C} - C \diagup^{SR} + RS^{\bullet} \quad (8.101)$$

The exothermic addition step (8.100) is reversible and the value of k_{-1} can be large compared to k_1 and k_2. This reversibility has been demonstrated in many cases, such as the isomerization of *cis* 2-butene (8.102). Reaction (8.101), which is also exothermic because BDE(C–H) is generally greater than BDE(S–H), explains why thiol addition reactions are easy and occur without telomer formation.

$$\underset{H}{\overset{CH_3}{\diagdown}}C=C\underset{H}{\overset{CH_3}{\diagup}} \xrightarrow{+CH_3S^{\bullet}} \underset{H}{\overset{CH_3}{\diagdown}}\overset{\bullet}{C}-\underset{H}{\overset{CH_3}{\underset{SCH_3}{\diagup}}}C \xrightarrow{-CH_3S^{\bullet}} \underset{H}{\overset{CH_3}{\diagdown}}C=C\underset{CH_3}{\overset{H}{\diagup}} \quad (8.102)$$

On the other hand, thiyl radicals, like alkoxyl radicals, give a β-displacement reaction with phosphines:

$$(CH_3O)_3P \xrightarrow{+CH_3S^{\bullet}} (CH_3O)_3\overset{\bullet}{P}-SCH_3 \xrightarrow{-{}^{\bullet}CH_3} (CH_3O)_3P=S \quad (8.103)$$

Sulfinyl radicals

Sulfinyl radicals are very unreactive, only giving dimers which after rearrangement lead to thiosulfones.

$$2\ \text{Ar–}\overset{\bullet}{\text{S}}\text{=O} \longrightarrow \text{Ar–S–O–S–Ar} \longrightarrow \text{Ar–}\underset{\underset{\text{O}}{\|}}{\text{S}}\text{–S–Ar} \quad (8.104)$$

Sulfonyl radicals

Sulfonyl radicals add to double bonds (8.105).

$$\text{RSO}_2^\bullet + \overset{\diagdown}{\underset{\diagup}{\text{C}}}=\overset{\diagup}{\underset{\diagdown}{\text{C}}} \rightleftharpoons \overset{\text{RSO}_2}{\underset{}{\text{C–C}^\bullet}} \quad (8.105)$$

They dimerize with formation of a S–O bond (8.106) rather than S–S (8.107).

$$2\ \text{ArSO}_2^\bullet \begin{array}{l} \xrightarrow{5\%} \text{ArS–SAr (with 4 O)} \quad (8.106) \\ \xrightarrow{95\%} \text{ArS–O–SAr} \longrightarrow \text{Ar}\overset{\bullet}{\text{SO}}_3 + \text{Ar}\overset{\bullet}{\text{SO}} \quad (8.107) \end{array}$$

8-6 HALOGEN ATOMS

8-6-1 Production

The most common source of halogen atoms is the molecular halogen itself (8.108).

$$X_2 \xrightarrow{\Delta\ \text{or}\ h\nu} 2\ X^\bullet \quad (8.108)$$

The energy required for such a scission (Table 8.4) demands fairly high temperatures. Thus, at 120°C in the absence of light chlorine does not react with ethane. Since molecular halogens absorb in the visible photolysis is the preferred method.

X	F	Cl	Br	I
BDE(X–X)	38	58	46	36
$\lambda_{max}(X_2)$		335	425	525
$\Delta H(X^\bullet + \text{Et–H})$	–36	–3	12	29
$\Delta H(\text{Et}^\bullet + X–X)$	–72	–26	–24	–20

Table 8.4 – *BDE, reaction enthalpy (kcal/mol) and visible absorption (nm).*

Hypohalites such as ROCl and ROBr which have low BDE(RO–X) (47 and 44 kcal/mol, respectively) readily undergo homolysis by thermolysis or photolysis and are, therefore, sources of halogen atoms. Halogenations by hypohalites follow a chain mechanism in which the radical RO$^\bullet$ abstracts hydrogen:

initiation $\quad$ RO–X $\quad\xrightarrow{\Delta \text{ or h}\nu}\quad$ RO$^\bullet$ + X$^\bullet$ $\quad$ (8.109)

chain $\quad \Big| \begin{array}{l} \text{RO}^\bullet + \text{R}^1\text{–H} \quad\longrightarrow\quad \text{RO–H} + \text{R}^{1\bullet} \quad (8.110) \\ \text{R}^{1\bullet} + \text{RO–X} \quad\longrightarrow\quad \text{R}^1\text{–X} + \text{RO}^\bullet \quad (8.111) \end{array}$

8-6-2 Reactions

Hydrocarbon halogenation is typical of this class of free radicals. These reactions follow a chain mechanism, (8.112) and (8.113), the efficiency of which depends markedly on the enthalpy of each of the two steps (Table 8.4). With fluorine both steps are strongly exothermic and the reaction can be explosive if the mixture is not diluted with an inert solvent.

chain $\quad \Big| \begin{array}{l} \text{X}^\bullet + \text{R–H} \quad\longrightarrow\quad \text{X–H} + \text{R}^\bullet \quad (8.112) \\ \text{R}^\bullet + \text{X}_2 \quad\longrightarrow\quad \text{R–X} + \text{X}^\bullet \quad (8.113) \end{array}$

Initiation is considered to occur by a molecule-molecule reaction (8.114) which is endothermic to the extent of only 7 kcal/mol for methane and becomes exothermic as soon as BDE(R–H) is lower than 98 kcal/mol, which is the case for most C–H bonds.

F–F + R–H $\quad\longrightarrow\quad$ F$^\bullet$ + H–F + R$^\bullet$ $\quad$ (8.114)

With chlorine the reaction is initiated photochemically. Since the two steps of the chain are reasonably exothermic, chain lengths are of the order of 10^6. With bromine and iodine, reaction (8.112) is endothermic and reversible at high temperature and low halogen concentration. Consequently, brominations and iodinations are slow. Since halogen radicals are markedly electrophilic, reaction (8.112) is significantly regioselective.

PART II

REACTIONS : CLASSIFICATIONS AND MECHANISMS

9 – PRODUCTION OF FREE RADICALS

Whatever the objective, physical or physicochemical studies, reaction mechanism or synthesis, the first step is the production of radicals. There are many methods which, depending on the viewpoint, are by no means equivalent. The energy required to perform this first homolysis step can be provided by photolysis, thermolysis or a redox system.

9-1 GENERAL PRINCIPLES

9-1-1 Photolysis

The light energy absorbed by a molecule can cause, amongst other processes, homolytic fission of a bond. This energy is related to the wavelength λ of the light by the following equation, where h is Planck's constant and c is the speed of light.

$$E = hc / \lambda \quad \text{or} \quad E \text{ (kcal/mol)} = 28\,600 / \lambda \text{ (nm)}$$

In theory, the complete absorption of a quantum at 270 nm is equivalent to 106 kcal/mol, which is enough to break the majority of the bonds in organic chemistry. In practice not all the energy absorbed is used in bond breaking. The primary quantum yield, Φ_1, which is the fraction of molecules homolyzed per photon absorbed, is between 0 and 1. The quantum yield, Φ_2, defined as the number of product molecules formed per photon absorbed, can be very great (rup to 10^6); in this case the radicals produced by the primary photolysis act have initiated a chain reaction; the chain length depends on Φ_2 and the nature of the reactants.

Photolysis can often be used to bring about homolytic fissions which would require high temperatures if performed by thermolysis. Thus, halogenation by chlorine, a chain process, is initiated by simple exposure to sunlight, whereas it is necessary to use temperatures of the order of 200°C to homolyze chlorine in the dark. Many molecules can be used in this way to generate radicals by photolysis.

$$\text{Cl–Cl} \xrightarrow[\Delta H = 59 \text{ kcal/mol}]{h\nu} 2 \text{ Cl}^\bullet \qquad (9.1)$$

Peroxides and azoalkanes

$$Y^1\text{O–O}Y^2 \xrightarrow[\Delta H = 30 \text{ to } 50 \text{ kcal/mol}]{h\nu} Y^1\text{O}^\bullet + Y^2\text{O}^\bullet \qquad (9.2)$$

$$Y^1, Y^2 = \text{H, R, RC(O), ArC(O), ROC(O), etc.}$$

$$\text{R–N=N–R} \xrightarrow[\Delta H = 35 \text{ to } 50 \text{ kcal/mol}]{h\nu} N_2 + 2 \text{ R}^\bullet \qquad (9.3)$$

Nitrites and hypochlorites

$$\text{RO–NO} \xrightarrow[\Delta H = 41 \text{ kcal/mol}]{h\nu} \text{RO}^\bullet + \text{NO}^\bullet \quad (9.4)$$

$$\text{RO–Cl} \xrightarrow[\Delta H = 48 \text{ kcal/mol}]{h\nu} \text{RO}^\bullet + \text{Cl}^\bullet \quad (9.5)$$

Polyhalomethanes

$$\text{Br–CCl}_3 \xrightarrow[\Delta H = 56 \text{ kcal/mol}]{h\nu} \text{Br}^\bullet + {}^\bullet\text{CCl}_3 \quad (9.6)$$

Thiols

$$\text{RS–H} \xrightarrow[\Delta H = 88 \text{ kcal/mol}]{h\nu} \text{RS}^\bullet + {}^\bullet\text{H} \quad (9.7)$$

N-Chloroamines

$$\text{R}_2\text{N–Cl} \xrightarrow[\Delta H = 45 \text{ kcal/mol}]{h\nu} \text{R}_2\text{N}^\bullet + \text{Cl}^\bullet \quad (9.8)$$

Organometallic compounds

$$\text{R}_3\text{Sn–SnR}_3 \xrightarrow[\Delta H = 63 \text{ kcal/mol}]{h\nu} 2\ \text{R}_3\text{Sn}^\bullet \quad (9.9)$$

$$(\text{PhCH}_2)_2\text{Hg} \xrightarrow{h\nu} 2\ \text{PhCH}_2^\bullet + \text{Hg} \quad (9.10)$$

Radical formation can be mediated by a photochemical sensitizer. Benzophenone, for example, gives a triplet when irradiated at about 320 nm; this has the characteristics of a biradical and reacts in much the same way as an alkoxyl radical (9.11) - (9.12). Mercury also can be used as a sensitizer.

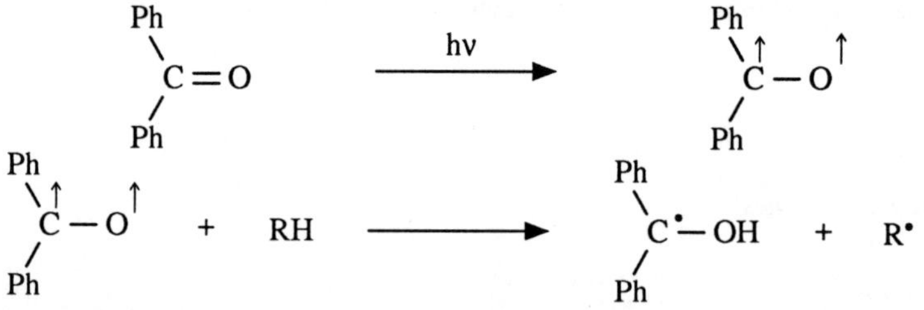

$$\text{Ph}_2\text{C=O} \xrightarrow{h\nu} \text{Ph}_2\overset{\uparrow}{\text{C}}-\text{O}^\uparrow \quad (9.11)$$

$$\text{Ph}_2\overset{\uparrow}{\text{C}}-\text{O}^\uparrow + \text{RH} \longrightarrow \text{Ph}_2\overset{\bullet}{\text{C}}-\text{OH} + \text{R}^\bullet \quad (9.12)$$

9-1-2 Radiolysis

When organic molecules are exposed to high energy radiation, such as X-rays or

General principles 107

γ-radiation, free radicals are produced. Because of the high energy supplied the reactions are often complex and unselective. The basic mechanism involves electron abstraction and the formation of a radical cation. Water, when irradiated, is a source of very reactive HO• radicals, which can be used in turn to produce other radicals.

$$H_2O \xrightarrow{\gamma} H_2O^{\bullet+} + e^-_{aq} \quad (9.13)$$

$$H_2O^{\bullet+} + H_2O \longrightarrow H_3O^+ + HO^{\bullet} \quad (9.14)$$

$$HO^{\bullet} + RH \longrightarrow H_2O + R^{\bullet} \quad (9.15)$$

Nitrous oxide, N_2O, is added to the solution to eliminate e^-_{aq} (9.16).

$$N_2O + e^-_{aq} \longrightarrow N_2 + HO^{\bullet} + HO^- \quad (9.16)$$

Because of its important biological and medical implications, much work has been devoted to the radiolysis of aqueous solutions.

The direct action of γ-radiation on an organic molecule generally leads to free radicals. $BrCCl_3$, for example, is a source of •CCl_3 radicals.

$$BrCCl_3 \xrightarrow{\gamma} [BrCCl_3]^{+\bullet} \longrightarrow \longrightarrow CCl_3^{\bullet} \quad (9.17)$$

There are important differences between γ-radiation or X-rays and UV:

– the primary process is electron abstraction instead of the formation of an excited state;
– creation of a cluster of radicals along the trajectory of the γ-radiation instead of a homogeneous distribution of excited states as in UV;
– bond breaking is rather unselective.

γ-Radiation is much used in order to perform polymerizations and polymer grafts. Its application in the laboratory remains uncommon in view of the investment required. Nevertheless, it is very useful for creating radicals in solids and at the surface of films.

9-1-3 Thermolysis

Raising the temperature increases the vibrational energy of a molecule. It can lose this internal energy by collision and dispersion in the medium or by homolytic fission of one or several bonds. At high temperature (800 ~ 1000°C) all bonds may be broken; this is the range of pyrolysis and combustion. Obtaining radicals at temperatures below 150°C requires weak bonds, with energies of the order of 30 to 40 kcal/mol. For this purpose peroxides, certain azoalkanes, N-hydroxy-2-thiopyridone derivatives, organometallic compounds, etc. are used. In what follows we shall examine the mechanisms of radical formation from these compounds, often referred to as chemical initiators.

9-2 METHODOLOGY OF RADICAL PRODUCTION

If one wants, for example, to create a carbon radical by homolysis of a C–H bond the energy required is much too great for this to be achieved directly (thermolysis or photolysis). One resorts to different systems involving molecules with a weak bond (*peroxides, azo compounds or organometallics*) likely to provide radicals easily. These, by energetically favorable processes, initiate chain, transfer or addition reactions.

9-2-1 Peroxides

The O–O bonds of peroxides, whose dissociation energies are between 25 and 40 kcal/mol, are easily broken thermally (between 50 and 150°C) or photolytically. The oxyl radicals formed then perform an *exothermic* hydrogen transfer reaction:

$$t\text{-BuO–O}t\text{-Bu} \xrightarrow{\Delta H = 38 \text{ kcal/mol}} 2\ t\text{-BuO}^\bullet \qquad (9.18)$$

$$t\text{-BuO}^\bullet + \text{R–H} \xrightarrow{\Delta H = -5 \text{ kcal/mol}} \text{R}^\bullet + t\text{-BuOH} \qquad (9.19)$$

Peroxides are generally referred to as *initiators*. This name comes from the fact that they constituted the first family of products used to *initiate* chain reactions in polymerization. The kinetics and mechanisms of their thermal or photochemical decomposition have been thoroughly investigated (Table 9.1). If the general formula is written as Y^1–O–O–Y^2, the homolysis of the O–O bond and the reactivity of the radicals $Y^1O^\bullet$ and $Y^2O^\bullet$ depends greatly on the nature of Y^1 and Y^2. Peroxides are characterized by:

– the activation energy, E_a, of reaction (9.2) for the breaking of the O–O bond and, therefore, the dissociation energy of this bond;
– the first-order rate constant of reaction (9.2) measured at the beginning of the reaction;
– the half-life ($t_{1/2}$). The experimental $t_{1/2}$ values are lower than those expected from the first-order rate constant. This is due to induced reactions which accelerate the disappearance rate of peroxide. In the case of AIBN, there no induced reaction: first-order rate constant and half-life are comparable.

These data, presented in Table 9.1, are important if peroxides are to be used intelligently in terms of the stated objectives. The range of temperatures which can be used in synthesis is limited and depends on the structure of the peroxide. It is generally considered that the ideal temperature is that for which the $t_{1/2}$ is of the order of a few hours. The rate of radical formation can be too slow at lower temperatures or too great at higher temperatures compared to the overall rate of the chain reaction to be initiated. All peroxides are potentially dangerous products, to a greater or lesser degree, depending on their structure. They can also undergo decomposition reactions induced by metals. They must always be handled and stored with caution. It is *always* wise, when the chain reaction is finished and the reactants have been consumed, to check that the peroxide initiator has disappeared completely. This precaution is particularly necessary when at the end of the experiment the solvent is evaporated by heating and the reaction product is distilled: the concentrated and heated peroxides will then lead frequently to an explosion. This precaution must also be taken every time one distils a substance which can be autoxidized and is likely to produce hydroperoxides, especially if it is used as a solvent. From this point of view, ether and particularly tetrahydrofuran, but also other substances, such as conjugated dienes, are so dangerous that their use is prohibited in many industries. The oxidizing power of peroxides and hydroperoxides is used to monitor their presence, by adding a few drops of ferrous sulfate and potassium thiocyanate. In the presence of peroxide there is an immediate blood-red coloration due to ferricyanide. If this test is positive the peroxides must be eliminated by treatment with ferrous sulfate solution. The peroxides described in what follows are commercial products.

name formula	E_a	k (T°C)	$t_{1/2}$ (T°C)	radicals obtained
benzoyl peroxide $(PhC(O)O-)_2$	30	1.95×10^{-6} (60) 3×10^{-5} (80)	7h (70) 1h (95)	$PhCO_2^\bullet$, $Ph^\bullet$
acetyl peroxide $(MeC(O)O-)_2$	31	0.5×10^{-5} (60) 1.3×10^{-5} (80)	8h (70) 1h (85)	$MeCO_2^\bullet$, $Me^\bullet$
diisopropylperoxydicarbonate $(i\text{-}PrO-C(O)O-)_2$	28	2.3×10^{-5} (50)	5h (50)	$i\text{-}PrOCO_2^\bullet$, $i\text{-}PrO^\bullet$
acetylperoxybenzoate $PhC(O)O-OC(O)Me$	30	2×10^{-5} (70)	5h (60)	$PhCO_2^\bullet$, $Ph^\bullet$, $MeCO_2^\bullet$, $Me^\bullet$
tert-butylperoxybenzoate $Ph-C(O)O-Ot\text{-}Bu$	34	1×10^{-4} (120)	1h (125)	$PhCO_2^\bullet$, $Ph^\bullet$, $t\text{-}BuO^\bullet$, $Me^\bullet$
tert-butylperoxyoxalate $(t\text{-}BuO-OC(O)-)_2$	25	7×10^{-5} (35) 1×10^{-3} (55)	1mn (60)	$t\text{-}BuO^\bullet$, $Me^\bullet$
di-*tert*-butylperoxide $t\text{-}BuO-Ot\text{-}Bu$	37	1.4×10^{-8} (70)	218h (100) 10h (126) 1h (150)	$t\text{-}BuO^\bullet$, $Me^\bullet$
tert-butylhydroperoxide $t\text{-}BuO-OH$	43	1×10^{-5} (150)	5h (100)	$t\text{-}BuO^\bullet$, $HO^\bullet$, $t\text{-}BuOO^\bullet$, $Me^\bullet$
azobisisobutyronitrile $(Me_2C(CN)-N=)_2$	31	1.5×10^{-4} (80)	5h (70) 1h (85)	$Me_2C^\bullet(CN)$

Table 9.1 – *Chemical initiators: E_a activation energy in kcal/mol, k first-order rate constant in s^{-1}; $t_{1/2}$, half-life.*

Diacyl peroxides RC(O)O–OC(O)R

The two most used diacyl peroxides are benzoyl peroxide (R = Ph) and, to a lesser extent, acetyl peroxide (R = Me). The second, which is dangerous when pure, is kept in solution. Benzoyl peroxide, a crystalline product (m.p. 105~6°C), is stable for several months in the refrigerator. Nevertheless, it decomposes violently when heated to 100°C and it can be dangerous to recrystallize it from hot chloroform; it is preferable to dissolve it in chloroform at room temperature and to cause it to crystallize by adding methanol.

Initiation

The initial step in the decomposition of a diacyl peroxide is the homolysis of the

O–O bond which results in the formation of two acyloxyl radicals (9.20). The activation energies, E_a, for benzoyl and acetyl peroxides are about the same (Table 9.1).

$$R-C(=O)-O-O-C(=O)-R \xrightarrow{R = Ph \text{ or } CH_3} 2\ R-C(=O)-O^\bullet \qquad (9.20)$$

Acyloxyl radicals can then decarboxylate (9.21) to give a radical, $Ph^\bullet$ in one case and $Me^\bullet$ in the other. However, $MeCO_2^\bullet$ decarboxylates about 10^3 times faster than $PhCO_2^\bullet$ (Table 26.18). For this reason acetyl peroxide is used as a source of $Me^\bullet$ radicals; these are very reactive since $BDE(Me-H) = 104$ kcal/mol.

$$R-C(=O)-O^\bullet \xrightarrow{R = Ph \text{ or } CH_3} R^\bullet + CO_2 \qquad (9.21)$$

Benzoyl peroxide gives two very reactive radicals, $PhCO_2^\bullet$ and $Ph^\bullet$ (BDE $(PhCO_2-H) = 110$ kcal/mol; $BDE(Ph-H) = 110$ kcal/mol). Their relative concentrations depend on the temperature $(E_a(9.20) > E_a(9.21))$ but also on the nature of the solvent. Decarboxylation can compete with hydrogen transfer from the solvent, $H-\Sigma$.

$$PhCO_2^\bullet + H-\Sigma \longrightarrow PhCO_2H + \Sigma^\bullet \qquad (9.22)$$

In the case of cyclohexane large amounts of cyclohexanol are formed because of autoxidation by dissolved oxygen (9.23).

$$\Sigma^\bullet \xrightarrow{+O_2} \Sigma OO^\bullet \xrightarrow[-O_2]{+\Sigma OO^\bullet} 2\ \Sigma O^\bullet \xrightarrow{+H-\Sigma} \Sigma OH + \Sigma^\bullet \qquad (9.23)$$

The $PhCO_2^\bullet$ radical can also add to double bonds or aromatic systems. Thus, in benzene, a poor hydrogen-donor, benzoyl peroxide decomposes mainly to CO_2 and Ph–Ph with the formation of small amounts of $PhCO_2Ph$ and $PhCO_2H$.

The decomposition of diacyl peroxides is not limited to these reactions; other processes intervene and modify both the decomposition kinetics and the availability of the radicals produced in reactions (9.20) and (9.21), that is, the efficiency of the peroxide as an initiator. These reactions are induced decomposition, cage recombination and non-radical processes.

Induced decomposition
At low concentration in an inert solvent, the decomposition of a peroxide follows a first-order rate law corresponding to reaction (9.20). If the concentration is increa-

$$PhCO_2^\bullet + EtOCH_2CH_3 \longrightarrow PhCO_2H + EtO\overset{\bullet}{C}HCH_3 \qquad (9.24)$$

$$EtO\overset{\bullet}{C}HCH_3 + Ph-C(=O)-O-O-C(=O)-Ph \longrightarrow PhCO_2^\bullet + PhCO_2CH(CH_3)OEt \qquad (9.25)$$

sed or the solvent is a hydrogen-donor the rate increases and the kinetic order is greater than unity. This behavior is characteristic of induced decomposition, that is, *the occurrence of a bimolecular reaction between a radical and the peroxide.* For example, benzoyl peroxide has a half-life of 5 mn in diethyl ether at 80°C (sealed tube). This acceleration is due to the following reactions, (9.24) and (9.25).

Induced decomposition also results from the non-radical attack of a reactive nucleophile such as an amine on the O–O bond. This very rapid reaction can become explosive with aniline or triethylamine. Induced decomposition may also be caused by a metallic reducing agent.

The extent of the induced reaction depends also on the nature of the peroxide; thus, acetyl peroxide gives less induced decomposition than benzoyl peroxide.

Cage recombination
Along with other decomposition products, a certain amount of ester, RCO_2R, is formed. This arises in part from the recombination of radicals within the cage of solvent molecules before diffusion into the medium (9.26).

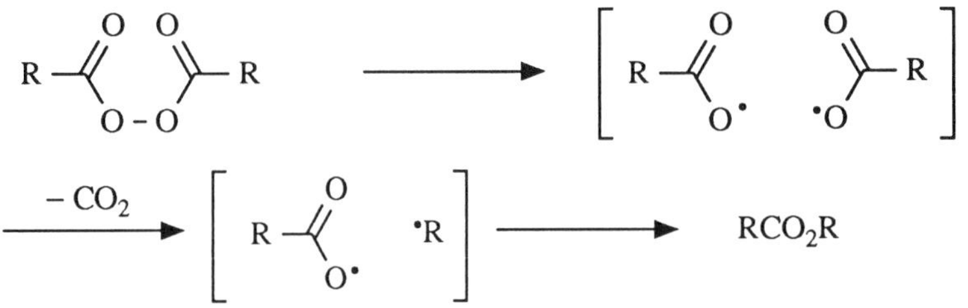

$$R-\overset{O}{\underset{O-O}{C}}-\overset{O}{C}-R \longrightarrow \left[R-\overset{O}{\underset{O^\bullet}{C}} \quad {}^\bullet O-\overset{O}{C}-R \right]$$

$$\xrightarrow{-CO_2} \left[R-\overset{O}{\underset{O^\bullet}{C}} \quad {}^\bullet R \right] \longrightarrow RCO_2R \qquad (9.26)$$

This cage recombination, which can be studied by CIDNP, depends on the solvent viscosity, the temperature and other factors. For example, benzoyl peroxide when photolyzed in benzene gives 13% of phenyl benzoate but only traces when thermolyzed at 80°C. This difference is due mainly to the increase in the rate of decarboxylation with the temperature (second step of reaction (9.26)).

Carboxy-inversion
Carboxy-inversion gives the ester, RCO_2R, by a non-radical process (9.27). The contribution of this reaction depends on the solvent polarity and the nature of R.

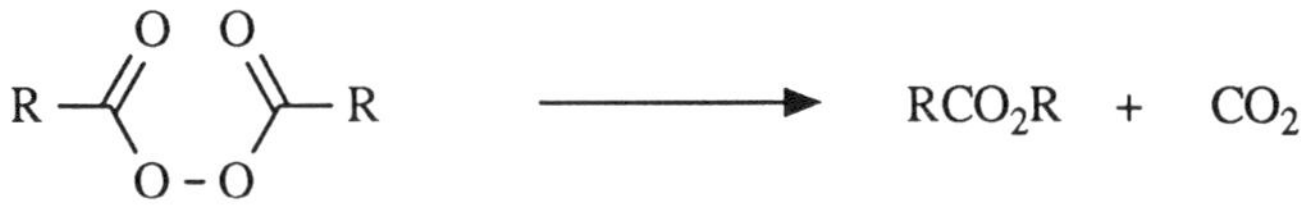

$$R-\overset{O}{\underset{O-O}{C}}-\overset{O}{C}-R \longrightarrow RCO_2R + CO_2 \qquad (9.27)$$

Sources of alkyl radicals
Diacyl peroxides can be efficient sources of alkyl radicals. Thus, dodecanoyl peroxide gives long-chain primary alkyl radicals, since the decarboxylation (9.28) is very fast ($k \approx 10^{10}$ s^{-1}).

$$CH_3(CH_2)_9CH_2CO_2^\bullet \longrightarrow CH_3(CH_2)_9CH_2^\bullet + CO_2 \qquad (9.28)$$

Peroxydicarbonates $ROC(O)O-OC(O)OR$
These are interesting because they decompose at moderate temperatures and

provide *(alkoxycarbonyl)oxy* radicals, $ROCO_2^•$, the decarboxylation of which is 10 and 10^5 times slower than that of $PhCO_2^•$ and $RCO_2^•$, respectively. The most commonly used are those where R = isopropyl, *tert*-butyl or cyclohexyl. Their low activation energy (28 to 30 kcal/mol) explains their easy decomposition; cyclohexyl peroxydicarbonate decomposes at 50°C at the same rate as benzoyl peroxide at 80°C. Their disadvantage is their instability. They are employed industrially in the polymerization of vinyl chloride.

$$RO-C(=O)-O-O-C(=O)-OR \longrightarrow 2\ RO-C(=O)-O^• \quad (9.29)$$

Peresters RC(O)O–OR'

tert-Butyl perbenzoate is the most widely used, decomposing a little more slowly than benzoyl peroxide. It produces $PhCO_2^•$, $Ph^•$, t-$BuO^•$ and $Me^•$ radicals at relative concentrations which depend on the temperature and the solvent (reactions (9.30) - (9.32)). It can be a source of $Me^•$ radicals but it is more interesting from this viewpoint to use acetyl peroxide, since the decarboxylation of $MeCO_2^•$ is 15 000 times as fast as reaction (9.32).

$$Ph-C(=O)-O-O-t\text{-}Bu \longrightarrow PhCO_2^• + t\text{-}BuO^• \quad (9.30)$$

$$PhCO_2^• \longrightarrow Ph^• + CO_2 \quad (9.31)$$

$$t\text{-}BuO^• \longrightarrow CH_3COCH_3 + {}^•CH_3 \quad (9.32)$$

Di-*tert*-butyl peroxalate is an interesting source of t-$BuO^•$ radicals since it decomposes at moderate temperatures about 10^3 and 10^5 times as fast as benzoyl peroxide and *tert*-butyl peroxide, respectively.

$$t\text{-}Bu-O-O-C(=O)-C(=O)-O-O-t\text{-}Bu \longrightarrow 2\ t\text{-}BuO^• + 2\ CO_2 \quad (9.33)$$

Dialkyl peroxides ROOR

Because of its stability and ease of handling, di-*tert*-butyl peroxide is, with benzoyl peroxide, the peroxide most often employed as an initiator, its only disadvantage being that it is necessary to work at a relatively high temperature to obtain an acceptable decomposition rate. At 80°C its half-life is about 1 000 h whereas that of benzoyl peroxide is only 7 h. In practice, it is used between 120 and 150°C. Di-*tert*-butyl peroxide is a source of t-$BuO^•$ radicals (9.34) but also of $Me^•$ radicals by β-fragmentation (9.32). However, to obtain a good yield of $Me^•$ it is necessary to raise the temperature, as the activation energy for (9.32) is 6 ~ 8 kcal/mol higher than that of hydrogen abstraction by t-$BuO^•$. Induced reactions are relatively unimportant except in the presence of alcohols or amines.

$$t\text{-}BuO-Ot\text{-}Bu \longrightarrow 2\ t\text{-}BuO^• \quad (9.34)$$

9-2-2 Azo compounds R–N=N–R

The mechanism for the decomposition of azoalkanes implies the breaking of two C–N bonds and the release of a stable nitrogen molecule (9.35). The activation energy for the decomposition of azo compounds is low and depends markedly on the nature of the radicals formed; it is 51 kcal/mol for azomethane (R = Me) but only 31 kcal/mol for azobisisobutyronitrile (AIBN) (R = C(CN)Me$_2$). AIBN is widely used as an initiator because of its moderate decomposition temperature; its half-life of 5 h at 70°C is reduced to 1 h at 85°C (Table 9.1). It is therefore shorter than that of benzoyl peroxide. Its decomposition can also be initiated photochemically.

$$\begin{array}{c} R \\ \diagdown \\ N=N \\ \diagdown \\ R \end{array} \longrightarrow \left[\begin{array}{c} R \\ \diagdown \\ N=N \\ \diagdown \\ R \end{array} \right]^* \longrightarrow 2\,R^\bullet + N\equiv N \quad (9.35)$$

Regardless of the solvent there is no induced reaction and the kinetics are therefore first-order. On the other hand, the rather unreactive 2-cyano-2-propyl radical ($^\bullet$(CN)Me$_2$) (BDE(H–C(CN)Me$_2$) = 86 kcal/mol) is only able to abstract weakly bonded hydrogen (allylic or benzylic hydrogen, thiol, tin hydride). Moreover, when AIBN decomposes it gives appreciable amounts, never less than 20%, of cage recombination products, which significantly reduces its efficiency as an initiator.

9-2-3 Organometallic compounds

Trialkyltin hydrides
The use of trialkyltin hydrides to produce radicals, discovered in 1960, has expanded enormously in recent years. The method consists in reacting the Bu$_3$Sn$^\bullet$ (Bu = *n*-butyl) radical, for example, with a halide, R–X; the radical is obtained by reaction of AIBN with the corresponding hydride, Bu$_3$Sn–H. Both reactions (9.36) and (9.37) are exothermic but the rate constant for reaction (9.37) varies considerably with the halogen; very slow and therefore useless with fluorides, when R is primary it is 2.5 x 10^9, 5 x 10^7 and 8.5 x 10^2 M^{-1}s^{-1} for X = I, Br and Cl, respectively.

$$\text{Bu}_3\text{Sn–H} \xrightarrow{\text{AIBN}/80°C} \text{Bu}_3\text{Sn}^\bullet \quad (9.36)$$

$$\text{Bu}_3\text{Sn}^\bullet + \text{R–X} \longrightarrow \text{R}^\bullet + \text{Bu}_3\text{Sn–X} \quad (9.37)$$

Instead of halides one can also use compounds such as phenylsulfides (9.38) or phenylselenides (9.39) which undergo easy displacement reactions.

$$\text{Bu}_3\text{Sn}^\bullet + \text{R–SPh} \longrightarrow \text{R}^\bullet + \text{Bu}_3\text{Sn–SPh} \quad (9.38)$$

$$\text{Bu}_3\text{Sn}^\bullet + \text{R–SePh} \longrightarrow \text{R}^\bullet + \text{Bu}_3\text{Sn–SePh} \quad (9.39)$$

The reactivity order is:

$$\text{RI} > \text{RSePh} \gg \text{RBr} > \text{RCl} > \text{RSPh} \quad (9.40)$$

Alkyl radicals can be obtained by addition-fragmentation reactions with secondary or tertiary nitro derivatives (9.41) or with xanthates (9.42). By means of this

AIBN-tin hydride couple a wide variety of carbon radicals and, in particular, aryl and vinyl radicals can be generated. Because tin salts are toxic, Bu_3SnH can be replaced by silicon compounds such as $(Me_3Si)_3SiH$ in which the Si–H bond strength is close to that of Sn–H.

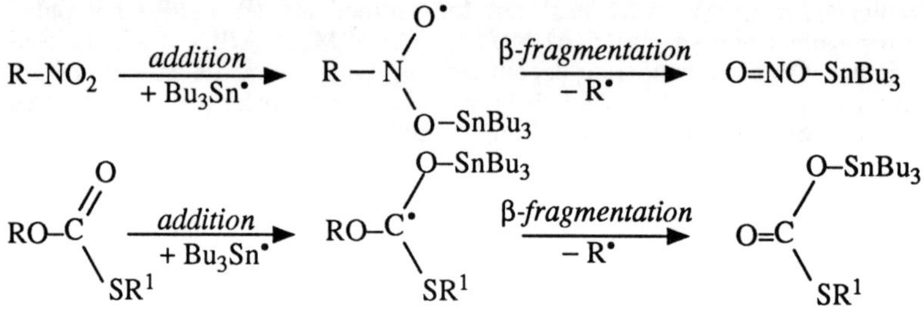

$$R-NO_2 \xrightarrow[+Bu_3Sn^\bullet]{addition} R-N{\overset{O^\bullet}{\underset{O-SnBu_3}{}}} \xrightarrow[-R^\bullet]{\beta\text{-}fragmentation} O=NO-SnBu_3 \quad (9.41)$$

$$RO-C{\overset{O}{\underset{SR^1}{}}} \xrightarrow[+Bu_3Sn^\bullet]{addition} RO-C{\overset{O-SnBu_3}{\underset{SR^1}{}}}^\bullet \xrightarrow[-R^\bullet]{\beta\text{-}fragmentation} O=C{\overset{O-SnBu_3}{\underset{SR^1}{}}} \quad (9.42)$$

Mercury and cobalt compounds

In general C–metal bonds have low BDE values and their homolysis is therefore easy. Thus, alkylmercury salts, which are easily obtained by solvomercuration of olefins, once reduced to hydrides become excellent hydrogen-donors and good radical precursors (9.43). The same is true of alkylcobalt complexes, such as alkylcobaloximes, where BDE(C–Co) is about 30 kcal/mol.

$$RHgX \xrightarrow[Bu_3SnH]{NaBH_4 \text{ or}} RHgH \xrightarrow[-H-X]{X^\bullet} RHg^\bullet \xrightarrow{-Hg} R^\bullet \quad (9.43)$$

$$RCo^{III}(dmgH)_2py \xrightarrow[dmg\,=\,dimethylglyoxime]{h\nu \text{ or } \Delta} R^\bullet + {}^\bullet Co^{II}(dmgH)_2py \quad (9.44)$$

9-2-4 Addition-fragmentation processes

Derivatives of thiohydroxamic acids, such as N-hydroxy-2-thiopyridone, are used to initiate and propagate chain reactions; the example of carboxylic esters shows how the reaction proceeds (Scheme 9.1).

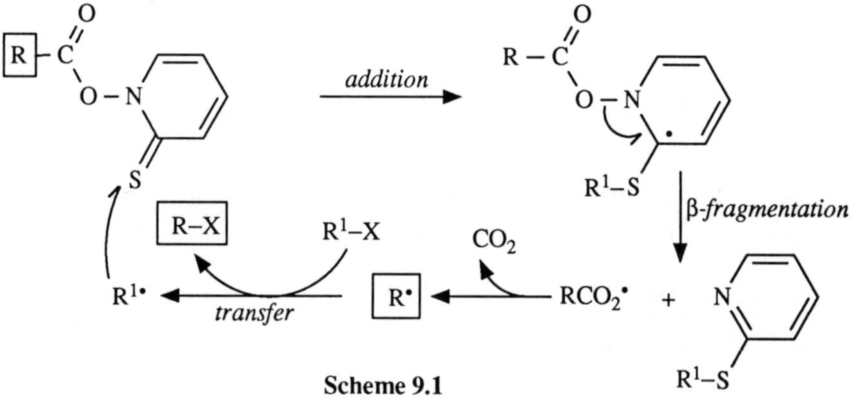

Scheme 9.1

The radical R• can perform transfer and addition reactions before reacting with the thiopyridone. Addition of $R^{1\bullet}$ to the C=S double bond is exothermic and may be reversible; the subsequent β-fragmentation is easy, since it is a unimolecular reaction (entropically favored) which gives rise to irreversible rearomatization (energy increase).

9-2-5 Redox systems

Redox reactions (oxidation or reduction) consist formally of two steps, electron transfer and fragmentation of the radical ion obtained. These reactions are carried out by means of a metal salt or electrochemically.

$$\text{oxidation} \quad \text{R–X} \xrightarrow{-e^-} |\text{R–X}|^{\bullet+} \longrightarrow \text{R}^\bullet + \text{X}^+ \quad (9.45)$$

$$\text{reduction} \quad \text{R–X} \xrightarrow{+e^-} |\text{R–X}|^{\bullet-} \longrightarrow \text{R}^\bullet + \text{X}^- \quad (9.46)$$

Oxidation
Carboxylic acids
Directly, by means of an oxidizing salt (9.47) - (9.48),

$$\text{RCO}_2\text{H} + \text{M}^{2+} \xrightarrow{\text{M}^{2+} = \text{Ce}^{4+}, \text{Pb}^{4+}, \text{Co}^{3+}} \text{RCO}_2^\bullet + \text{M}^+ + \text{H}^+ \quad (9.47)$$

$$\text{RCO}_2^\bullet \longrightarrow \text{R}^\bullet + \text{CO}_2 \quad (9.48)$$

by use of a chemical mediator (9.49) - (9.50)

$$\text{S}_2\text{O}_8^{2-} + 2\,\text{Ag}^+ \longrightarrow 2\,\text{Ag}^{2+} + 2\,\text{SO}_4^{2-} \quad (9.49)$$

$$\text{RCO}_2\text{H} + \text{Ag}^{2+} \longrightarrow \text{R}^\bullet + \text{CO}_2 + \text{Ag}^+ + \text{H}^+ \quad (9.50)$$

or by electrochemical oxidation of carboxylates (Kolbe reaction (9.51)).

$$\text{RCO}_2^- \xrightarrow{-e^-} \text{RCO}_2^\bullet \longrightarrow \text{R}^\bullet + \text{CO}_2 \quad (9.51)$$

Alkylaromatics
Oxidation can be chemical or electrochemical.

$$\text{ArCH}_3 \xrightarrow{-e^- \text{ or } \text{Ce}^{4+}} |\text{ArCH}_3|^{\bullet+} \xrightarrow{\text{Nu}} \text{ArCH}_2^\bullet + \text{NuH}^+ \quad (9.52)$$

Enolizable keto compounds

$$\text{CH}_2(\text{CO}_2\text{CH}_3)_2 \xrightarrow[\text{Nu}]{\text{Mn}^{3+}} {}^\bullet\text{CH}(\text{CO}_2\text{CH}_3)_2 + \text{NuH}^+ \quad (9.53)$$

Reduction
Diazonium salts

$$ArN_2^+ + Cu^+ \longrightarrow ArN_2^\bullet + Cu^{2+} \quad (9.54)$$

$$ArN_2^\bullet \longrightarrow Ar^\bullet + N_2 \quad (9.55)$$

Halides

The electrochemical reaction involves two steps for aromatic halides (9.56) but only one for aliphatic halides, R–X (9.57). The problem in this type of reaction is to avoid the reduction of the radical to an anion.

$$Ar-X \xrightarrow{+e^-} |Ar-X|^{\bullet-} \longrightarrow Ar^\bullet + X^- \quad (9.56)$$

$$R-X \xrightarrow{+e^-} R^\bullet + X^- \quad (9.57)$$

Chemical reduction can be achieved by means of a good electron donor, such as samarium iodide (SmI_2).

$$R-X + SmI_2 \longrightarrow |R-X|^{\bullet-} \longrightarrow R^\bullet + XSmI_2 \quad (9.58)$$

Peroxides

Fenton's classical reaction (9.59)

$$H_2O_2 + Fe^{2+} \longrightarrow HO^\bullet + HO^- + Fe^{3+} \quad (9.59)$$

is applicable to many other peroxides and reducing metal salts (9.60). In this type of reaction half the peroxide is lost as a radical source since it is converted to anion (9.61).

$$t\text{-BuOOH} + Fe^{2+} \longrightarrow t\text{-BuO}^\bullet + HO^- + Fe^{3+} \quad (9.60)$$

$$(PhCO_2)_2 + M^+ \xrightarrow{M^+ = Cu^+, Cr^{2+}, Ti^{3+}, Fe^{2+}} PhCO_2^\bullet + PhCO_2^- + M^{2+} \quad (9.61)$$

Chloroamines

$$\begin{array}{c} R \\ \diagdown \\ NH\text{-}Cl \\ \diagup \\ R \end{array} + M^+ \xrightarrow{M^+ = Ti^{3+}, Fe^{2+}} \begin{array}{c} R \\ \diagdown \\ N\text{--}H \\ \diagup \\ R \end{array}^{\bullet+} + Cl^- + M^{2+} \quad (9.62)$$

9-2-6 Molecule-molecule reactions

Radicals are sometimes generated by the interaction of two or more molecules. This type of reaction may be very fast, as in the case of the polymerization, without initiator, of styrene or methyl methacrylate or the reaction of *tert*-butyl hypochlorite with styrene (9.63), which gives the radical addition product in the dark at 0°C. The addition of N-chloroamines to olefins is probably the result of an analogous process. The mechanism of molecule-molecule interactions which lead to radical formation is not yet well understood.

$$t\text{-BuOCl} + \text{Ph–CH=CH}_2 \longrightarrow \text{Ph–CHCl–CH}_2\text{–O}t\text{-Bu} \quad (9.63)$$

9-3 CONCLUSIONS

There are many ways of producing radicals but they are not all equivalent; their principal differences can be summarized as follows:

– *temperature*. Redox systems can be used at room temperature whereas peroxides require temperatures which are more or less high (80 to 150°C) depending on their structure; photolysis, however, sometimes makes it possible to operate at lower temperatures.

– *reaction medium*. Certain methods (chloroamine, decarboxylation of acids by $Ag^+/S_2O_8^{2-}$) imply an acidic medium.

– *nature of the radical*. Peroxide decomposition gives essentially electrophilic radicals, acyloxyl or oxyl, leading to strong O–H bonds, whereas acid decarboxylation gives nucleophilic carbon radicals. Moreover, enthalpic factors and the polar character can have an important bearing on the outcome of the reactions.

Table 9.2 summarizes the various methodologies and the radicals formed.

Starting material	Methodology	Radical formed
R–H	peroxide, AIBN	R$^\bullet$ (alkyl)
R–Hal	peroxide, AIBN Bu$_3$SnH/AIBN addition/fragmentation	
R–SeR, R–SR	Bu$_3$SnH/AIBN	
R–CO$_2$H	addition/fragmentation	
Ar–Hal	Bu$_3$SnH/AIBN redox (reduction)	Ar$^\bullet$
Ar–SePh	Bu$_3$SnH/AIBN	
Ar–N$_2^+$Cl$^-$, ArCO$_2$H	redox (oxidation)	
RONO, ROCl, ROOR^1	hv, redox (reduction)	RO$^\bullet$
R$_2\overset{+}{\text{N}}$HCl	hv, redox (reduction)	R$_2\overset{\bullet+}{\text{N}}$H

Table 9.2 – *Radical production*.

10 – RADICAL-RADICAL REACTIONS

Radical-radical reactions are very important since they cause radical species to disappear. They are particularly involved in the termination steps of chain reactions. There are two types of radical-radical reaction:

- *recombination* (10.1) which leads to a single molecule;
- *disproportionation* (10.2) which gives, by transfer of a hydrogen from one radical to another, two molecules, one of which is saturated and the other unsaturated.

$$R^1-CH_2-CH_2^\bullet + {}^\bullet R^2 \quad \begin{array}{l} \xrightarrow{k_r \text{ recombination}} R^1-CH_2-CH_2-R^2 \quad (10.1) \\ \xrightarrow{k_d \text{ disproportionation}} R^1-CH=CH_2 + R^2-H \quad (10.2) \end{array}$$

10-1 RECOMBINATION

Recombination reactions are also referred to as *coupling* reactions, or *dimerization* if the two radicals are identical. There have been many studies both in the gas phase and in solution.

The recombination of two alkyl radicals to form a C–C bond releases about 80 kcal/mol of energy. Since recombination reactions are exothermic the activation energies are practically zero and the rate constants close to the diffusion rate. Nevertheless, it must not be forgotten in interpreting the results that the instantaneous radical concentrations are often *very* small.

10-1-1 Gas phase

The limitations of the reaction are the rate of displacement, the frequency and the efficiency of collisions. In the case of methyl radicals the rate constant for recombination (k_r) is about 10^{10} $M^{-1}s^{-1}$ and reaction occurs every 4 ~ 6 collisions. Rate constant k_r is independent of the temperature, which means that $E_a \approx 0$, and decreases slightly with the size of the radical, by a factor of about 10 on going from Me$^\bullet$ to t-Bu$^\bullet$.

10-1-2 Solution

Two coupling processes are to be considered:

- radicals produced separately must diffuse in the solvent before meeting. The recombination rate is then limited by the rate of radical diffusion;
- radicals produced in pairs from a single precursor (then called *geminate*) recombine in the solvent cage before diffusion.

The identical products formed inside and outside the solvent cage can be distinguished by chemically induced nuclear polarization (Chap. 2). The effect of different factors on the recombination rate has been studied.

Solvent

The recombination reaction is controlled by the diffusion rate of the two radicals. The second-order rate constant which can be calculated from the laws of diffusion is related to the viscosity of the solvent (equation (10.3)).

$$k_{diffusion} = 8\ RT / (3 \times 10^3\ \eta)\ M^{-1}s^{-1} \qquad (10.3)$$

Since the solvents most commonly used (cyclohexane, benzene, CCl_4, water, etc.) have low viscosities, the variations observed are small. Nonetheless, at 30°C, *tert*-butylperoxyacetate gives, by cage recombination, *tert*-butylmethylether with, respectively, yield of 27% in pentane, 40% in decalin and 75% in nujol. In certain solvents, benzene in particular, the radical can be solvated.

Scheme 10.1

In general, for small radicals the diffusion rate constants are of the order of 10^{10} $M^{-1}s^{-1}$ and those of recombination about $10^9\ M^{-1}s^{-1}$ or less. For Me$^{\bullet}$ k_r is about 5 times smaller in solution than in the gas phase ($4.5 \times 10^9\ M^{-1}s^{-1}$ as against 2.4×10^{10} $M^{-1}s^{-1}$).

Temperature

As in the gas phase, the temperature has little effect upon the value of k_r. Therefore E_a is close to zero.

Structure

The data in Table 10.1 indicate that k_r is close to $10^9\ M^{-1}s^{-1}$ for lightly substituted carbon radicals.

radical	solvent	$k_c \times 10^{-9}\ M^{-1}s^{-1}$
$^{\bullet}CH_3$	cyclohexane	4.5
$C_6H_{13}^{\bullet}$	cyclohexane	1.4
cy-$C_6H_{11}^{\bullet}$	cyclohexane	1.4
cy-$C_6H_{11}^{\bullet}$	benzene	0.83
$PhCH_2^{\bullet}$	cyclohexane	1.0
$PhCH_2^{\bullet}$	benzene	0.9
$(CH_3)_3C^{\bullet}$	cyclohexane	1.1
$^{\bullet}CCl_3$	CCl_4	2.5

Table 10.1 – *Rate constants (± 10~25%) for radical dimerization in solution at 25°C.*

Steric hindrance or the nature of the substituents at the radical center can lead to large variations in E_a and k_r. Thus, the di-*tert*-butylmethyl radical **1** and 2,4,6-tri-*tert*-butylphenyl radical **2** which have half-lives of 6 and 0.1 s, respectively, are persistent

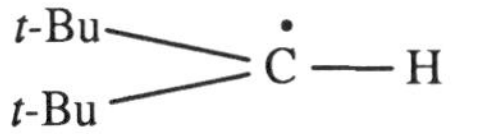

Recombination

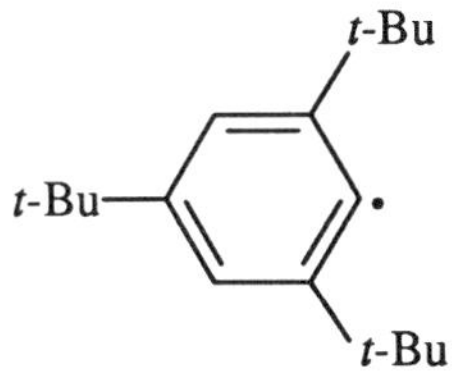

Scheme 10.2

(Chap. 4) since steric hindrance markedly increases E_a ($\approx$ 20 kcal/mol for **1**) and decreases the rate of recombination. The reverse reaction, homolysis, is very sensitive to steric crowding, which weakens the C–C bond of the dimer (Table 10.2).

R–R $\xrightarrow{\Delta H}$ 2 R•			
R	CH$_3$–	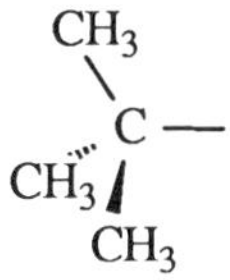 (CH$_3$)$_3$C–	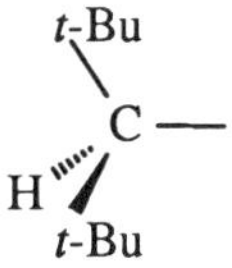 (t-Bu)$_2$CH–
ΔH (kcal/mol)	88.0	67.4	36.4

Table 10.2 – *Enthalpy of reaction for homolysis of saturated hydrocarbons.*

Because of steric hindrance the triphenylmethyl radical **3** does not recombine to hexaphenylethane but gives a product of addition to the aromatic nucleus **4** (10.4).

$$2 \text{ Ph}_3\text{C}^\bullet \longrightarrow \mathbf{4} \qquad (10.4)$$

Thermodynamic stabilization provided by the substituents can also be relevant. For example, the activation energy of 6.4 kcal/mol for recombination of the dicyanoaminomethyl radical (10.5) is not due to steric hindrance but to captodative stabilization of the radical.

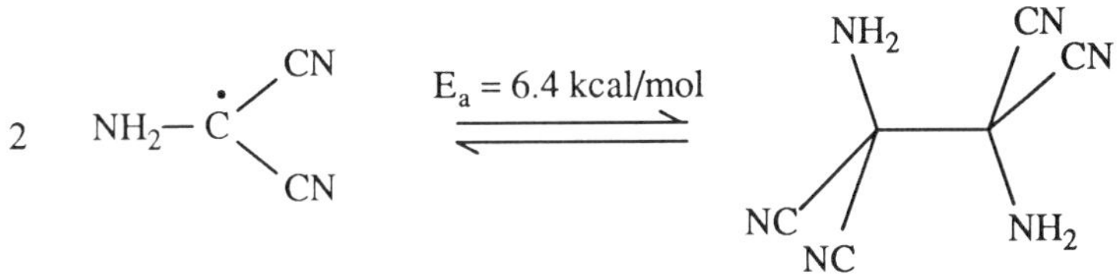

$$2 \text{ NH}_2-\overset{\bullet}{\text{C}}(\text{CN})_2 \xrightleftharpoons[]{E_a = 6.4 \text{ kcal/mol}} (\text{NC})_2(\text{NH}_2)\text{C}-\text{C}(\text{NH}_2)(\text{CN})_2 \qquad (10.5)$$

Stereoselectivity

The recombination of two radicals is stereoselective if one of the substituents on the radical carbon is sufficiently bulky. In reaction (10.6) the *dl*/meso ratio is 1 when R = Me or Et and 1.66 when R = *t*-Bu.

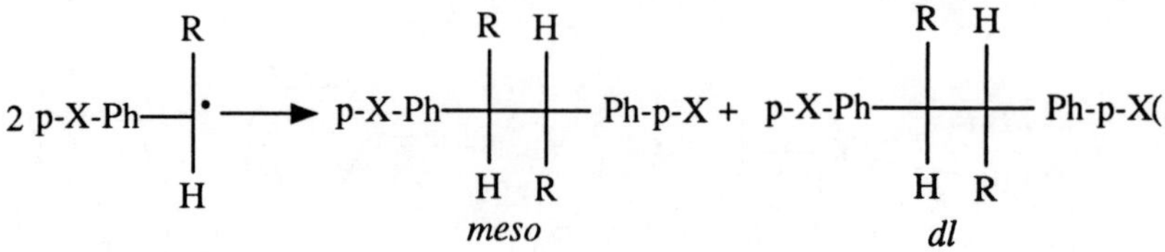

$$2 \text{ p-X-Ph} \underset{H}{\overset{R}{-}}\!\!\cdot \longrightarrow \text{p-X-Ph}\underset{H\ \ R}{\overset{R\ \ H}{-\!\!|\!\!|\!\!-}}\text{Ph-p-X} + \text{p-X-Ph}\underset{H\ \ R}{\overset{R\ \ H}{-\!\!|\!\!|\!\!-}}\text{Ph-p-X} \quad (10.6)$$

meso dl

Cage recombination can favor retention of stereochemistry. The thermal decomposition of the optically active azo compound (Scheme 10.3) takes place with 17% retention. This indicates that inversion of the radical is about 15 times faster than recombination.

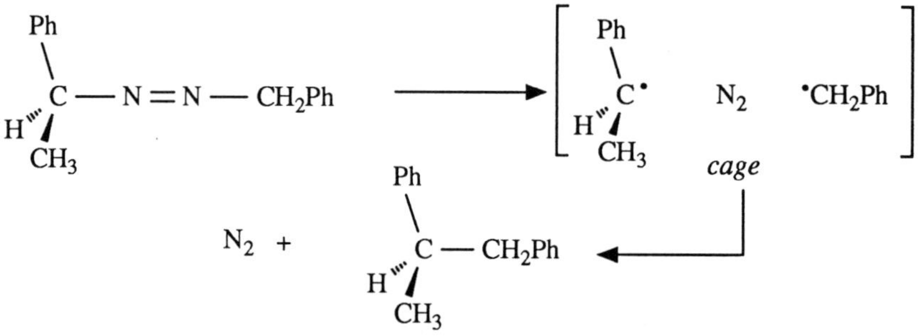

Scheme 10.3

10-1-3 Peroxyl radicals

Autoxidation reactions proceed by a radical chain mechanism (Chap. 16-2) in which termination is mainly the duplication of two peroxyl radicals to give *non-radical products*. The mechanism by which these products are formed is complex but it is generally accepted that a tetroxide is formed in equilibrium with the peroxyl radicals. With tertiary peroxyl radicals this tetroxide decomposes to alkoxyl radicals in the solvent cage; these may or may not lead to a termination reaction (Scheme 10.4). However, if the temperature is high enough the peroxide, ROOR, can again produce RO• radicals capable of initiating a chain.

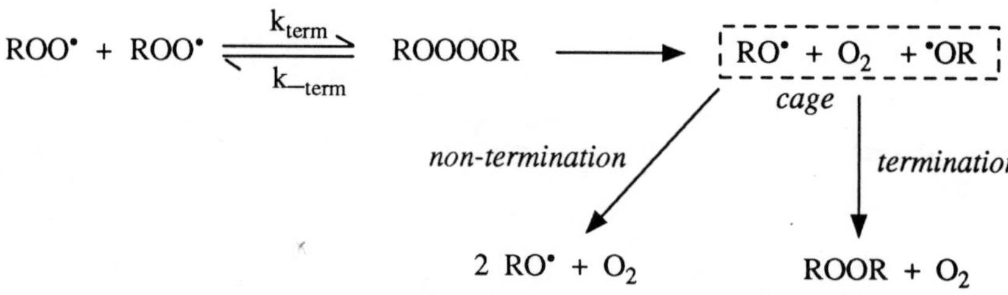

$$\text{ROO}^\bullet + \text{ROO}^\bullet \underset{k_{-term}}{\overset{k_{term}}{\rightleftharpoons}} \text{ROOOOR} \longrightarrow [\text{RO}^\bullet + \text{O}_2 + {}^\bullet\text{OR}]_{cage}$$

non-termination → $2\,\text{RO}^\bullet + \text{O}_2$

termination → $\text{ROOR} + \text{O}_2$

Scheme 10.4

In the case of primary and secondary peroxyl radicals an electrocyclic tetroxide decomposition reaction (10.7) has been proposed to explain the formation of non-radical products. The tetroxide formation equilibrium depends on the temperature and the solvent. Since the activation energy for the decomposition of the tetroxide is generally greater than that of the reverse reaction, k_{-d}, raising the temperature favors decomposition to products. The rate constant, $2 k_d$, ranges from 10^4 to 10^9 $M^{-1}s^{-1}$, being lower for tertiary peroxyl radicals than for secondary and primary.

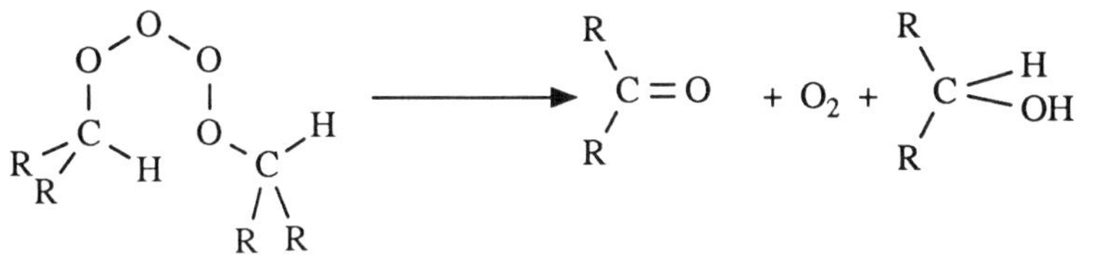

$$\underset{R}{\overset{R}{C}}\text{(tetroxide)} \longrightarrow \underset{R}{\overset{R}{C}}=O + O_2 + \underset{R}{\overset{R}{C}}\overset{H}{\underset{OH}{\diagdown}} \quad (10.7)$$

10-2 DISPROPORTIONATION

Disproportionation consists in the transfer of a hydrogen β to a radical center to another radical to give a saturated product and an unsaturated molecule (10.8).

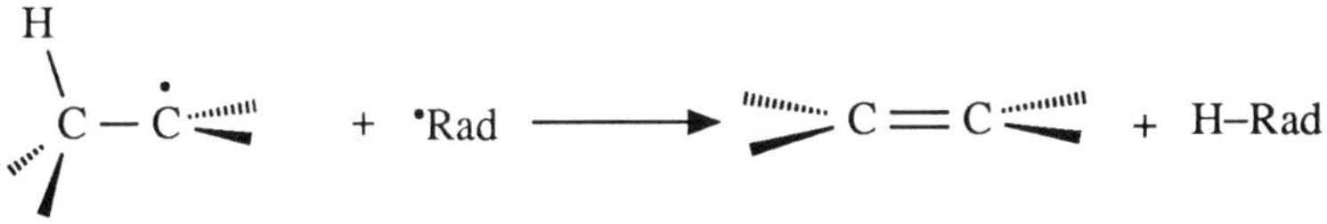

$$\overset{H}{\underset{}{\diagdown}}C-\overset{\cdot}{C}\diagup + \;^{\bullet}\text{Rad} \longrightarrow \diagdown C=C\diagup + \text{H–Rad} \quad (10.8)$$

The probability of this reaction occurring increases with the number of β hydrogens. The tri-*tert*-butylmethyl radical **3**, because there are no β hydrogens, gives no disproportionation product and, since dimerization is very slow because of steric hindrance, this radical is persistent ($t_{1/2}$ = 15 s). Phenyl radicals **4** give only recombination products, since disproportionation would produce benzyne which is much too unstable.

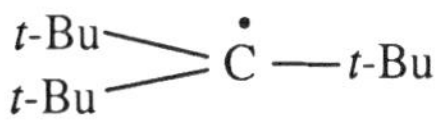

$$\underset{t\text{-Bu}}{\overset{t\text{-Bu}}{\diagdown}}\overset{\cdot}{C}-t\text{-Bu} \quad \mathbf{3} \qquad \qquad \mathbf{4}$$

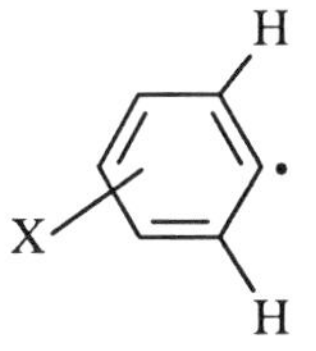

Scheme 10.5

Compared to hydrogen transfer from a molecule to a radical (10.9), disproportionation has a much higher pre-exponential factor (10^{11} as against 10^8) and a much lower activation energy (close to zero as against 10 kcal/mol). The BDE of a C–H bond β to a radical center is much reduced. It can be concluded that the transition state of a disproportionation reaction has breaking and forming bonds much longer than those in a molecule-radical hydrogen transfer.

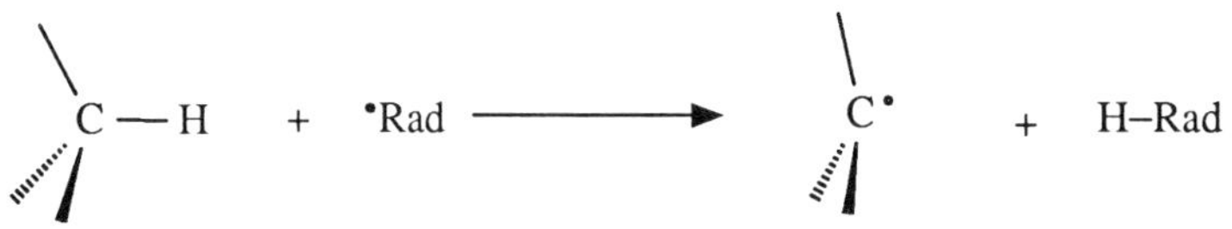

$$\diagdown C-H + \;^{\bullet}\text{Rad} \longrightarrow \diagdown \overset{\bullet}{C} + \text{H–Rad} \quad (10.9)$$

Disproportionation can be slowed by steric and conformational effects. The tri-isopropylmethyl radical **5** is persistent due to steric hindrance but also because the β hydrogens are in the plane orthogonal to the p orbital containing the unpaired electron, which reduces the disproportionation rate (Scheme 10.6).

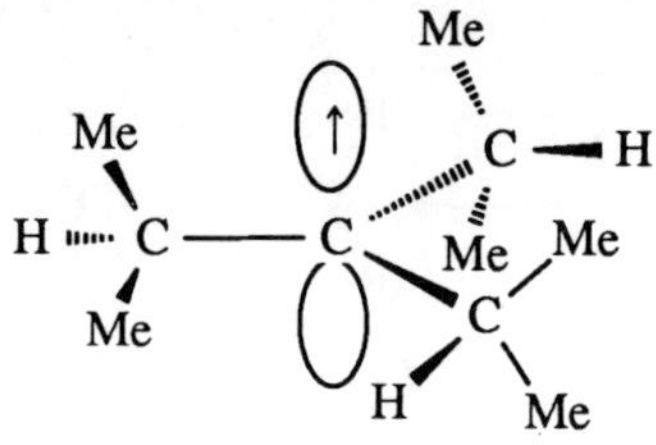

Scheme 10.6

10-3 DISPROPORTIONATION/RECOMBINATION RATIO

It is interesting to know the relative importance of the two types of radical-radical reaction. This is expressed by the rate constant ratio, k_d/k_r, and has been determined for a good number of radicals. In Table 10.3 some results are given for the reactions of identical radicals.

Radicals	k_d/k_r
$CH_3CH_2^{\bullet}$	0.15
$CH_3^{\bullet}CHCH_3$	1.2
$(CH_3)_3C^{\bullet}$	4.5
$cy\text{-}C_6H_{11}^{\bullet}$	1.1
cumyl$^{\bullet}$	0.054
$CH_3CH_2O^{\bullet}$	12

Table 10.3 – *Ratio k_d/k_r for various radicals in solution.*

The ratio k_d/k_r is about the same in the gas phase as in solution. It is little affected by the temperature and the nature of the solvent. On the other hand, the structure of the radicals, particularly steric effects, can be all-important. The ratio increases on going from the ethyl to the *tert*-butyl radical because the number of β hydrogens increases. The ratio is high for alkoxyl radicals since the O–O bond in the dimer is weaker than the O–H and C=O bonds formed by disproportionation.

11 - SUBSTITUTION REACTIONS

In general, in *bimolecular homolytic substitution* reactions, S_H2 (11.1) a part, Rad^2, of a molecule, Rad^2A, is replaced by an entering radical, $Rad^{1\bullet}$, with the formation of a new radical, $Rad^{2\bullet}$. These reactions are particularly important since they occur as an elementary step in chain mechanisms, which enter into very many radical processes.

$$Rad^{1\bullet} \; + \; Rad^2A \; \longrightarrow \; Rad^1A \; + \; Rad^{2\bullet} \quad (11.1)$$

This type of reaction manifests the greatest differences between radical and ionic chemistry. Substitution reactions *on carbon*, frequently observed in ionic chemistry (nucleophilic substitution at sp^3 carbon or sp^2, electrophilic substitution) essentially because of the polarizability of the breaking bond, are very rare in radical chemistry. Conversely, substitution at a monovalent atom (hydrogen or halogen), which is found in ionic chemistry in carbanion formation or the first stage of elimination reactions, plays a major part in radical chemistry.

11-1 DIFFERENT TYPES OF S_H2

11-1-1 Transfer or abstraction reactions

If A is a monovalent atom (hydrogen or halogen) the reaction is known as *transfer* or *abstraction*. It goes through a linear, concerted transition state (a bond is formed at the same time as another breaks) with frontal attack on the radical.

$$R^{1\bullet} \; + \; A-R^2 \; \longrightarrow \; [\, R^1 \cdots A \cdots R^2 \,]^{\bullet *} \; \longrightarrow \; R^1-A \; + \; R^{2\bullet} \quad (11.2)$$
$$A = H \text{ or } Hal$$

11-1-2 Displacement reactions

If A is a multivalent atom there are two types of mechanism:

— for a peroxide bond the reaction is a one-step process (11.3) exactly as above (11.2);

$$R^{\bullet} \; + \; \underset{R^1}{O-OR^2} \; \longrightarrow \; R-O-R^1 \; + \; {}^{\bullet}OR^2 \quad (11.3)$$

— if A is an atom such as phosphorus the *displacement* reaction occurs in two steps: formation, by addition, of a hypervalent radical intermediate (expansion of the valence shell of the atom attacked), then α-fragmentation of this intermediate.

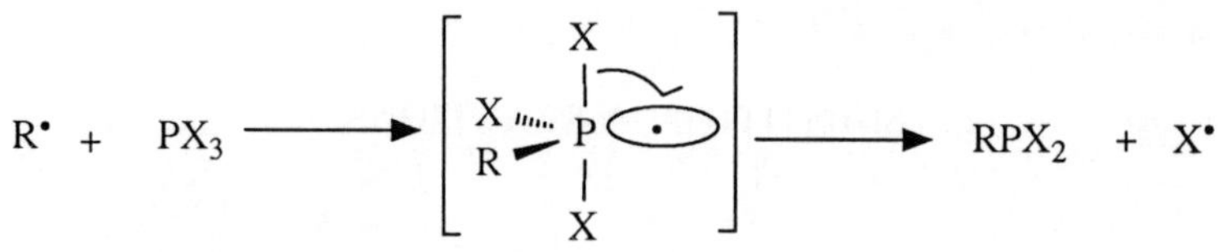

$$R^{\bullet} + PX_3 \longrightarrow \left[\begin{array}{c} X \\ | \\ X_{\cdots} P \\ R \nearrow | \\ X \end{array} \right] \longrightarrow RPX_2 + X^{\bullet} \quad (11.4)$$

11-1-3 Intramolecular reactions

The transfer reaction can be intramolecular, in which case it is denoted S_Hi: *intramolecular homolytic substitution*.

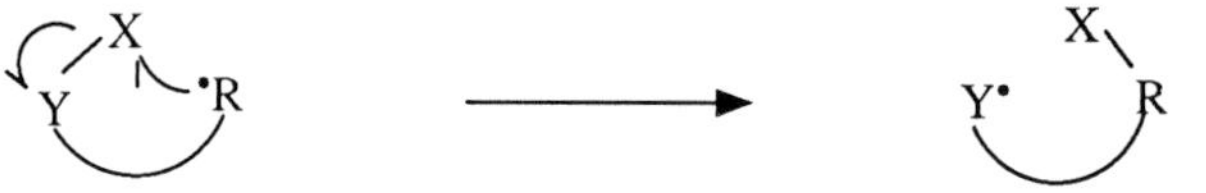

$$\text{(11.5)}$$

11-2 S_H2 ON MONOVALENT ATOMS

11-2-1 Hydrogen transfer

A radical reacts with a C–H bond, via a linear transition state (11.2) by frontal attack on the hydrogen atom. There are two consequences of this fundamental difference from the S_N2 reaction, where the nucleophile reacts at carbon:

– lack of stereoselectivity at the carbon of the C–H bond;
– weak steric effects, e.g. no neopentyl retardation.

Rate constant

Since it is often difficult to measure the absolute rate constant a relative rate constant or *relative reactivity* is defined by comparing:

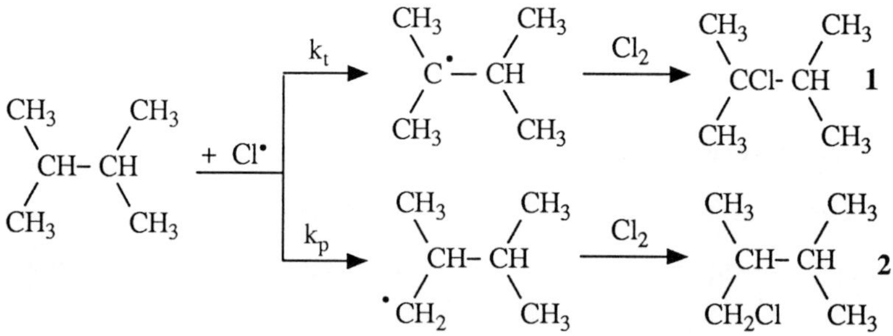

Scheme 11.1

– either several substrates, one of which is taken as a reference, by determining the relative amounts of products formed. In a competitive system, where toluene is mixed with a substituted toluene, the relative reactivity in hydrogen abstraction from methyl is expressed by taking toluene as the reference;
– or between two reaction centers in a single molecule. Given the number of

hydrogens in the molecule (and their type), a *relative selectivity per hydrogen* is defined. Thus, in the chlorination of 2,3-dimethylbutane (Scheme 11.1) the ratio [1]/[2] of the tertiary and primary chlorides, corrected for a statistical factor of 6 (2 tertiary and 12 primary hydrogens), gives:

$$RS_{t/p} = 6\ k_t/k_p = 6\ [1]/[2] \qquad (11.8)$$

The data in Table 11.1 show that the transfer of primary, secondary and tertiary hydrogens is largely controlled by enthalpy.

radical	T(°C)	parameters for primary H			relative selectivity		
		E_a	logA	ΔH	prim.	sec.	tert.
$\cdot CH_3$	150	12	9	−6	1	10	80
$F\cdot$	25	0.3	11	−36	1	1	2
$Cl\cdot$	25	1	11	−5	1	4	7
$Br\cdot$	150	14	12	+11	1	80	1600
$t\text{-BuO}\cdot$	25	8	8	−4	1	8	36
$\cdot CCl_3$	150	14	9	+8	1	80	2300
$\cdot CF_3$	150	9	8	−7	1	8	90

Table 11.1 – E_a *(kcal/mol), logA ($M^{-1}s^{-1}$) and ΔH (kcal/mol) for primary hydrogen transfer by different radicals. Selectivity relative to primary hydrogen for a secondary or tertiary hydrogen.*

Pre-exponential factor, logA

The pre-exponential factor, logA, has about the same value for all monoatomic radicals (11 ~ 12) but is different from that of polyatomic radicals (8 ~ 9). This explains why chemical reactivity is often expressed in terms of E_a only.

Activation energy, E_a

The most important result is that the more the reaction is enthalpy-favored, the smaller the E_a the lower the selectivity. Polyani's relationship (Chap. 6, eq. (6.6)), which correlates E_a with ΔH, is only valid for reactions of the same type, without marked polar effects; for example, Me$\cdot$ abstracting a primary, secondary or tertiary hydrogen from an alkane. If not, there is no general relationship between E_a and ΔH. The reactions of Cl$\cdot$ and Me$\cdot$ radicals have the same ΔH but very different E_a because the polar effect is involved (Chap. 6). The sequence is always $H_{prim} < H_{sec} < H_{tert}$ and the less favorable the reaction enthalpy, the greater the selectivity (cf. Br$\cdot$ and $\cdot CCl_3$). In these last cases the stability of the radical formed is very important as, in the transition state, which is late on the reaction path (Hammond postulate), the C–H bond is almost completely broken and the character of the incipient radical very pronounced.

Polar effect

Table 11.2 presents a few examples where this effect has an important bearing upon the regioselectivity of hydrogen abstraction from a substituted alkane by different radicals. When the substituent X is electron-attracting the C–H bonds at atoms 1 and 2 are electron-deficient and less reactive towards an electrophilic radical, and vice versa when the substituent is electron-donating (MeO, t-Bu). The more the radical is electrophilic ($Me_2NH^{\cdot+}$), the more the polar effect is marked. In the case where X = Ph, hydrogen abstraction from the 1 position is preferred for reasons of enthalpy. This latter effect is accentuated with Br$\cdot$ since, the reaction being much less exothermic, the transition state resembles the products (Hammond postulate) and the regioselectivity is determined by the stability of the benzylic radical formed. The selectivity for abstraction of secondary benzylic and aliphatic hydrogens is 0.75 for Cl• and 7 300 for Br•.

X	radical	1 XCH$_2$—	2 CH$_2$—	3 CH$_2$—	4 CH$_3$
H	Cl•	1	3.6	3.6	1
H	Br•	1	80	80	1
Cl	Cl•	0.8	2.1	3.7	1
Cl	Br•	34	32	80	1
Cl	(CH$_3$)$_2$NH•+	0	2	24	1
CF$_3$	Cl•	0.04	1.2	4.3	1
CH$_3$O	Cl•	3.5	0.7	4,4	1
(CH$_3$)$_3$C	Cl•	2.9	3.7	5.3	1
C$_6$H$_5$	Cl•	5.9	2.7	4.0	1
CH$_2$=CH	Cl•	4.4	1.1	3.6	1
CN	Cl•	0.2	1.7	3.9	1
CN	Br•	20	8	80	1

Table 11.2 – *Relative selectivity, $RS_{s/p}$, for various radicals on 1-substituted butanes.*

Solvent effect

In solvents such as benzene or carbon disulfide, the Cl• radical gives charge-transfer complexes (Scheme 11.2). The complexed Cl• radicals are less reactive and much more selective than when *free* in the gas phase or in carbon tetrachloride. Thus, in the chlorination of 2,3-dimethylbutane the selectivity, $R_{t/p}$, is about 4 in CCl$_4$, 11 in benzene and 15 in CS$_2$ (concentration = 2 M). For the latter solvents $R_{t/p}$ increases with increase in the concentration of the solvating solvent.

Scheme 11.2

11-2-2 Halogen transfer

A radical reacts with a C–X bond, via a linear transition state (11.2) by frontside attack on the halogen atom. This is a fundamental difference from the S$_N$2 reaction, where the nucleophile attacks a C–X bond at carbon.

Carbon radicals

The reactivity for transfer of a halogen bonded to carbon (C–X bond) by carbon radicals follows the order of the C–X bond dissociation energies, (11.9). Fluorine is almost never transferred but iodine transfer is so rapid that it can serve as a carbon

$$>\!\!\overset{\bullet}{C}\!\!- \;+\; X-\overset{|}{\underset{\backslash}{C}}\!\!\! \xrightarrow{F<Cl<Br<I} \;>\!\!\overset{}{C}\!\!-X \;+\; >\!\!\overset{\bullet}{C}\!\!- \qquad (11.9)$$

radical trap. This reaction is often used to obtain bromo and chloro derivatives by means of solvents such as $CBrCl_3$ and CCl_4.

When a compound has a hydrogen and a halogen on the same carbon the reaction is usually *chemoselective*. Hydrogen is transferred preferentially for chloro derivatives (11.10) but Br or I are transferred (11.11) in the case of bromo and iodo derivatives. Nevertheless, in every case hydrogen or halogen transfer is approximately thermoneutral. This is explained by correlation diagrams (Chap. 6).

$$\ce{>\dot{C}-} + \ce{>C<^H_X} \longrightarrow \begin{cases} X = Cl: \ce{>C-H} + \ce{>\dot{C}-X} & (11.10) \\ X = Br, I: \ce{>C-X} + \ce{>\dot{C}-H} & (11.11) \end{cases}$$

Si, Ge or Sn-centered radicals

Except for fluoro compounds which do not react, it is always the halogen atom which is transferred in preference to hydrogen, because of enthalpic control (Table 11.3). This is, moreover, the usual method both for producing radicals centered on Si, Ge or Sn (Chap. 9) and for reducing halides (Chap. 16). The reaction is selective as regards the halogen and the radical.

$$\ce{M^{\bullet}} + \ce{X-C} \xrightarrow[\ce{M^{\bullet}(Si^{\bullet} > Ge^{\bullet} \approx Sn^{\bullet})}]{X (I > Br > Cl > H)} \ce{M-X} + \ce{>\dot{C}-} \quad (11.12)$$

ΔH		X			
		I	Br	Cl	H
M	Si	−29	−32	−32	+15
	Ge	−15	−40	−36	+27
	Sn	−16	−21	−21	+31

Table 11.3 − ΔH (kcal/mol) for reaction (11.12).

Gem-dihalo compounds can be reduced selectively by Bu_3SnH.

$$\text{cyclopropane-Br,Cl} \xrightarrow{Bu_3SnH} \text{cyclopropane-H,Cl} \quad (11.13)$$

$$\text{norcarane-Cl,F} \xrightarrow{Bu_3SnH} \text{norcarane-H,F} \quad (11.14)$$

11-2-3 Stereoselectivity

In most cases a carbon radical is practically planar (Chap. 3). The transfer reaction can occur on one or other of the faces of a carbon radical (11.15) - (11.16). Configurational, steric, conformational and stereoelectronic effects can make these faces non-equivalent (heterotopic), which induces a certain stereoselectivity. Generally speaking, the stereoselectivity decreases when the temperature is raised.

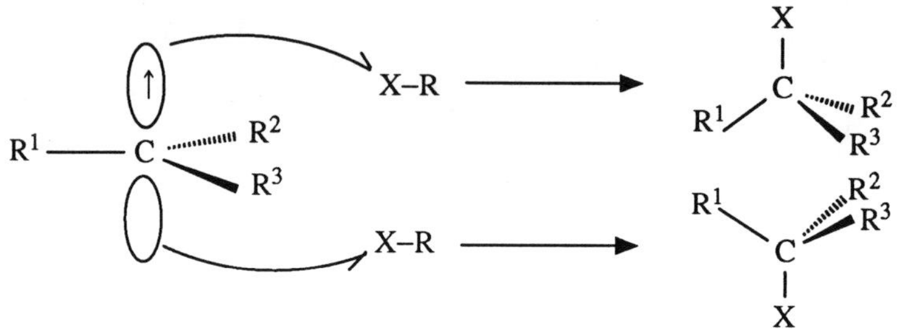

Configurational effect

In pyramidal radicals with a significant inversion barrier transfer can occur before inversion. Such is the case for cyclopropyl radicals with an electron-attracting group like halogen at the 1 position, (reactions (11.13), (11.14) and (11.17)), and for phosphonyl radicals (11.18).

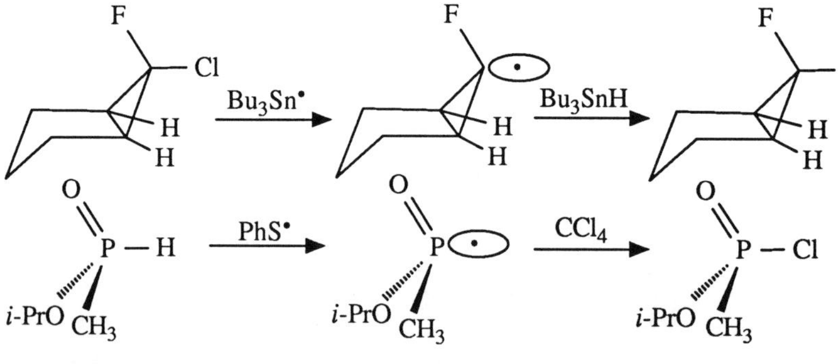

Steric effect

In cyclic systems the steric effect determines to a large extent the approach of the reactant to the radical. In the case, for example, of the radical obtained by addition of R• to methylmaleic anhydride, *anti* transfer, to the less crowded face, leads to the *cis* product and *syn* transfer to the *trans*. The *cis/trans* ratio goes from 62/38 (R = *n*-hexyl) to 94/6 (R = *t*-Bu), the increase in steric hindrance disfavoring *syn* attack (Scheme 11.3).

Conformational effect

Examples of stereoselectivity in acyclic systems are somewhat rare because of free rotation about the C–C bonds. However, stereoselectivity is possible when the radical reacts in a preferred conformation and steric effects disfavor attack on one of the faces.

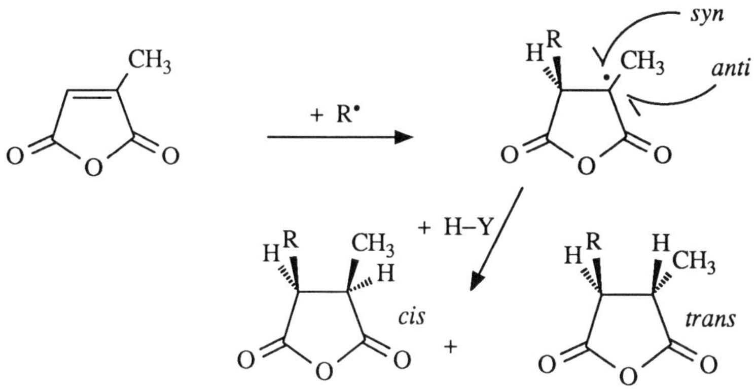

S_H2 on monovalent atoms

Scheme 11.3

(11.19)

The photochemical reaction of excess DBr on *cis* and *trans* 2-butenes is stereospecific at −70°C; *cis* 2-butene gives *erythro* 2-deuterio-3-bromobutane and *trans* 2-butene leads to the *threo* stereoisomer. The first step is the reversible addition of Br• to the double bond (11.20).

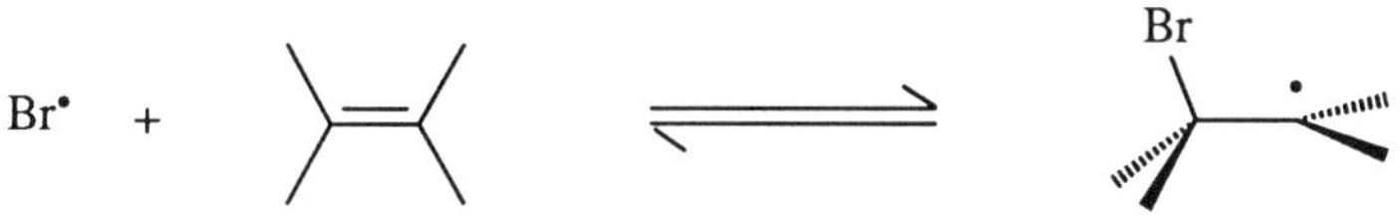

(11.20)

The second step is the transfer reaction where D enters *anti* to bromine (reactions (11.21) and (11.22)). Stereoselectivity is lost when the temperature is raised. This shows that the activation energies for reactions (11.21) and (11.22) are smaller than those for the elimination of Br•, (11.20), and for the interconversion of radicals **3** and **4** by rotation about the C–C bond. This rotation followed by Br• elimination explains the *cis* ⇌ *trans* isomerization of the double bonds.

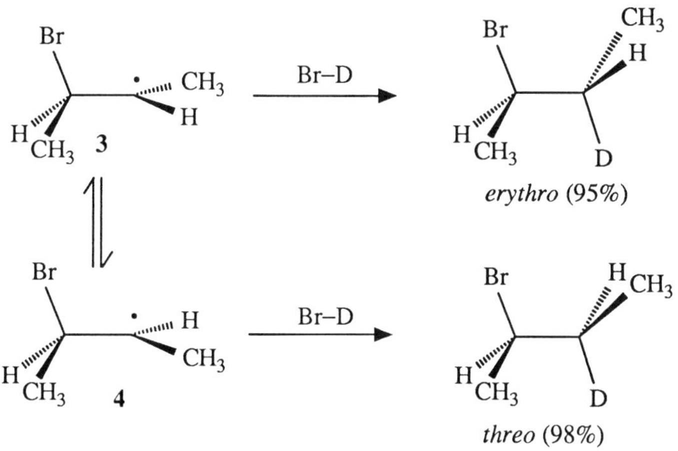

(11.21)

erythro (95%)

(11.22)

threo (98%)

In certain cases this free rotation can be slowed or prevented by the interaction of two large α and β substituents, favoring a conformation where steric effects control the stereoselectivity. For example, in radical **5**, obtained by addition of R• to diethyl citronate, the two ester groups are on opposite sides of the molecule. Consequently, hydrogen transfer from RHgH, *anti* or *syn* with respect to R, leads to the *threo* (11.24) and *erythro* (11.23) derivatives, respectively, in a ratio of 40 when R = *t*-Bu and only 2 for a less bulky group, R = *n*-hexyl.

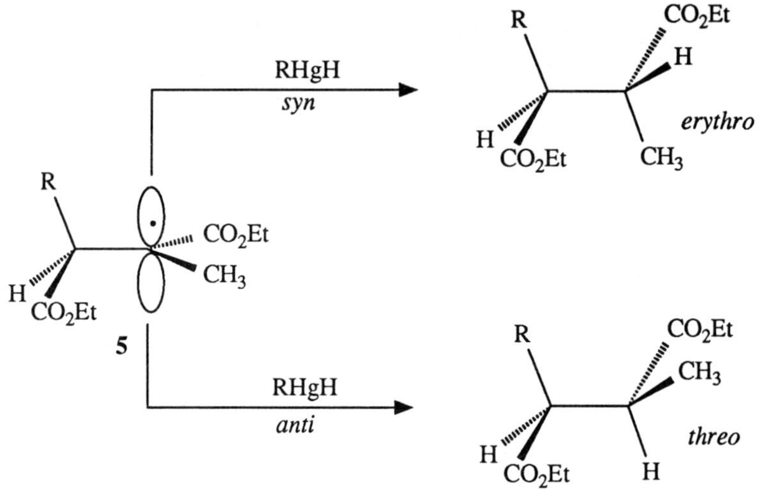

(11.23)

(11.24)

Stereoelectronic control

In the radical chemistry of sugars, the presence of an oxygen α to the radical center induces stereoelectronic control (antiperiplanar effect between one of the orbitals containing an oxygen lone pair and the p orbital of the radical carbon) which favors the transfer reaction at the axial face, despite the fact that it is the most crowded.

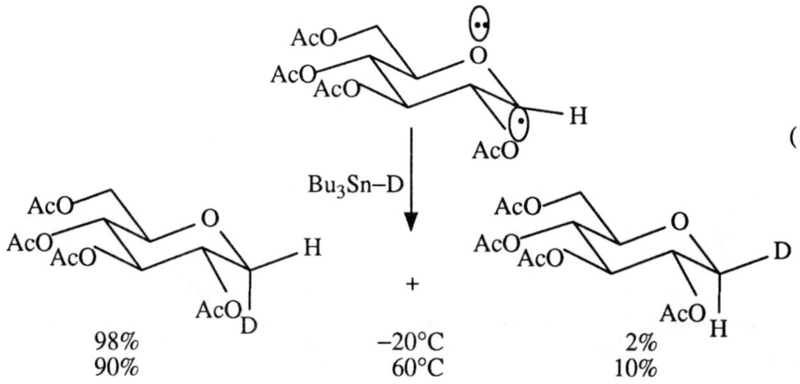

(11.25)

11-3 S_H2 ON MULTIVALENT ATOMS

Displacement reactions can occur by different mechanisms, depending on the nature of the multivalent atom M:

– either directly, via a concerted transition state (11.26). This reaction is usually only observed if the bond which forms (R–M) is significantly stronger than that which is breaking (M–X) (exothermic reaction).

$$R^\bullet + M-X \longrightarrow [R\cdots M\cdots X]^{\bullet *} \longrightarrow R-M + X^\bullet \quad (11.26)$$

– or via a radical intermediate resulting from expansion of the valence shell of M (11.27).

$$R^\bullet + M-X \longrightarrow [R-\overset{\bullet}{M}-X] \longrightarrow R-M + X^\bullet \quad (11.27)$$

11-3-1 Direct or concerted S_H2

This concerns essentially Group 14 (M = C, Si, Sn) and 16 (M = O, S, Se) atoms.

Group 14 atoms (C, Si, Sn)
S_H2 Reactions on carbon are extremely rare. One of the few examples known is the halogenation of strained cyclic compounds such as cyclopropane (11.28).

$$Cl^\bullet + \triangle \longrightarrow Cl-CH_2-CH_2-CH_2^\bullet \xrightarrow{Cl_2} Cl-(CH_2)_3-Cl \quad (11.28)$$

On the other hand, Si–C and especially Si–Si and Sn–Sn bonds are easily broken by radical attack.

$$^\bullet CF_3 + (CH_3)_4Si \longrightarrow CF_3Si(CH_3)_3 + {}^\bullet CH_3 \quad (11.29)$$

$$I^\bullet + (CH_3)_3Si-Si(CH_3)_3 \longrightarrow (CH_3)_3Si-I + (CH_3)_3Si^\bullet \quad (11.30)$$

$$t\text{-BuO}^\bullet + Bu_3Sn-SnBu_3 \longrightarrow t\text{-BuO-SnBu}_3 + Bu_3Sn^\bullet \quad (11.31)$$

Group 16 atoms (O, S, Se)
S_H2 Reactions of carbon radicals on *peroxide* (11.32) and *disulfide* (11.33) bonds are common.

$$R^\bullet + \overset{R^1}{\underset{}{\diagdown}}O-OR^2 \longrightarrow R-O-R^1 + {}^\bullet OR^2 \quad (11.32)$$

$$R^\bullet + \overset{R^1}{\underset{}{\diagdown}}S-SR^2 \longrightarrow R-S-R^1 + {}^\bullet SR^2 \quad (11.33)$$

With the peroxide bond, which is electrophilic, the more nucleophilic $R^\bullet$, the faster the reaction. This explains the acceleration of the decomposition of peroxides in solvents such as diethyl ether or alcohols, which generate nucleophilic radicals by hydrogen abstraction (Chap. 9). Benzoyl peroxide in ethanol, for example, gives reaction (11.34).

$$CH_3\overset{\bullet}{C}HOH + Ph-\underset{O-O}{\overset{O\quad\;\; O}{\overset{\|\quad\;\;\|}{C\;\;\;\;\;C}}}-Ph \longrightarrow \underset{\underset{HO}{CH-O}}{\overset{CH_3}{\underset{|}{\overset{|}{C}}}}\overset{O}{\overset{\|}{C}}-Ph + PhCO_2^\bullet \quad (11.34)$$

The S_H2 reaction of carbon radicals occurs also with hydroperoxides and peracids. In particular, aliphatic peracids under certain conditions are transformed into alcohols by the mechanism:

$$R-\underset{O-O}{\overset{O}{\overset{\|}{C}}}\overset{H}{\underset{}{}} \longrightarrow R-\underset{O^\bullet}{\overset{O}{\overset{\|}{C}}} + {}^\bullet OH \quad (11.35)$$

$$RCO_2^\bullet \longrightarrow R^\bullet + CO_2 \quad (11.36)$$

$$R^\bullet + H\underset{O-O}{\overset{O}{\overset{\|}{C}}}-R \longrightarrow ROH + RCO_2^\bullet \quad (11.37)$$

In the case of ethers R^1–O–R^2, as opposed to peroxides, the S_H2 reaction is only observed with radicals centered on atoms less electronegative than carbon, such as Si and Sn, which bond more strongly to oxygen than carbon does.

$$R-CH_2-O-CH_2-R + {}^\bullet SiCl_3 \longrightarrow R-CH_2-O-SiCl_3 + R-CH_2^\bullet \quad (11.38)$$

With sulfides and selenides the reaction works well only with radicals like $R_3Sn^\bullet$ and provided that the leaving radical is relatively stable.

$$Bu_3Sn^\bullet + Ph-X-CH_2CO_2Et \xrightarrow{X = S \text{ or } Se} Bu_3Sn-X-Ph + {}^\bullet CH_2CO_2Et \quad (11.39)$$

This reaction is used very generally to produce radicals; it should be noted that selenides are much more reactive than sulfides.

11-3-2 Stepwise S_H2

The cases where the formation of a radical intermediate is most clearly established are those of S_H2 reactions on mercury, boron and phosphorus.

Group 12 atoms (Mg, Hg)
Carbon radicals react with mercuric halides as shown in reaction (11.40).

With organomagnesium compounds, t-BuO• radicals give two competing reactions one of which corresponds to a displacement mechanism (11.41) while the other is a hydrogen abstraction (11.42).

$$t\text{-BuO}• + \underset{R^2}{\overset{R^1}{\underset{|}{C}}}\text{—MgX} \begin{array}{c} \nearrow \; t\text{-BuO–MgX} + \underset{R^2}{\overset{R^1}{}}\dot{C}\text{—H} \quad (11.41) \\ \\ \searrow \; t\text{-BuOH} + \underset{R^2}{\overset{R^1}{}}\dot{C}\text{—MgX} \quad (11.42) \end{array}$$

Group 13 atoms (B)
The addition of alkoxyl radicals to organoboranes gives a radical intermediate which then fragments.

$$B(i\text{-Pr})_3 \xrightarrow{+\; CH_3O•} [CH_3\dot{O}B(i\text{-Pr})_3] \xrightarrow{-\; i\text{-Pr}•} CH_3OB(i\text{-Pr})_2 \quad (11.43)$$

Group 15 atoms (N, P)
S_H2 Reactions on nitrogen are rare, whereas those on phosphorus compounds are very common and give a radical intermediate by valency increase. Only the α-fragmentation (11.44) of the phosphoranyl radical corresponds to a displacement reaction. The relative importance of α- and β-fragmentations depends mainly on the strengths of the P–X and P–O bonds.

$$RO• + PX_3 \longrightarrow [RO\dot{P}X_3] \begin{array}{c} \overset{\alpha\text{-}fragmentation}{\nearrow} RO\text{–}PX_2 + X• \quad (11.44) \\ \\ \underset{\beta\text{-}fragmentation}{\searrow} O{=}PX_2 + R• \quad (11.45) \end{array}$$

The displacement reaction (11.46) is primarily observed with phosphines. At high temperature and in polar solvents only β-fragmentation occurs.

$$PEt_3 \xrightarrow{+\; t\text{-BuO}•} [t\text{-Bu}O\dot{P}Et_3] \xrightarrow{-\; Et•} t\text{-BuOPEt}_2 \quad (11.46)$$

11-4 INTRAMOLECULAR SUBSTITUTION, S_Hi

11-4-1 Hydrogen transfer

Homolytic 1,5 and 1,6 (n = 3 and 4) hydrogen transfers are very often observed, whether $\Sigma^\bullet$ is $CH_2^\bullet$, $O^\bullet$ (Barton reaction) or $^{\bullet+}NH_2$ (Hofmann-Löffler-Freytag reaction). Calculations and experimental results show that the activation energy of the reaction is a minimum when the C–H–Σ angle is close to 180° (linear transition state, reaction (11.47)).

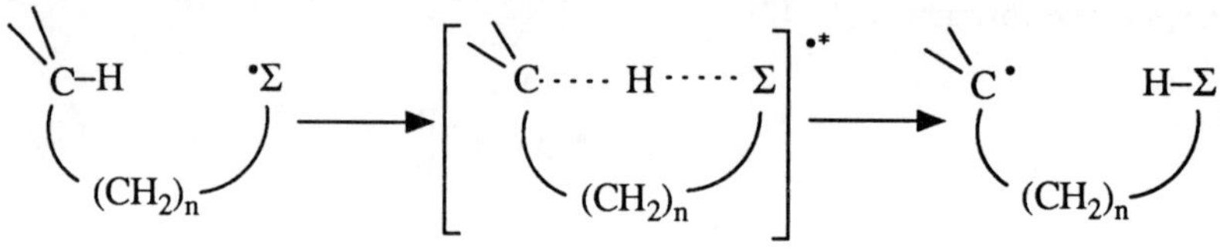

$$\begin{array}{c}\diagdown\\ C-H\\ \diagup\diagdown\\ (CH_2)_n\end{array}\;\;\Sigma^\bullet \longrightarrow \left[\begin{array}{c}\diagdown\\ C\cdots H\cdots \Sigma\\ \diagup\diagdown\\ (CH_2)_n\end{array}\right]^{\bullet *} \longrightarrow \begin{array}{c}\diagdown\\ C^\bulletH-\Sigma\\ \diagup\diagdown\\ (CH_2)_n\end{array} \quad (11.47)$$

The rate constant ratio 1,5/1,6 is 3.3 when Σ is CH_2 and 10 for Σ = O or $^+NH_2$. The reaction is more exothermic, therefore faster, with protonated oxyl and aminyl radicals than with carbon radicals. The difference in the activation energies for 1,5 and 1,6 transfers is small. The preference for 1,5 transfers is due essentially to the entropy term.

When n is less than 3 the C–H–Σ angle departs from the ideal value of 180° and the activation energy increases; for this reason 1,4 transfer is slow ($E_a \approx$ 15 kcal/mol) while 1,3 and 1,2 transfers are almost unknown. Beyond n = 4 the entropy term rules out the reaction.

This need to approach linear geometry has stereochemical consequences in the case of cyclic molecules. The alkoxyl radical **6** readily undergoes intramolecular hydrogen transfer, but not its stereoisomer **7**.

Scheme 11.4

11-4-2 Halogen transfer

Unlike hydrogen, long-distance transfers of halogens are rarely observed. Halogen atoms, especially chlorine, undergo 1,2 transfers much more frequently. Reaction (11.48), where the chlorine atom belongs to the CCl_3 group, is extremely fast ($k_{1,2} > 10^8$ s^{-1}). It is the chlorine which is transferred, not the bromine. This is doubtless due to the fact that the radical formed is stabilized by two chlorine atoms; bromine transfer would give a primary radical.

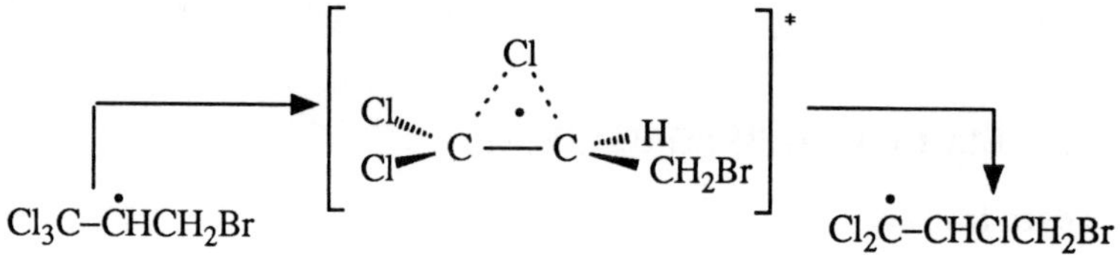

$$Cl_3C-\overset{\bullet}{C}HCH_2Br \longrightarrow \left[\begin{array}{c}Cl\\ \vdots\\ Cl_{\cdots\cdots}C-C\overset{H}{\underset{CH_2Br}{\cdots\cdots}}\\ Cl^{\nearrow}\end{array}\right]^* \longrightarrow Cl_2\overset{\bullet}{C}-CHClCH_2Br \quad (11.48)$$

Intramolecular substitution, S_Hi

Spectroscopic studies and calculations show that the β-chloroethyl radical prefers conformation **8** (Scheme 11.5) in which the C–Cl bond eclipses the p orbital bearing the unpaired electron, favoring a dissymmetric bridged structure for the 1,2 transfer (11.48).

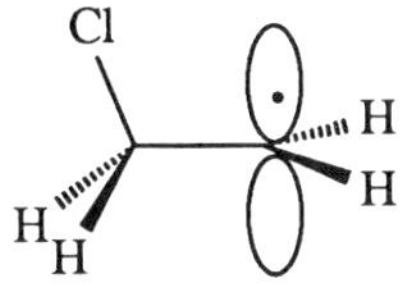

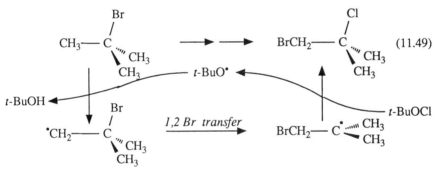

Scheme 11.5

Since the C–F bond strength is higher, fluorine transfer is never seen. Since C–Br and C–I bonds are weaker than C–Cl, 1,2 transfers for Br and I are faster ($k_{1,2} \approx 10^{11} s^{-1}$ for Br as against $10^8 s^{-1}$ for Cl) but, for this reason, there is also competition with elimination which can become the dominant reaction. With bromine one can observe a direct 1,2 transfer, as in the case of the chlorination of *tert*-butyl bromide by *t*-BuOCl (11.49).

Scheme 11.6

or, in other cases, by an elimination-addition process (11.50).

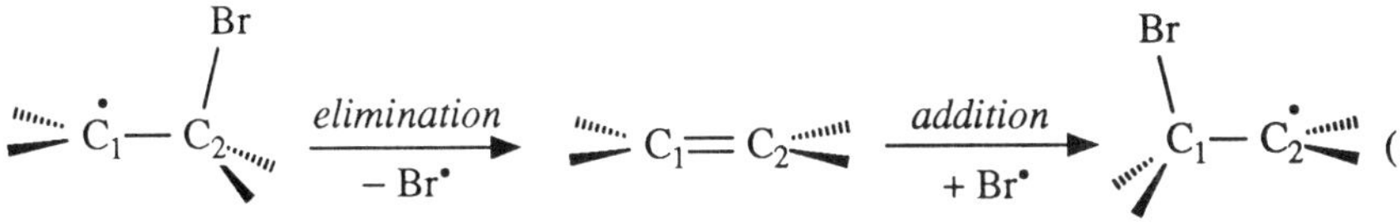

12 – ADDITION AND FRAGMENTATION REACTIONS

Radical additions constitute one of the commonest methods for making C–C or C–heteroatom bonds and for obtaining a wide variety of multifunctional products. In general, the reaction involves a chain process one step of which is *the addition of a radical to an unsaturated system*, C=C or C=heteroatom.

12-1 ADDITION TO A >C=C< DOUBLE BOND

The rate of addition depends on the substituents on the radical and the double bond.

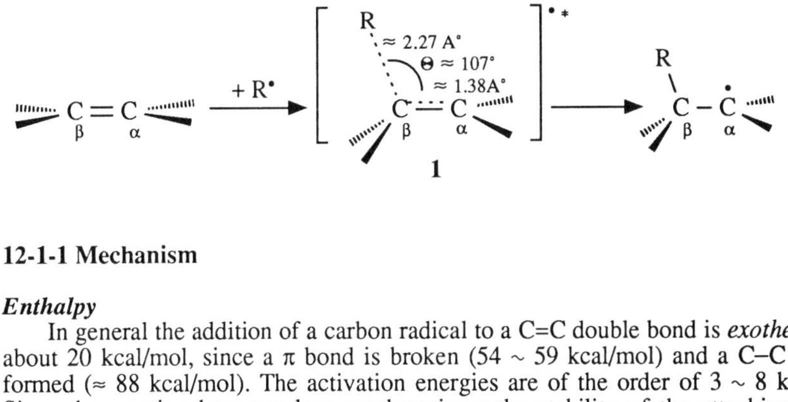

$$\overset{\beta}{C}=\overset{\alpha}{C} \quad + R^{\bullet} \longrightarrow \left[\begin{array}{c} R \\ \approx 2.27\,\text{Å} \\ \theta \approx 107° \\ \approx 1.38\,\text{Å} \\ \underset{\beta}{C} \cdots \underset{\alpha}{C} \end{array} \right]^{\bullet *} \longrightarrow \underset{\beta}{C}-\underset{\alpha}{\overset{R}{C}} \quad (12.1)$$

1

12-1-1 Mechanism

Enthalpy
In general the addition of a carbon radical to a C=C double bond is *exothermic* by about 20 kcal/mol, since a π bond is broken (54 ~ 59 kcal/mol) and a C–C bond is formed (≈ 88 kcal/mol). The activation energies are of the order of 3 ~ 8 kcal/mol. Since the reaction becomes less exothermic as the stability of the attacking radical increases, the reaction may become reversible.

Transition state structure
The consequences of the exothermicity of the reaction on the position of the transition state (Hammond postulate) are confirmed by theoretical calculations which show that:

– the structure of the transition state is dissymmetric and non-centrosymmetric: the angle of attack (≈ 107°) is close to the value for an sp^3 carbon (109.5°);
– the bond which forms is relatively long (≈ 2.27 Å as against ≈ 1.54 Å in the product propyl radical);
– the bond which breaks is relatively intact (≈ 1.38 Å as against ≈ 1.34 Å in the reacting ethylene.

This dissymmetry, added to the fact that bond formation is little advanced, has two consequences:

– steric effects of substituents on carbon C_β are relatively weak and those on

carbon C_α even weaker. Nevertheless, these effects to a certain extent control the regioselectivity (para. 12-1-4);
— the stability of the carbon C_α-centered radical formed is relatively unimportant.

Depending on the nature of the substituents, one can obtain a development of charges of opposite sign (polar effect) which reduces the activation energy and increases the reaction rate.

Orbital interpretation

The existence of a transition state close to the initial products allows us to describe the reactivity in terms of frontier orbital interactions (SOMO of the radical, HOMO and LUMO of the double bond, i.e. the π and π^* orbitals). The HOMO is stabilized by interaction with the SOMO. This is destabilized by the HOMO and stabilized by the LUMO. To a first approximation one gains the stabilization of 2 electrons in the HOMO. The closer the energies of the orbitals, the greater this stabilization.

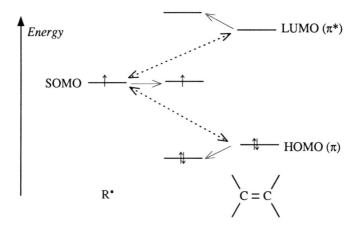

Figure 12.1 – *SOMO ↔ HOMO and SOMO ↔ LUMO orbital interactions.*

The SOMO ↔ LUMO interaction is most important for radicals with electron-donating substituents (high SOMO) and an electron-poor olefin (low LUMO). Since the energy levels are close the reactivity is high; the radical then behaves as a nucleophile. Conversely, with electron-attracting substituents on the radical (low SOMO) and donors on the double bond (high HOMO and LUMO), the SOMO ↔ HOMO interaction is dominant and the radical behaves as an electrophile.

12-1-2 Effects of substituents on the double bond

Substituents on C_α

The more the substituent Z is electron-attracting, the greater the reactivity with a nucleophilic radical such as cyclohexyl, since the lowering of the LUMO increases the SOMO ↔ LUMO interaction (Table 12.1).

Addition to a >C=C< double bond

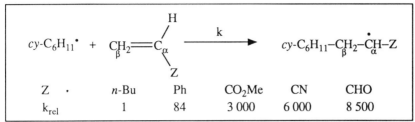

Table 12.1 – *Effects of an electron-donating substituent at C_α on the reactivity.*

The stability of the radical formed is relatively unimportant compared to the polar effect, since when Z = Ph (stabilized benzyl radical) the reaction is about 100 times slower than with Z = CHO (more electron-attracting substituent).

It can be seen from the reaction in Table 12.2 that the retardation due to the steric effect of Z is small. Moreover, part of this reduction in the rate can be attributed to the donor character of the alkyl substituents.

cy-C$_6$H$_{11}$· + CH$_2$=C(Z)(CO$_2$Me) → cy-C$_6$H$_{11}$–CH$_2$–CZ·–CO$_2$Me			
Z	H	Me	*t*-Bu
k$_{rel}$	1	0.75	0.26

Table 12.2 – *Steric effect of a substituent at C_α on the reactivity.*

In the addition of the Me• radical to alkenes (Table 12.3), the polar effect is negligible. The data show that the reactivity is controlled mainly by the steric effect, which would give the order (with respect to the radical formed): primary > secondary > tertiary, and partially by the stability of the radical formed, which would predict the reverse order.

olefin	=	=/	=/⌐	⫽<	/⌐\	/=\	⩥<	⨳<
k$_{rel}$	1	0.7	0.6	1.1	0.23	0.33	0.32	0.20

Table 12.3 – *Relative rate of addition of Me• radical to different olefins.*

Substituents on C_β

Steric effects are much more important at C_β than at C_α. Thus, a *t*-Bu substituent reduces the rate of addition of a cyclohexyl radical by a factor of only 4 when it is at C_α and 20 000 at C_β.

12-1-3 Effects of substituents on the radical

Polar effect

With an electron-poor olefin the reactivity of a radical with one or two donor substituents is greater than that of a radical with an attracting substituent (Table 12.4). In terms of orbital interactions, the presence of a donor substituent raises the energy level of the SOMO and increases the SOMO ↔ LUMO interaction and, hence, the reactivity. If the substituent is an electron-attractor the SOMO energy is lowered and the reactivity is lower.

$$R^{\bullet} + CH_2=C{<}^{H}_{CN} \xrightarrow{k} R-CH_2-\overset{\bullet}{C}H-CN$$

	$R^{\bullet}_{prim}$	$R^{\bullet}_{sec}$	$R^1-\overset{\bullet}{C}H-CN$
$R^{\bullet}$			
k_{rel}	1	7.3	0.0015

Table 12.4 – *Reactivity effect of a substituent on the radical.*

Different substituents on the radical and the olefin give rise to synergistic and antagonistic effects. Thus, the selectivities of malonyl (electrophile) and cyclohexyl (nucleophile) radicals are reversed on going from an olefin with a donor substituent (Me) to an attractor (CO_2Me).

	$CH_2=C{<}^{Ph}_{Me}$	$CH_2=C{<}^{Ph}_{CO_2Me}$
$^{\bullet}CH(CO_2Et)_2$	3.7	1
$cy\text{-}C_6H_{11}^{\bullet}$	1	42

Table 12.5 – *Effect of synergy or antagonism on the selectivity (k_{rel}).*

The impact of polar effects on the outcome of radical addition reactions is very significant. Thus, the very favored addition of a nucleophilic radical to acrylonitrile gives an electrophilic radical, which is unreactive towards acrylonitrile. This means that hydrogen transfer from the solvent is then preferred to polymerization.

$$R^{\bullet} \xrightarrow[fast]{+CH_2=CHCN} R-CH_2-\overset{\bullet}{C}HCN \xrightarrow[slow]{+CH_2=CHCN} RCH_2C(CN)H-CH_2-\overset{\bullet}{C}HCN \quad (12.2)$$

Steric effects

A $t\text{-}Bu^{\bullet}$ radical reacts 12 times faster than a primary radical on an electrophilic double bond; we can conclude that the polar effect outweighs the steric effect. However, the latter is not unimportant, and radical **2** for example, reacts 260 times more slowly with methyl fumarate when $X = t\text{-}Bu$ than when $X = Me$. In this case the steric effect of $t\text{-}Bu$ is more important than its polar contribution.

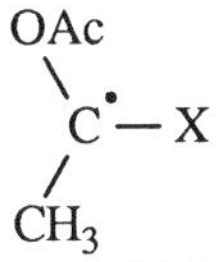

Scheme 12.1

12-1-4 Regioselectivity

In the case of a disubstituted double bond the radical can attack either the α or the β carbon (12.3). The regioselectivity is guided essentially by the difference in the steric crowding of Y and Z. As all substituents are bigger than hydrogen, addition is favored at the less substituted carbon. In particular, for olefins with a methylene group, >C=CH$_2$, it is steric hindrance which orients addition to the terminal carbon. The regioselectivity is controlled by the steric effect rather than by the stability of the radical formed. This is in agreement with the fact that the reaction is exothermic (Hammond postulate) and with calculation, which indicates that the transition state structure is closer to the reactants than to the products (para. 12-1).

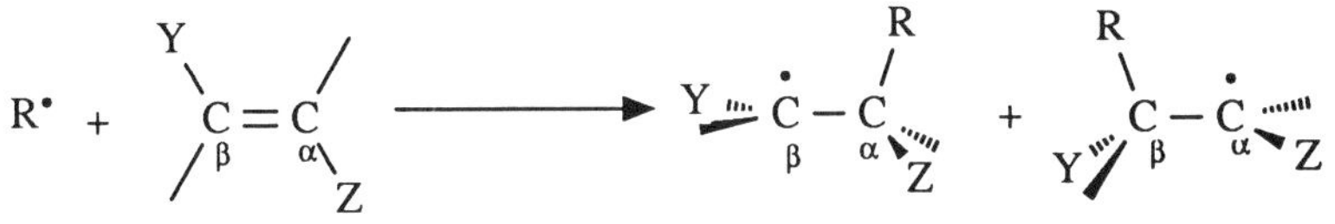

$$R^\bullet + \underset{Z}{\overset{Y}{\underset{\beta}{}\!\!C=C\!\!\underset{\alpha}{}}} \longrightarrow Y\cdots\underset{\beta}{C}-\underset{\alpha}{\overset{R}{C}}\cdots_{Z} + \underset{Y}{\overset{R}{C}}\cdots\underset{\beta}{C}-\underset{\alpha}{C}\cdots_Z \quad (12.3)$$

Nevertheless, polar effects can determine regioselectivity. The cyclohexyl radical, which is nucleophilic, reacts preferentially at the most electrophilic carbon of a double bond. In terms of orbital interactions, it reacts at the carbon with the highest atomic coefficient in the LUMO, provided that it is not too encumbered. Thus, the high regioselectivity observed in the following cases, (Scheme 12.2), is explained by the fact that a donor substituent (mesomeric effect of F) increases the coefficient of the LUMO on C$_\alpha$ and that an attractor substituent increases that on C$_\beta$. Since CN is a more powerful attractor than CO$_2$Me, the cyclohexyl radical attacks the carbon bonded to CO$_2$Me despite its greater bulk.

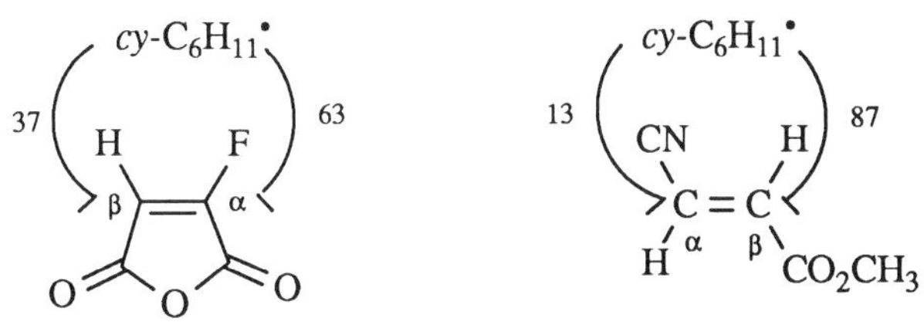

Scheme 12.2

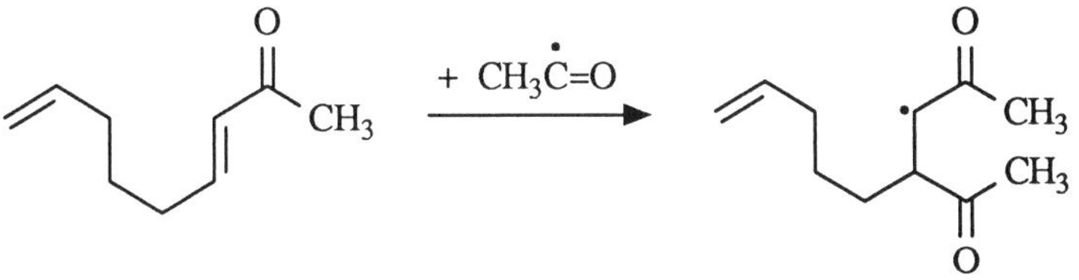

In the reaction (12.4) the selectivity of the nucleophilic acetyl radical with respect to the two double bonds depends on the polar effect, which in this case outweighs the steric effect.

12-1-5 Stereoselectivity

The non-equivalence of the two faces (Scheme 12.3), both of the attacking radical and of the olefin, leads to four stereoisomers. The stereoselectivity depends on the temperature but also on the reactivity of the species. According to the Hammond postulate, the less a system is reactive, the more the transition state resembles the products and the greater the steric effects. An unreactive radical can therefore be very stereoselective.

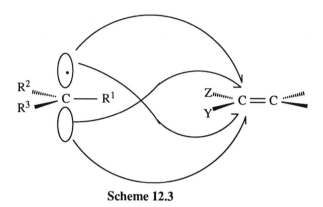

Scheme 12.3

In the case of a cyclopentyl radical **3** which is practically planar, the presence of a substituent such as Me favors *anti* attack (12.5).

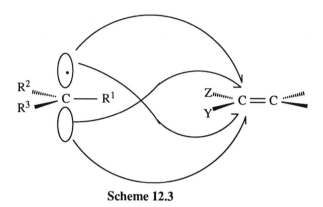

For the cyclohexyl radical **4** the stereoselectivity is a compromise between *steric interactions*, of the attacking radical with the axial 3 substituents (axial approach), and *torsional interactions* between the forming bond and the equatorial bonds to the 2 carbon (equatorial approach).

Addition to a >C=C< double bond

[Scheme showing axial/equatorial radical addition to CH$_2$=CH-CN, reactions (12.7) and (12.8), giving 45% and 55% respectively]

Increasing the torsional interactions, for example, by replacing the axial 2 hydrogen by an OH group, leads to greater stereoselectivity (12.9) - (12.10).

[Scheme showing axial/equatorial radical with OH substituent addition to CH$_2$=CH-CN, reactions (12.9) and (12.10), giving 73% and 27% respectively]

In carbohydrates the stereoselectivity is much greater than that in cyclohexanes (12.11) - (12.12).

[Scheme showing carbohydrate radical (with OAc, AcO, OCH$_3$ groups) addition to CH$_2$=CH-CN, reactions (12.11) and (12.12), giving 15% and 85% respectively]

When the radical carbon is in the anomeric position, the ring oxygen affects both the conformation of the ring and the direction of attack (axial) (12.14). Attack occurs antiperiplanar with respect to the axial lone pair on the oxygen (anomeric stereoelectronic effect).

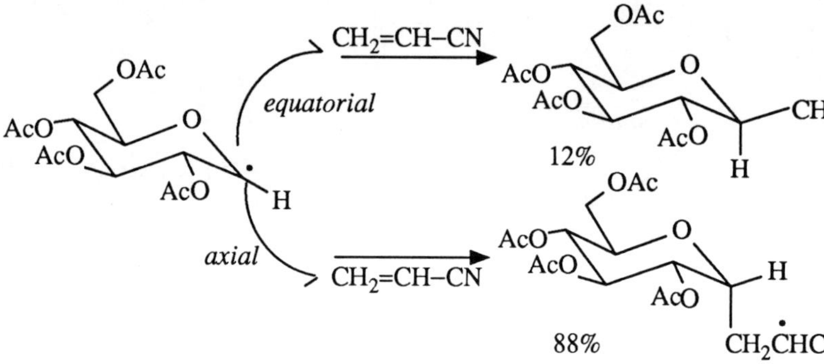

$$\text{(12.13)}$$
$$\text{(12.14)}$$

Dissymmetry in an olefin also causes stereoselective addition. In the case of norbornene, radical addition goes mainly *exo*, the *endo* face being less accessible.

$$\text{(12.15)}$$

12-1-6 Reversibility

When the bond formed by addition is weak the reaction becomes appreciably reversible at normal temperatures. This is particularly true for radicals $X^{\bullet}$ = $Br^{\bullet}$, $RS^{\bullet}$, $R_3Sn^{\bullet}$ or $RSO_2^{\bullet}$ for which the X–C dissociation energies are 70, 70, 60 and 60 kcal/mol, respectively. One of the consequences of the reversibility is the *cis* ⇌ *trans* isomerization of a disubstituted double bond, by rotation about the C–C bond of the radical adduct, as we have already seen in Chapter 11. Higher temperatures strongly favor elimination over addition ($E_{elim} > E_{add}$) (Scheme 12.4).

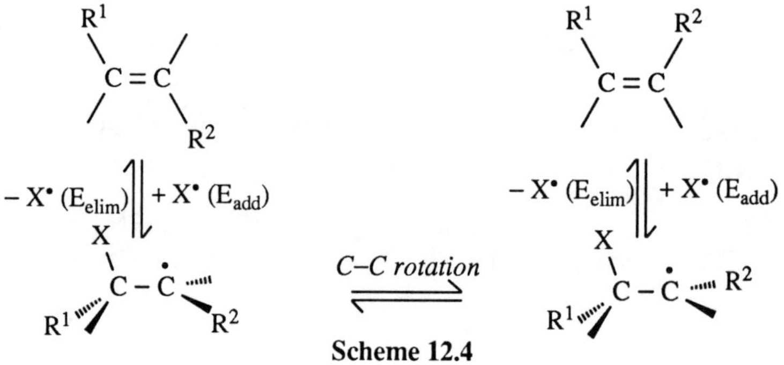

$$\text{(12.16)}$$

Scheme 12.4

12-2 ADDITION TO VARIOUS UNSATURATED SYSTEMS

Acetylenic bond $-C\equiv C-$

Carbon radicals add to triple bonds more slowly than to double bonds. The vinyl radical obtained is more reactive than a saturated carbon radical for reasons of enthalpy; this may be interesting in the subsequent reactions.

$$R^{1\bullet} + R^2-C\equiv C-R^3 \longrightarrow \underset{R^2}{\overset{R^1}{}}C=C\overset{\bullet}{\underset{R^3}{}} \qquad (12.17)$$

Ketonic bond $>C=O$

Intermolecular additions to a carbonyl group are fairly uncommon. They are more difficult than on a C=C double bond because the bond energies of $\pi(C=O)$ and $\pi(C=C)$ are $73 \sim 78$ and $54 \sim 59$ kcal/mol, respectively. The regioselectivity of the addition depends on the polar character of the radical (a nucleophilic radical adds preferentially to oxygen) or the steric effect of R^2 and R^3 (steric hindrance favors (12.18)) and the rate of the reverse reaction. The β-fragmentation, k_{-c}, is known to be very fast (para. 12.3).

$$R^{1\bullet} + O=C\overset{R^2}{\underset{R^3}{}} \underset{k_{-o}}{\overset{k_o}{\rightleftarrows}} \underset{R^3}{\overset{R^1}{}}O-\overset{R^2}{\underset{}{C^{\bullet}}} \qquad (12.18)$$

$$\underset{k_{-c}}{\overset{k_c}{\rightleftarrows}} \overset{\bullet}{O}\cdots\underset{R^1}{\overset{R^2}{C}}\underset{R^3}{} \qquad (12.19)$$

Thioketonic bond $>C=S$

Addition of a radical to a C=S double bond (12.20) is very easy and more exothermic than addition to carbonyl, for example. The methodology for creating radicals by means of xanthates and hydroxythiopyridone (Chap. 9) takes advantage of this reaction.

$$R^{1\bullet} + S=C\overset{R^2}{\underset{R^3}{}} \longrightarrow \underset{R^3}{\overset{R^1}{}}S-\overset{R^2}{\underset{}{C^{\bullet}}} \qquad (12.20)$$

>C=N– and –C≡N bonds

Unlike heteroatom- and metal-centered radicals, carbon radical additions to the >C=N– double bond are rare as far as imines are concerned, but very common with nitrones, used as radical traps (12.21).

$$R^{1\bullet} + R^2-CH=\overset{+}{\underset{R^3}{N}}-O^- \longrightarrow \underset{R^3}{\overset{R^1-CH(R^2)}{N}}-O^\bullet \quad (12.21)$$

Nitriles are very unreactive. Isonitriles, however, are much more reactive and can be used to transform a carbon radical into a nitrile by an addition-fragmentation process (12.22), with $k_{add} \approx 10^5$ M^{-1}s^{-1} for a secondary radical.

$$:C=N-t\text{-Bu} \xrightarrow[k_{add}]{+R^{1\bullet}} R^1-\overset{\bullet}{C}=N-t\text{-Bu} \xrightarrow{-t\text{-Bu}^\bullet} R^1-C\equiv N \quad (12.22)$$

12-3 FRAGMENTATION

Fragmentation, reactions (12.23) and (12.24), is the reverse of addition. This process is often endothermic and is favored by raising the temperature. The main driving force is the entropy increase resulting from the formation of two entities from one. The factors which promote fragmentation are several and analogous to those described for intermolecular addition: reaction enthalpy (stability of the radical formed) and stereoelectronic control. The homolysis can involve a bond either α or β to the radical center.

$$X\overset{\bullet}{\underset{\alpha}{Y}}\overset{\beta}{\underset{A}{\frown}}B \quad \begin{array}{l} \xrightarrow{\alpha\text{-fragmentation}} X-Y: + {}^\bullet A-B \quad (12.23) \\ \\ \xrightarrow{\beta\text{-fragmentation}} X^{-Y}\!\!\!=\!\!A + B^\bullet \quad (12.24) \end{array}$$

Enthalpy balance

The formation of a strong π bond and the breaking of a weak σ bond are important factors. Thus, the formation of Me$^\bullet$ is less endothermic from the *tert*-butoxyl radical (12.25) than from neopentyl (12.26). Reaction (12.25) is therefore faster than (12.26).

$$\underset{CH_3}{\overset{CH_3}{\underset{|}{C}}}\!\!-\!\!O^\bullet \xrightarrow{\Delta H = 5 \text{ kcal/mol}} CH_3\overset{O}{\underset{\|}{C}}CH_3 + {}^\bullet CH_3 \quad (12.25)$$

$$\underset{\underset{CH_3}{\overset{CH_3}{|}}}{CH_3-\overset{|}{C}-\overset{\bullet}{C}H_2} \xrightarrow{\Delta H = 20 \text{ kcal/mol}} \underset{CH_3}{\overset{CH_2}{\overset{\|}{C}}} \overset{}{}_{CH_3} + \ ^{\bullet}CH_3 \quad (12.26)$$

The ease of β-fragmentation varies with the type of multiple bond formed, in the order: C=O > C≡N > C=C. It depends also on the strength of the σ bond broken and only works if the BDE of the breaking bond is about 70 kcal/mol or less. This is the case of C–SR, C–Br, C–SnR$_3$, C–I or C–CH(CN)(CO$_2$Me) bonds.

Formation of small stable molecules

The formation of a stable molecule by α-fragmentation, such as CO or SO$_2$, although endothermic (ΔH = 20 ~ 25 kcal/mol) is fast enough (k = 10^4 ~ 10^5 s^{-1}) to be experimentally significant. The two reactions are reversible.

$$R-\overset{O}{\overset{\|}{\underset{\bullet}{C}}} \quad \rightleftharpoons \quad R^{\bullet} \quad + \quad CO \quad (12.27)$$

$$R-\overset{\bullet}{\overset{\overset{O}{\|}}{\underset{\|}{S}}}_{O} \quad \rightleftharpoons \quad R^{\bullet} \quad + \quad SO_2 \quad (12.28)$$

The formation of CO$_2$ by β-fragmentation (12.29) is exothermic (ΔH ≈ −14 kcal/mol), irreversible and much faster (k = 10^6 ~ 10^{10} s^{-1} depending on whether R is a phenyl or an alkyl group). If the fragmentation process is accompanied by the formation of a stabilized radical such as PhCH$_2^{\bullet}$ the rate is increased by a factor of 10^3 ~ 10^4 compared to a phenyl radical.

$$R-\overset{O}{\overset{\|}{\underset{\underset{\bullet}{O}}{C}}} \quad \longrightarrow \quad R^{\bullet} \quad + \quad CO_2 \quad (12.29)$$

Stereoelectronic control

This can be important in β-fragmentation. Thus, the 2-butyl radical is in conformational equilibrium (12.30).

$$(12.30)$$

The conformer which gives elimination the more easily is that where the C–X bond eclipses the p orbital. If the rotation about the C–C bond, which equilibrates the two rotamers, is faster than β-fragmentation one obtains the more stable *trans* compound. This is the case at normal temperatures for X = Br, Bu$_3$Sn or PhSO$_2$. If elimination is very fast (X = PhSO) the reaction becomes stereoselective and reproduces the starting conformer. The temperature has a considerable effect upon the equilibrium (12.30).

Alkoxyl radicals

These merit special attention since they turn up everywhere; due to the formation of a strong C=O bond they very easily undergo β-fragmentation. The commonest example is that of the *tert*-butoxyl radical (12.25) for which $\Delta H = 5$ kcal/mol, $E_a = 13$ kcal/mol and $k = 10^5$ s^{-1} at 60°C.

When the aliphatic substituents at the carbon are different, β-fragmentation gives mainly the most stable radical, i.e. by breaking the weakest C–C bond.

$$\text{MeC(O)Et} + i\text{-Pr}^{\bullet} \quad (12.31)$$

$$\text{MeC(O)}i\text{-Pr} + \text{Et}^{\bullet} \quad (12.32)$$

$$\text{EtC(O)}i\text{-Pr} + \text{Me}^{\bullet} \quad (12.33)$$

(95%, 3%, <0.5%)

However, this is not always the case, and in reactions (12.34) and (12.35) the strongest bond breaks (C–SPh ≈ 57 kcal/mol and C–SiR$_3$ ≈ 76 kcal/mol). The polar effect (S is more electronegative than Si) no doubt has something to do with this orientation.

$$\text{R}^1\text{CH(PhS)–C(SiR}_3\text{)(SPh)–O}^{\bullet} \longrightarrow \text{R}^1\text{CH(PhS)–C(O)(SPh)} + {}^{\bullet}\text{SiR}_3 \quad (12.34)$$

$$\not\longrightarrow \text{R}^1\text{CH(PhS)–C(O)(SiR}_3\text{)} + {}^{\bullet}\text{SPh} \quad (12.35)$$

The fact that alkoxyl radicals are strongly solvated by polar and polarizable solvents affects the competition between β-fragmentation and hydrogen transfer.

13 – CYCLIZATIONS AND REARRANGEMENTS

Radical *cyclizations* usually consist of *intramolecular additions* to double or triple bonds. They are of great interest since they allow the synthesis of a wide variety of 5- and 6-membered cyclic compounds with very high regioselectivity and often very good stereoselectivity. In general, other things being equal, intramolecular addition is faster than intermolecular addition. Certain cyclizations proceed by *intramolecular homolytic substitution*.

We shall also discuss *ring opening*, which is the reverse of cyclization.

This chapter deals with *radical rearrangements* involving the transfer of polyatomic groups from carbon (or heteroatom) to carbon (or heteroatom), except for the transfer of monovalent atoms (hydrogen or halogen) which was described in Chapter 11. These rearrangements go through an intermediate cyclization step.

13-1 CYCLIZATION ON A >C=C< DOUBLE BOND

13-1-1 Mechanism

Of the two possible cyclizations, *exo* (13.1) and *endo* (13.2), generally the former is kinetically preferred.

$$\text{C=C} \xrightarrow{k_{exo}} \text{exo} \quad (13.1)$$

$$\text{X}^\bullet \xrightarrow{k_{endo}} \text{endo} \quad (13.2)$$

Many thermodynamic and theoretical studies on the cyclization of the hex-5-en-1-yl radical, which we shall refer to as 5-hexenyl (13.3), have shown that the substantially greater yield of product **1** (*exo* cyclization), the less stable isomer (the 5-membered ring species is a primary radical whereas the 6-membered one is a secondary radical), is a result of the favorable activation enthalpy (13.4) rather than the activation entropy (13.5).

$$\xrightarrow{} exo \; \mathbf{1}\,(98\%) \; + \; endo \; \mathbf{2}\,(2\%) \quad (13.3)$$

$$\Delta H^*_{1,6} - \Delta H^*_{1,5} = 1.7 \text{ kcal mol}^{-1} \quad (13.4)$$
$$\Delta S^*_{1,5} - \Delta S^*_{1,6} = 2.8 \text{ cal mol}^{-1}\text{K}^{-1} \quad (13.5)$$

This regioselectivity, under enthalpy control, is explained by conformational and electronic effects. In the transition state, **3**, of the cyclization to a 5-membered ring there is better SOMO ↔ LUMO overlap than in transition state, **4**, to give a 6-membered ring (Scheme 13.1).

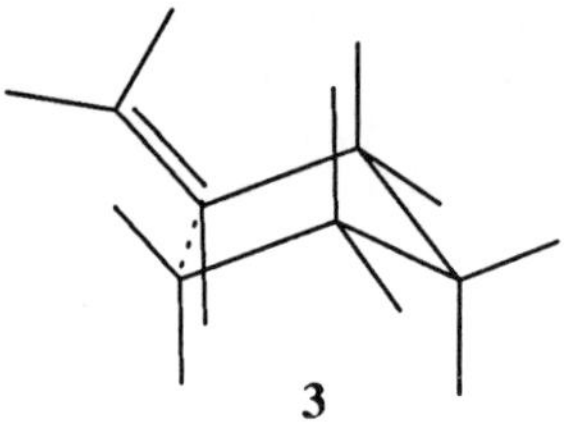

Scheme 13.1

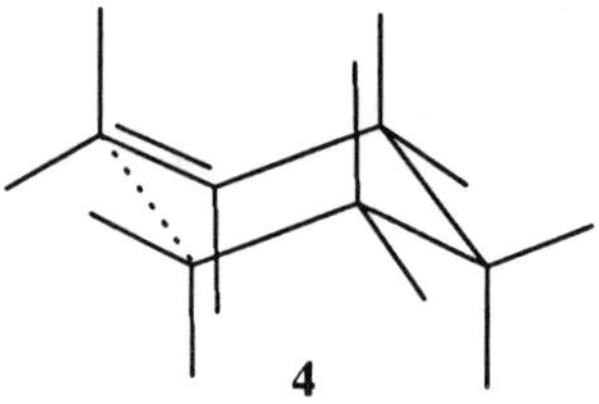

Cyclization reactions are sensitive to the same thermochemical, steric and polar effects as intermolecular additions, but they are subject to the additional constraints of ring formation. When the ring formed is 3-, 5- or 6-membered cyclization is generally faster than the analogous intermolecular addition.

13-1-2 Factors determining the regio- and stereoselectivity

Chain length
The *exo* or *endo* cyclization rate depends greatly on the chain length (Table 13.1).

$\diagup\!\!\diagdown(CH_2)_n\diagup\!\dot{C}H_2$	k_{exo}	k_{endo}	k_{exo}/k_{endo}	k_{-exo}
n = 1	1.8 x 10^4	not obs.	-	2.0 x 10^8
n = 2	1.0	not obs.	-	4.7 x 10^3
n = 3	2.3 x 10^5	4.1 x 10^3	58	-
n = 4	5.2 x 10^3	8.3 x 10^2	6	-
n = 5	< 70	1.2 x 10^2	< 0.6	-

Table 13.1 – *Cyclization and reopening of alkyl radicals (k in s^{-1} at 25°C).*

The reaction is fast for the butenyl radical (n = 1) and gives only *exo* cyclization to cyclopropane, but the reverse reaction, reopening, is even faster. Cyclization of the pentenyl radical (n = 2) is very difficult to observe. Cyclization is the fastest and the most *exo* regioselective for the hexenyl radical (n = 3). Beyond n = 3 the rate and the

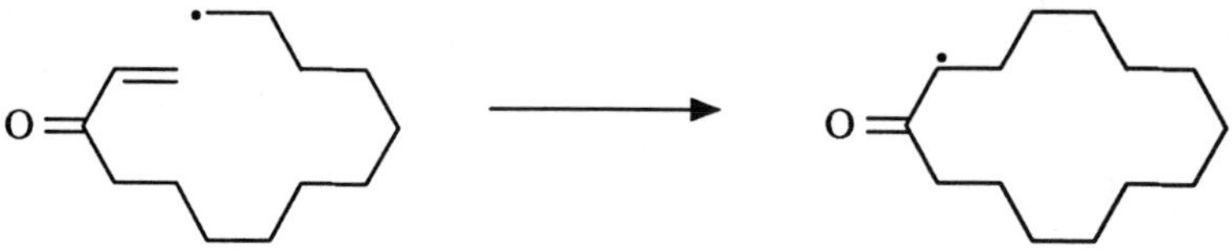

$$\quad (13.6)$$

Cyclization on a >C=C< double bond

regioselectivity of the reaction decrease and beyond n = 5 the *endo* mode is the most important and becomes exclusive. Despite an unfavorable entropy factor one can obtain macrocyclization (> 10-membered ring) in certain cases (nucleophilic carbon radical on an electron-poor double bond).

Chain substituents

A gem-dimethyl group accelerates cyclization (Thorpe-Ingold effect) compared to an unsubstituted radical. The reopening of this radical to a more stable tertiary radical is very fast. Since this cyclizes only slowly the overall reaction is the fast rearrangement of a primary radical to a tertiary radical.

$$k_c = 1.7 \times 10^7 \text{ s}^{-1} \quad k_o = 1.7 \times 10^9 \text{ s}^{-1}$$
$$k_o = 3 \times 10^8 \text{ s}^{-1} \quad k_c < 3 \times 10^4 \text{ s}^{-1}$$
$$k_{rearrangement} = 1.4 \times 10^7 \text{ s}^{-1} \quad (13.7)$$

In the much studied case of the hexenyl radical, a substituent such as Me has a different effect on the rate of cyclization, the regioselectivity (*exo* or *endo*) and the stereoselectivity (Table 13.2) depending on its position in the chain. The presence of a

radical		k_{exo}	k_{endo}	k_{exo}/k_{endo}
(6-5-4-3-2-1)	5	1	0.017	58
	6	1.4	0.02	70
	7	1.4	0.02	70
	8	1.0	<0.01	>100
	9	0.022	0.04	0.55
	10	24	0.27	85
	11	0.0024	0.005	0.5

Table 13.2 – *Cyclization of substituted hexenyl radicals (relative k at 65°C)*.

methyl group at the 1 or 6 position (radicals **6, 7, 8**) has little effect. On the other hand, in the 5 position (**9**) the rate of *exo* cyclization is greatly decreased, which makes *endo* cyclization less disfavored. The cyclization rate depends on the stability neither of the initial radical (**5, 6, 7**) nor of the radical formed (**5, 8**). A gem-dimethyl group, whether it be in the 2, 3 or 4 position, increases the cyclization rate, as in the case of the butenyl radical (reaction (13.7)), but by a factor of only 10 instead of 10^3. If the 3 carbon is replaced by another element (O, N, Si), both the bond lengths and angles in the transition state are modified, which leads to a variation of the rate and the regioselectivity (radicals **10, 11**).

initial radical	$\underset{k_o}{\overset{k_c}{\rightleftharpoons}}$	cyclized radical	k_c	k_o
			(s^{-1} at 25°C)	
			2.5×10^5	
			5×10^8	
			6×10^8	
			1×10^5	
			8.7×10^5	4.7×10^8
			1×10^6	1.1×10^7
			4×10^3	

Table 13.3 – *Rate constants for cyclization, k_c, and ring opening, k_o.*

Stereoselectivity

The major product has the *cis* configuration when the methyl is in the 1 position (13.8) or the 3 position, but *trans* for the other positions. This stereoselectivity is explained by a transition state in the chair conformation, the substituents being equatorial (**3**, Scheme 13.1).

$$\xrightarrow{1 \times 10^5 \text{ s}^{-1}} \quad cis \quad + \quad trans \quad (13.8)$$

Nature of radical

The structure of the radical is also relevant (Table 13.3). In the case of alkenyl-aryl radicals, the cyclization rate increases by about 1 000 because the aromatic ring causes a loss of flexibility, as compared with the hexenyl radical, and because of the high reactivity of the aryl radical; the *exo/endo* ratio stays the same at about 50.

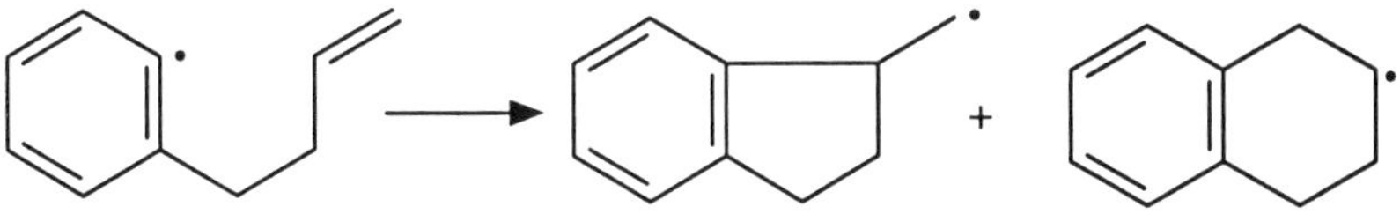

$$\text{(13.9)}$$

If the initial radical is stabilized, cyclization becomes reversible and the products are formed under thermodynamic control with a preference for the 6-membered ring (13.10) - (13.11).

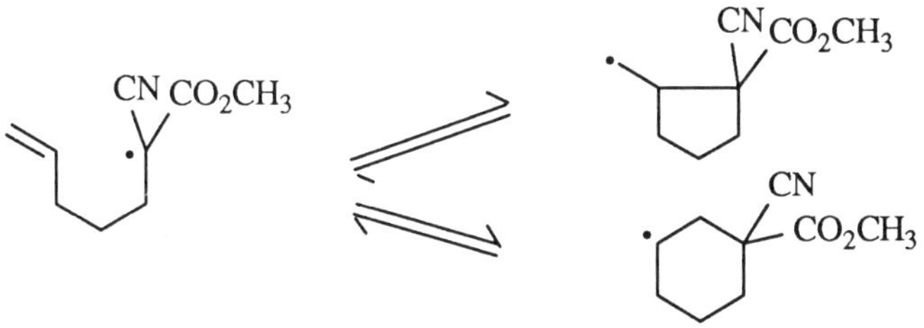

$$\text{(13.10)}$$

$$\text{(13.11)}$$

The cyclization of radicals centered on a heteroatom has been studied for different chain lengths and radicals. For thiyl radicals ($CH_2=CH(CH_2)_3-S^\bullet$) with a double bond in the δ position one obtains, because the cyclization is reversible, a mixture of 5- (*exo* cyclization) and 6-membered (*endo* cyclization) cyclic products, depending on the reaction conditions and particularly the temperature.

With oxyl and protonated or metal-complexed aminyl radicals, the C–O or C–N bond is shorter than C–C and the radical is more electrophilic than a primary carbon radical; these are the factors no doubt responsible for the speed, the irreversibility and the regioselectivity (*exo*) of cyclization. In the case of the oxyl radical the rate increases ($k > 10^8 \text{ s}^{-1}$ instead of $2 \times 10^5 \text{ s}^{-1}$ for the carbon radical) and *exo* cyclization predominates when there is a Me in the 5 position (compare with line 5 of Table 13.2).

$$\text{(13.12)}$$

For radicals centered on Si and P, such as **12** and **13**, the fact that the C–Si or

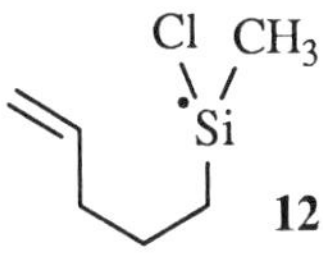

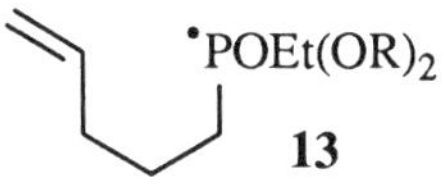

Scheme 13.2

C–P bond length is greater than C–C should lead to lesser regioselectivity. In fact, cyclization goes preferentially *exo* except when the radical atom (Si in particular) bears bulky substituents.

13-2 CYCLIZATION ON OTHER UNSATURATED SYSTEMS

Intramolecular addition of a radical occurs for many unsaturated systems other than the C=C double bond. We shall give some examples.

13-2-1 Triple bond

Cyclization on a triple bond is always *exo*. It is slower than a double bond, especially for nitriles (Table 13.3).

$$k_{exo} / k_{endo} \approx 50 \quad \longrightarrow \quad exo \quad + \quad endo \quad (13.13)$$

13-2-2 Carbonyl

Cyclization on a carbonyl group is much more frequently observed than intermolecular addition. It can take place either on the carbon (*exo*) or on oxygen (*endo*).

$$(13.14) \text{ exo}$$
$$(13.15) \text{ endo}$$

The regioselectivity depends mainly on the size of the ring. For $n = 2$ cyclization occurs exclusively on oxygen (*endo*) whereas for $n = 1, 3$ or 4 reaction is preferentially at carbon (*exo*). The reverse reaction, β-fragmentation, is usually faster than cyclization. Therefore, to obtain the cyclized alcohol (13.14) the equilibrium has to be displaced by trapping the radical with a good hydrogen-donor.

The rate of cyclization on carbonyl can, in certain cases, be greater than on a C=C double bond. In the following example, where C=O and C=C double bonds are disposed symmetrically, alcohol is obtained almost exclusively for $n = 2$ (6-membered ring cyclization on C=O) and a mixture of alcohol and aldehyde, with a slight preference for the latter (5-membered ring cyclization on C=C), when $n = 1$.

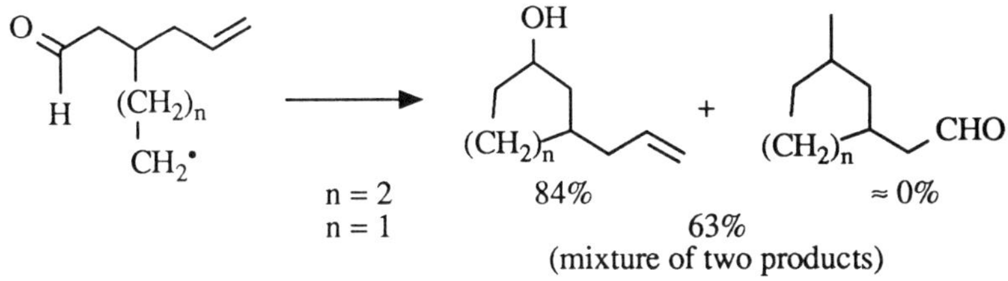

Cyclization by S_Hi

[Reaction scheme 13.16: aldehyde with (CH$_2$)$_n$ and CH$_2^\bullet$ radical giving cyclic alcohol with alkene (84% for n=2, 63% for n=1 mixture of two products) and acyclic CHO product (≈0%)]

(13.16)

n = 2, 84%
n = 1, 63% (mixture of two products)

13-3 CYCLIZATION BY S_Hi

Intramolecular substitution reactions (S_Hi) on multivalent atoms (displacement reaction), giving cyclized products, are most frequently observed with sulfur (sulfides) and oxygen. Substitution on sulfur appears to be an addition-elimination whereas on oxygen it is more probably a direct substitution.

13-3-1 Sulfur

For the reaction to be efficient it is necessary, as in the case of hydrogen, for the attacking radical, the sulfur atom and the leaving radical to be in a straight line. It is generally assumed that an intermediate sulfuranyl radical is formed.

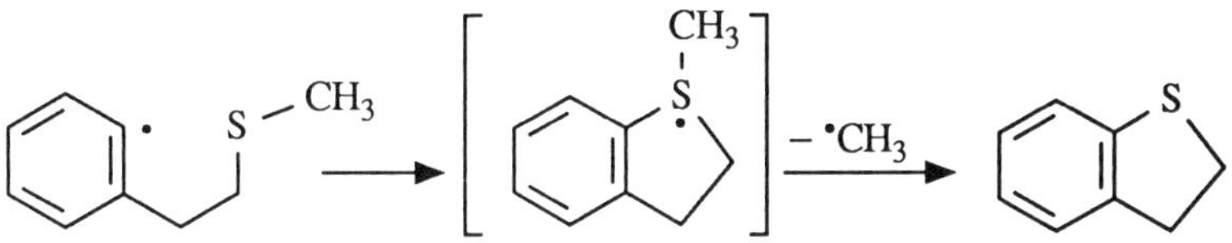

[Reaction scheme 13.17: benzyl radical with S–CH$_3$ → sulfuranyl intermediate → benzothiophene derivative + $^\bullet$CH$_3$]

(13.17)

[Reaction scheme 13.18: aryl radical with S–R and β-lactam ring → cyclic product, R = CH$_3$, t-Bu, Ph, –R$^\bullet$]

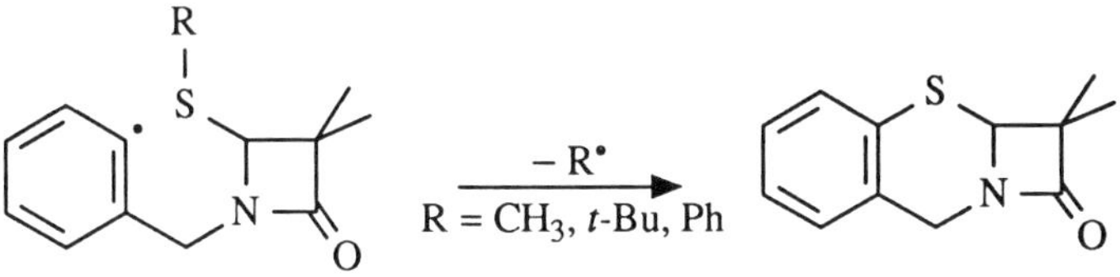

(13.18)

13-3-2 Oxygen

The intramolecular displacement reaction generally involves a weak peroxide bond. Depending on the position of the radical center and the nature of the O–O bond (dialkyl peroxide or perester) either cyclic ethers (13.19)

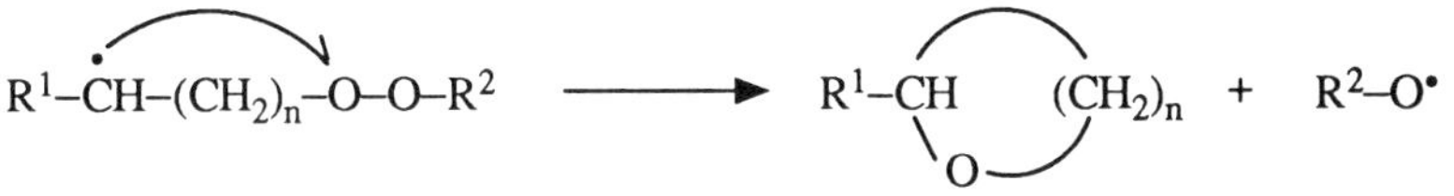

$$R^1\text{-}\overset{\bullet}{C}H\text{-}(CH_2)_n\text{-}O\text{-}O\text{-}R^2 \longrightarrow R^1\text{-}CH\underset{O}{\overset{}{\diagdown}}(CH_2)_n + R^2\text{-}O^\bullet \quad (13.19)$$

or lactones (13.20) are obtained.

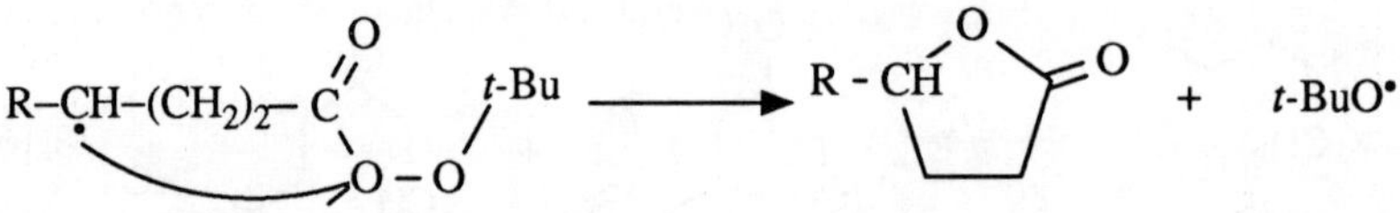

$$R-\underset{\bullet}{CH}-(CH_2)_2-\overset{O}{\underset{O-O}{C}}\diagdown_{t\text{-Bu}} \longrightarrow R-CH\overset{O}{\underset{}{\diagup}}\overset{}{\underset{}{\diagdown}}O + t\text{-BuO}^\bullet \quad (13.20)$$

Reaction (13.21), analogous to (13.19), explains the induced decomposition of di-*tert*-butyl peroxide.

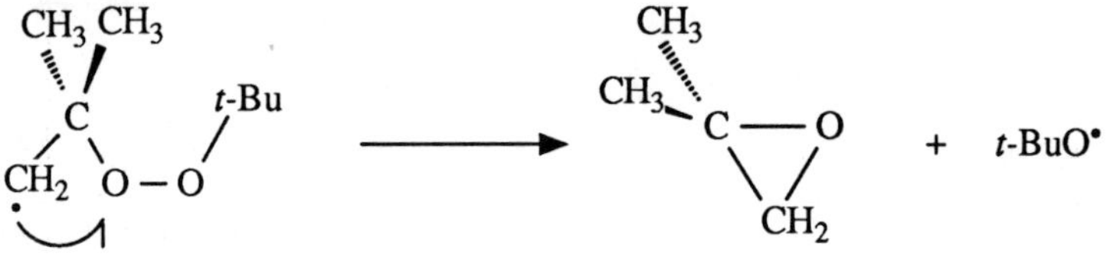

$$\begin{array}{c} CH_3\ CH_3 \\ \diagdown\!\!\diagup \\ C \\ \diagup\ \diagdown \\ \underset{\bullet}{CH_2}\ O-O \end{array}\!\!\!\!-t\text{-Bu} \longrightarrow \begin{array}{c} CH_3 \\ CH_3\!\!-\!\!C-O \\ \diagdown\ \diagup \\ CH_2 \end{array} + t\text{-BuO}^\bullet \quad (13.21)$$

When the radical center is β to an intracyclic O–O bond, the yield of $S_H i$ product depends on the ring size. In reaction (13.23) the –C•–C–O–O chain can be coplanar and allow better overlap between the p orbital containing the unpaired electron and the σ* orbital of the O–O bond, thus explaining why the yield of epoxide is higher than in the case of (13.22).

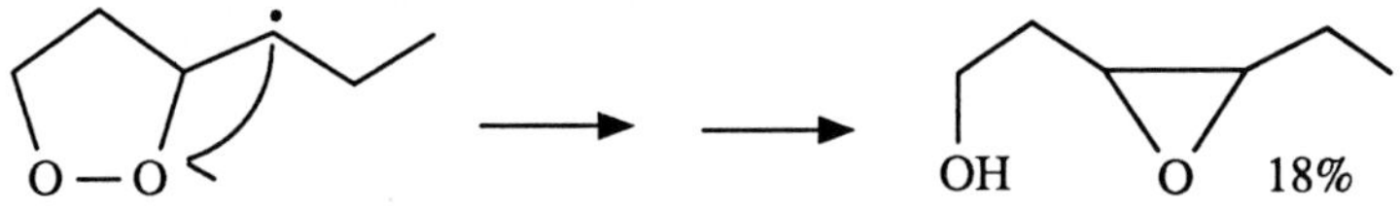

(13.22)

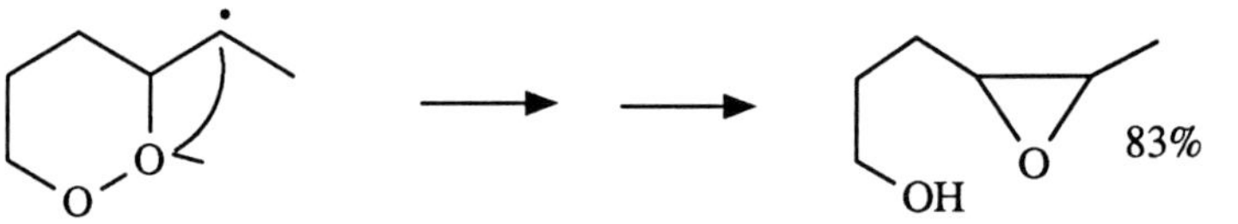

(13.23)

13-4 RING OPENING

Ring opening occurs by β-fragmentation. In this type of reaction ring strain is the most important thermodynamic factor.

13-4-1 Carbon radicals

Reactions (13.24) and (13.25) are exothermic for cyclopropyl and cyclobutyl radicals.

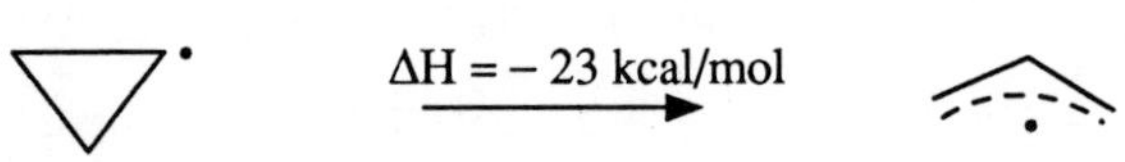

$$\triangle^\bullet \xrightarrow{\Delta H = -23\text{ kcal/mol}} \quad (13.24)$$

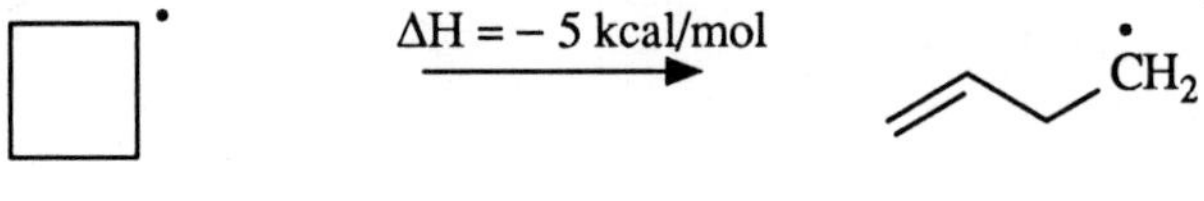

$$\square^\bullet \xrightarrow{\Delta H = -5\text{ kcal/mol}} \diagup\!\!\diagdown\!\!\diagup\overset{\bullet}{CH_2} \quad (13.25)$$

However, these reactions are slow and almost never observed under ordinary conditions, whereas those, (13.26) and (13.27), of cyclopropylmethyl and cyclobutylmethyl radicals are much faster and very common. Because it opens very rapidly the cyclopropylmethyl radical is used as a *radical clock* in competitive systems (Chap. 17).

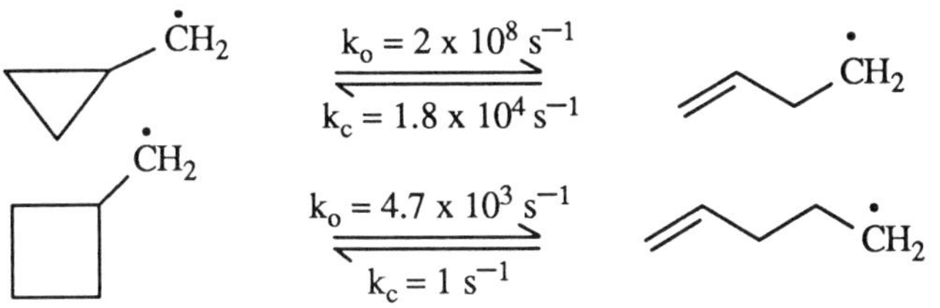

$$k_o = 2 \times 10^8 \text{ s}^{-1} \quad (13.26)$$
$$k_c = 1.8 \times 10^4 \text{ s}^{-1}$$

$$k_o = 4.7 \times 10^3 \text{ s}^{-1} \quad (13.27)$$
$$k_c = 1 \text{ s}^{-1}$$

The activation energy is reduced by a *stereoelectronic effect* when the p orbital of the radical center eclipses the antibonding σ* orbital of the bond which is to be broken.

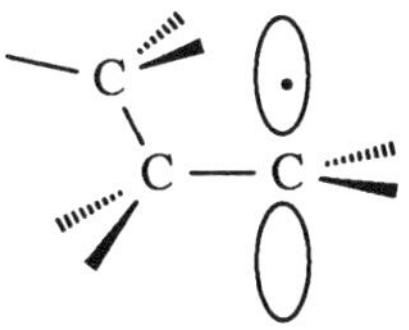

Scheme 13.3

The transition state is analogous to that of the reverse reaction (addition) which was described above (Chap. 12-1). Fragmentation is therefore oriented stereoelectronically. We shall give two examples:

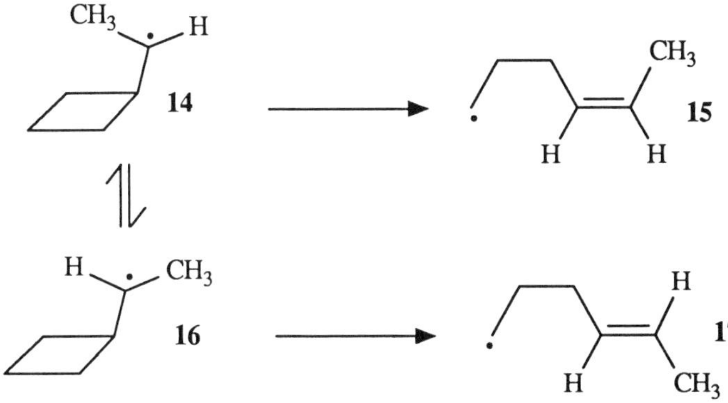

$$14 \longrightarrow 15 \quad (13.28)$$

$$16 \longrightarrow 17 \quad (13.29)$$

- the cyclobutylethyl radical **14** goes to the *trans* radical **17** and not to the *cis* radical **15** since the transoid form **16** is the more stable in the transition state;
- the stability of the radical formed is not always the driving force for β-fragmentation. Thus, for radicals **18** and **19** reactions (13.30) and (13.32) do not lead to the more stable radicals, secondary and benzylic, respectively. Breaking of an exo-

cyclic bond (13.32) is much preferred to fragmentation of an endocyclic bond (13.33) which would give, however, the more stable benzyl radical. This is a very fine example of a stereoelectronic effect (eclipsing of the breaking bond by the single occupied orbital).

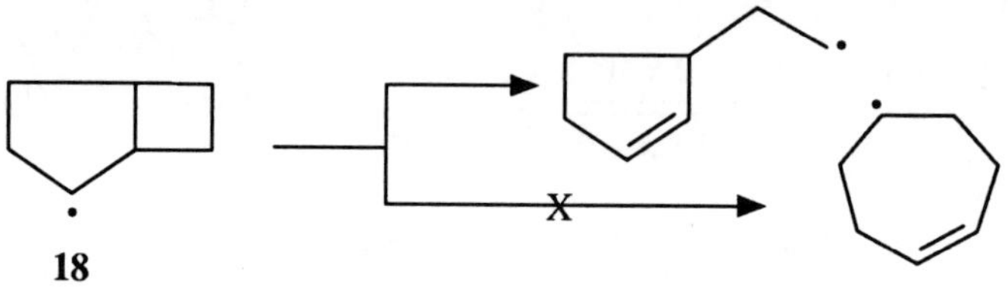

(13.30)

(13.31)

18

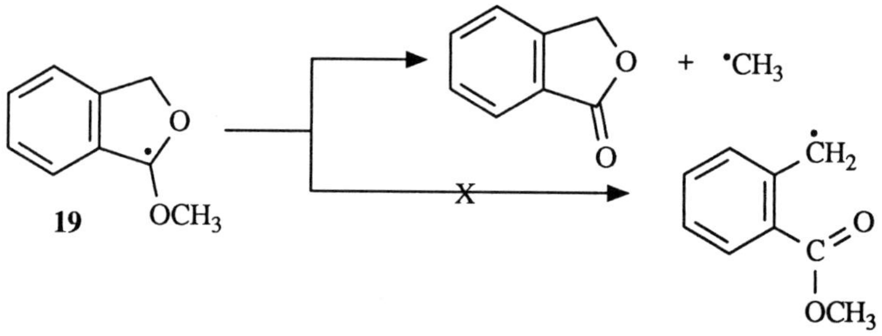

19 OCH$_3$

(13.32)   + •CH$_3$

(13.33)

Nevertheless, in many cases the *stability* of the radical formed is the controlling factor. For example, the opening of epoxides α to a radical center is a common reaction whose orientation, breaking of a C–O (13.34) or a C–C bond (13.35), is determined by the stability of the product radical.

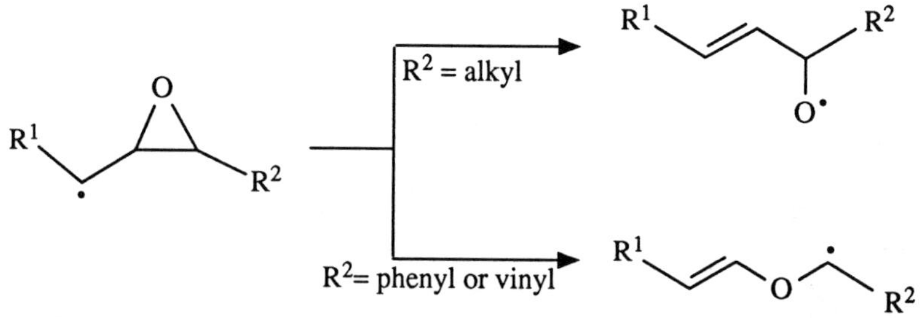

R^2 = alkyl (13.34)

R^2 = phenyl or vinyl (13.35)

In the same way, the rate of opening of radical **20** to the tertiary radical **21** explains why there are no products derived from **20** in the products of radical addition to β-pinene, unless one works with a transfer agent which can trap **20** very rapidly.

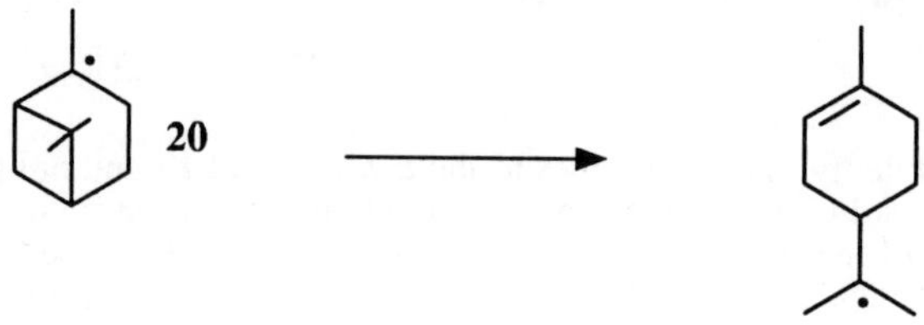

20 ⟶ **21**   (13.36)

13-4-2 Alkoxyl radicals

The ease of ring opening by β-fragmentation (13.37) depends on the ring size.

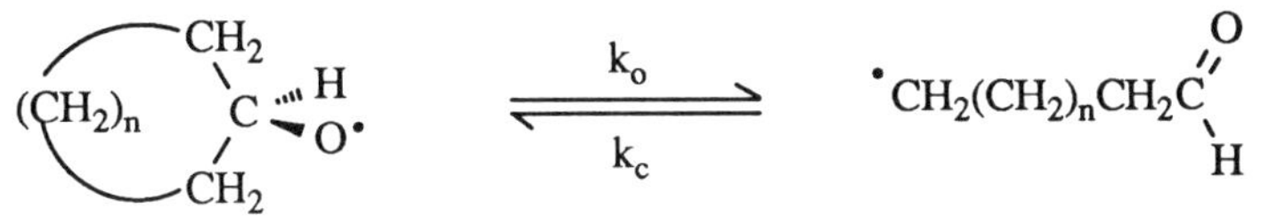

$$(CH_2)_n\underset{CH_2}{\overset{CH_2}{\diagdown}}C\underset{O\bullet}{\overset{H}{\diagup}} \underset{k_c}{\overset{k_o}{\rightleftharpoons}} {}^\bullet CH_2(CH_2)_nCH_2\underset{H}{\overset{O}{C}} \qquad (13.37)$$

The reaction is exothermic for n = O, 1, 2 or 4 and endothermic for n = 3. The rate varies as shown in (13.38). The reaction is reversible and in general $k_o > k_c$.

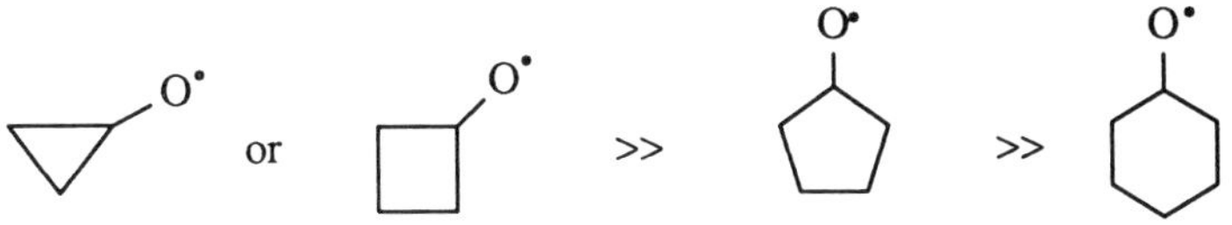

$$\text{cyclopropyl-O}^\bullet \text{ or cyclobutyl-O}^\bullet \gg \text{cyclopentyl-O}^\bullet \gg \text{cyclohexyl-O}^\bullet \qquad (13.38)$$

With bicyclic compounds β-fragmentation leads to ring enlargement, a reaction which is particularly useful in synthesis.

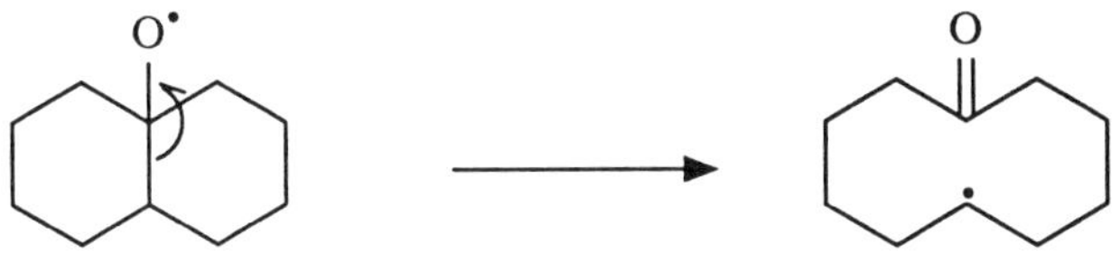

(13.39)

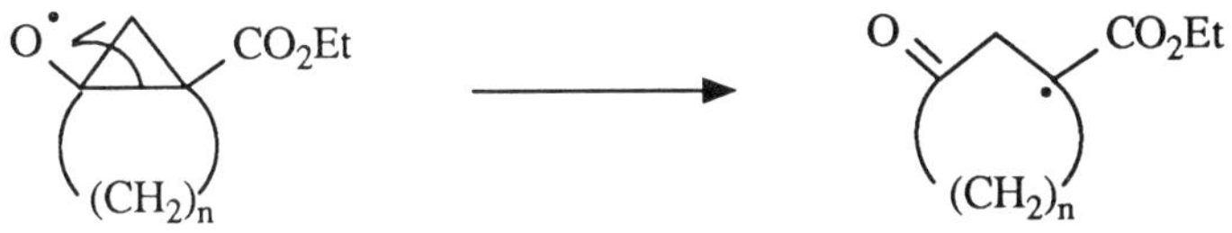

(13.40)

13-5 REARRANGEMENTS

We shall define rearrangements as intramolecular transfers of polyatomic groups. This type of reaction is much less common in radical chemistry than for carbocations. In most cases, for rearrangement to occur the group transferred must be unsaturated. The mechanism involves two steps: cyclization by intramolecular addition, then re-opening of the ring.

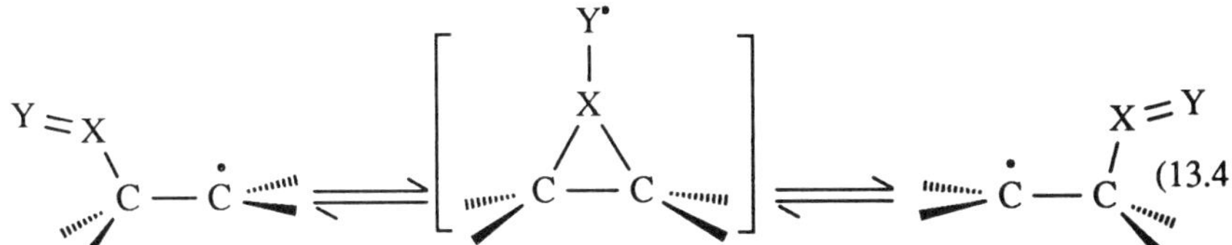

$$\underset{}{\overset{Y=X}{\diagdown}}C-\overset{\bullet}{C} \rightleftharpoons \left[\begin{array}{c}Y^\bullet\\|\\X\\C-C\end{array}\right] \rightleftharpoons \overset{\bullet}{C}-\underset{}{\overset{X=Y}{\diagup}}C \qquad (13.41)$$

The most frequently observed rearrangements concern aryl, vinyl and acetoxy groups. Generally speaking, thermodynamic factors, i.e. the relative stabilities of the initial and rearranged products, control the reaction. Intramolecular transfers of monovalent atoms (hydrogen or halogen) were described in Chapter 11.

13-5-1 Aryl transfer

1,2 Aryl transfers (neophyl rearrangements) take place by the formation of a cyclohexadienyl radical **22** which has been identified by UV spectroscopy in flash photolysis experiments. The lifetime of **22** is too short for it to be observed by EPR.

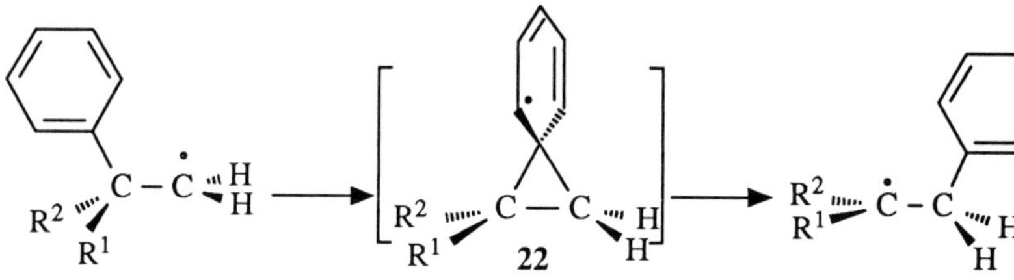

$$\text{(13.42)}$$

This reaction is slowed by electron-donating substituents on the phenyl, which raise the energy of the unoccupied orbitals of the aromatic nucleus and reduce the interaction with the singly occupied orbital (SOMO) of the radical center. Conversely, electron-attractors, which lower the energy of the unoccupied orbitals of the aromatic nucleus, accelerate the reaction. Thus, the relative rates for Ar = Ph, p-MeOC$_6$H$_4$ and p-CNC$_6$H$_4$ are 1, 0.36 and 35, respectively.

The stability of the radical formed and the overall enthalpy change related to the bonds formed and broken are important. Thus, the rate constant for reaction (13.42) is 59 s^{-1} when R^1 = R^2 = H and 3.6 x 10^5 s^{-1} when R^1 = R^2 = Ph.

Phenyl transfer from carbon to oxygen is faster than carbon to carbon, since BDE (Ph–O) > BDE(Ph–C); there is no β-fragmentation (13.44) as this is relatively slow.

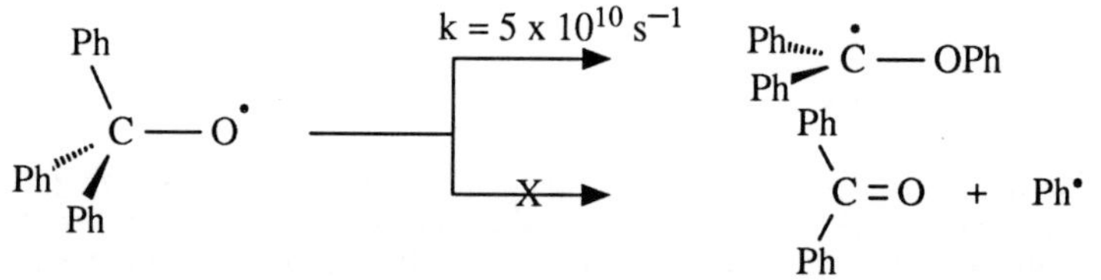

$$k = 5 \times 10^{10} \text{ s}^{-1} \quad \text{(13.43)}$$

$$\text{(13.44)}$$

While 1,3 aryl transfers do not exist, 1,4 and 1,5 (denoted Ar$_{1-5}$ and Ar$_{2-6}$) are easy. In the case of the 4-phenylbutyl radical **23**, there is competition between two cyclization reactions:

— Ar$_{1-5}$ which leads reversibly to the spiro radical **24** corresponds, after opening, to a 1,4 phenyl transfer (13.45);
— Ar$_{2-6}$ gives radical **25** irreversibly and then the cyclized product (13.46).

Radical **23** is about 2.5 kcal/mol more stable than **24** (spiro ring strain) and about 6.2 kcal/mol less stable than **25**.

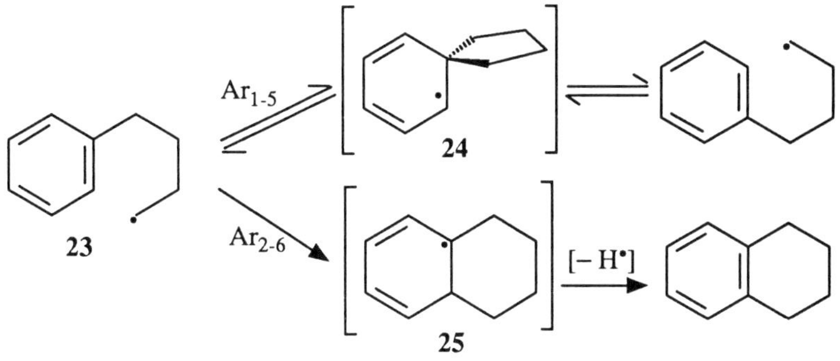

$$\text{23} \xrightarrow{\text{Ar}_{1\text{-}5}} [\textbf{24}] \rightleftharpoons \quad\quad (13.45)$$

$$\xrightarrow{\text{Ar}_{2\text{-}6}} [\textbf{25}] \xrightarrow{[-\text{H}^\bullet]} \quad\quad (13.46)$$

1,2 Phenyl transfer from Si to C does not occur since radical **26** is stabilized by interaction with the silicon orbitals (Scheme 13.5).

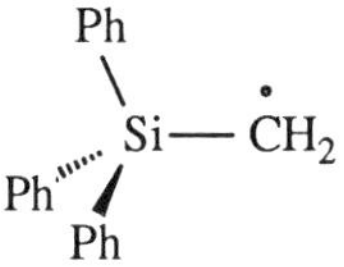

$$\text{Ph-Si(Ph)(Ph)-}\overset{\bullet}{\text{CH}}_2 \quad \textbf{26}$$

Scheme 13.4

However, 1,4 and 1,5 migrations (13.47) are observed.

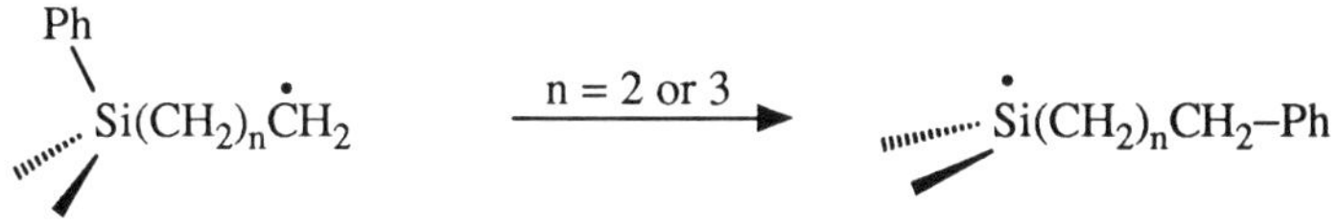

$$\text{Ph-Si(CH}_2\text{)}_n\overset{\bullet}{\text{CH}}_2 \xrightarrow{n = 2 \text{ or } 3} \text{Si(CH}_2\text{)}_n\text{CH}_2\text{-Ph} \quad (13.47)$$

1,2 Phenyl transfer from Si to O is enthalpy favored (13.48).

$$\text{Ph}_3\text{Si-O}^\bullet \longrightarrow \text{Ph}_2\overset{\bullet}{\text{Si}}\text{-OPh} \quad (13.48)$$

13-5-2 Vinyl transfer

Vinyl transfers (13.49) have been much less studied than those of aryl. There are several indications that 1,2 vinyl transfer in allylcarbinyl radicals proceeds by an

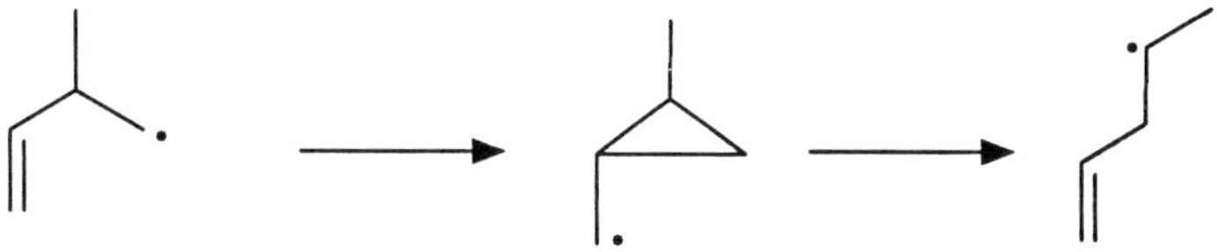

$$\text{(13.49)}$$

intermediate cyclopropylcarbonyl radical. The driving force of the reaction is the formation of a secondary radical from a primary.

In general, the ease of 1,2 rearrangement of unsaturated systems depends on the stability of the initial and final radicals but also on the difficulty of forming the intermediate 3-membered ring radical, as is shown by the data of Table 13.4 where E_a correlates with the calculated endothermicity for the formation of a 3-membered ring.

−X=Y	−CH=CH$_2$	$\underset{/}{\overset{t\text{-Bu}}{\text{C}}}$=O	−Ph	−C≡C-t-Bu	−C≡N
k (s^{-1} at 25°C)	10^7	1.7×10^5	7.6×10^2	93	0.9
E_a (kcal/mol)	5.7	7.8	11.8	12.8	16.4

Table 13.4 − *1,2 Transfer of unsaturated groups.*

13-5-3 1,2 Acetoxyl transfer

The mechanism of reaction (13.50) has been the subject of much work; depending on the structure, transfer occurs either by a concerted process with a delocalized radical **27** (path a) or via an intimate radical anion-cation pair **28** (path b). The process involving **28** appears to be faster (2 x 10^6 s^{-1}) than that via **27** (5 x 10^2 s^{-1}).

(13.50)

Whatever the case, there are several arguments against a localized radical intermediate **29**.

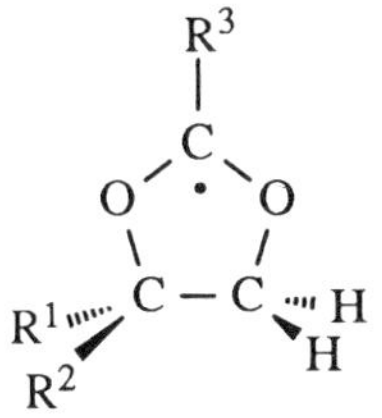

Scheme 13.5

This transfer is applied in the synthesis of 2-deoxy sugars (13.51).

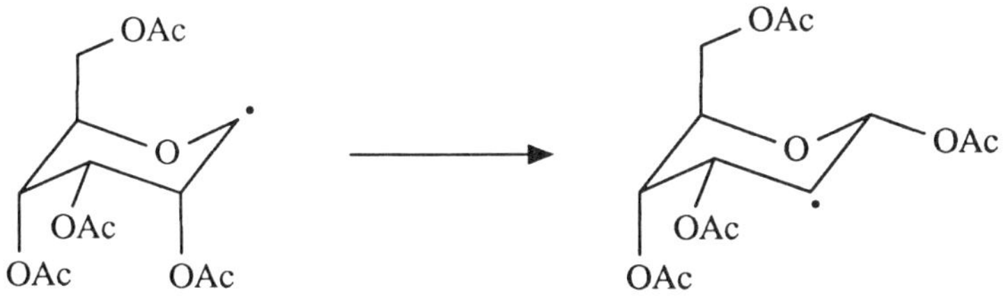

(13.51)

13-5-4 Peroxyl transfer

The rearrangement of an allyl peroxide consists of a concerted process analogous to path a of (13.50)

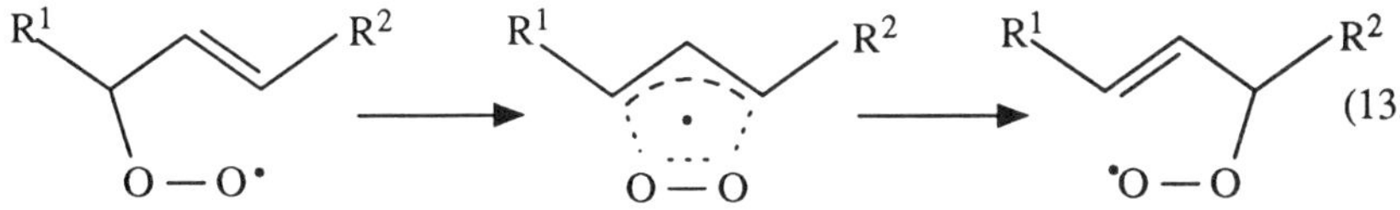

(13.52)

rather than one going by a localized radical intermediate **30**

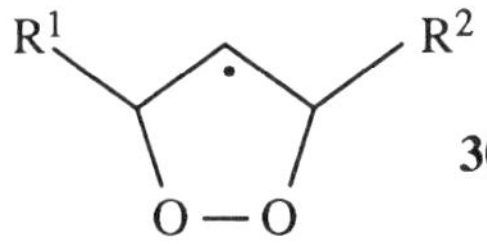

Scheme 13.6

but in some cases the reaction can be a dissociative process (13.53).

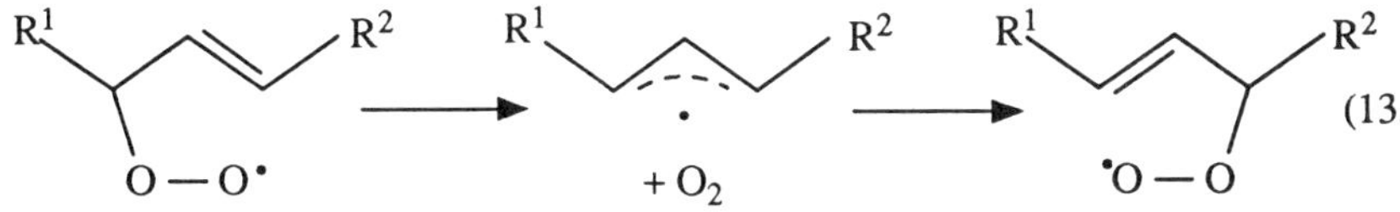

(13.53)

14 - AROMATIC SUBSTITUTION

Homolytic substitution reactions correspond to the following general scheme:

$$\text{Rad}^\bullet + X\text{-ArH} \xrightarrow{[-H^\bullet]} X\text{-Ar-Rad} \tag{14.1}$$

Arylation (Rad$^\bullet$ = Ar$^\bullet$ and various X) was studied by Hey as far back as 1937. The essential merit of the many results obtained was that they established the mechanism, but their synthetic interest is limited because of the low reactivity and regioselectivity of the reaction. Subsequently, by using pronounced polar antagonisms, Minisci developed an aromatic homolytic substitution chemistry, particularly with heteroaromatic bases, leading to very interesting results as regards both reactivity and regioselectivity.

14-1 MECHANISM

Aromatic substitution is an addition-elimination process (14.2) - (14.3).

$$\text{Rad}^\bullet + \text{X-ArH} \underset{k_{-ArH}}{\overset{k_{ArH}}{\rightleftharpoons}} \text{X-ArHRad}^\bullet \tag{14.2}$$

$$\text{X-ArHRad}^\bullet \xrightarrow{[-H^\bullet]} \text{X-ArRad} \tag{14.3}$$

Generally, addition does not involve a chain mechanism, unlike radical additions, since the ArHRad$^\bullet$ radical is too unreactive to perform a transfer reaction and regenerate Rad$^\bullet$. The existence of the ArHRad$^\bullet$ radical has been demonstrated by EPR and by the presence of dimerization and disproportionation products. The yield of the

$$2 \text{ X-ArHRad}^\bullet \xrightarrow{\text{dimerization}} \text{(H,Rad)-Ar-Ar-(H,Rad) with X} \tag{14.4}$$

$$\xrightarrow{\text{disproportionation}} \text{X-ArH}_2\text{Rad} + \text{X-ArRad} \tag{14.5}$$

Chapter 14: aromatic substitution

monosubstituted product, ArRad, depends on the factors controlling the addition (14.2) and elimination steps (rearomatization) (14.3).

14-2 ADDITION STEP

14-2-1 Reaction profile

The reaction profile for the addition of a radical to an aromatic system goes successively through a π^* transition state, a π complex, a σ^* transition state and a σ complex. This σ complex, the result of the addition (14.2), is the cyclohexadienyl radical, ArHRad$^\bullet$. It corresponds to the Wheland or Meisenheimer complexes of ionic aromatic substitutions. The existence of the intermediate π complex in addition to an aromatic nucleus is not sure. Generally, the addition step (14.2) is reversible and its transition state is σ^* type (Figure 14.1).

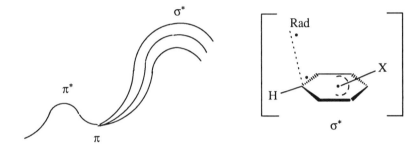

Figure 14.1 – *Reaction profile and transition states for addition of a radical to an aromatic nucleus.*

In the case of a substituted benzene there are therefore three σ^* transition states (*ortho, meta, para*) and the rate of formation of each of the corresponding σ complexes is different. The sum of the three rate constants, k_o, k_m and k_p, of the *ortho, meta* and *para* positions defines an overall rate constant, k_{Ar} (14.6).

$$\frac{d[ArRad_{ortho}]}{dt} + \frac{d[ArRad_{meta}]}{dt} + \frac{d[ArRad_{para}]}{dt} = (k_o + k_m + k_p)[ArH][Rad^\bullet]$$

$$k_{ArH} = k_o + k_m + k_p \qquad (14.6)$$

In the case of pyridines it is thought that the transition state for the addition step

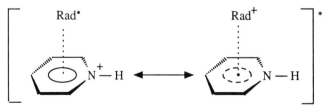

Figure 14.2 – π^* *type transition state.*

is π^* type. This can be described by two canonical forms, one of which corresponds to electron transfer from the radical to the substrate (Figure 14.2). Nevertheless, the carbocation structure, Rad$^+$, is only a canonical form and there is no skeletal rearrangement which would result if a carbocation really existed. In this case equation (14.6) is not applicable and k_{ArH} represents the rate of formation of the π or charge transfer complex.

14-2-2 Absolute rate constants

The addition reaction (14.2) is exothermic but, because of the loss of aromaticity, less than for addition to a C=C double bond. The rate constant depends on several factors.

Reaction enthalpy

The strength of the bond formed explains why Ph$^•$ reacts faster with benzene than does n-Bu$^•$ (Table 14.1).

R$^•$	PhH	4-CNPyH$^+$
t-Bu$^•$	a	6.3×10^7
$^•$CH$_2$OH	a	$\approx 10^7$
Ph$^•$	1.0×10^6	6.0×10^6
t-BuC$^•$=O	a	$\approx 10^6$
n-Bu$^•$	3.8×10^2	8.9×10^5
PhCO$_2$$^•$	1.4×10^2	-
HO$^•$	5×10^9	-

Table 14.1 – *Rate constants for addition, k_{ArH} ($M^{-1}s^{-1}$) of radicals to benzene and protonated 4-cyanopyridine. a) No reaction.*

substrate	k_{ArH}
PhH	3.8×10^2
PhOCH$_3$	1.3×10^3
4-Me pyridine	1.5×10^3
4-Me pyridine H$^+$	1.1×10^5
4-CN pyridine H$^+$	1.8×10^6
quinoxaline H$^+$	2.7×10^7

Table 14.2 – *Rate constants for addition, k_{ArH} ($M^{-1}s^{-1}$) of primary alkyl radical to aromatic substrates.*

The extent to which the reaction is reversible increases with the stability of the entering radical (14.7).

$$\text{Ph}^• < \text{R}^•_{prim} < \text{R}^•_{sec} < \text{R}^•_{tert} \ll \text{PhCH}_2^• \tag{14.7}$$

Polar character

Generally, making the aromatic or heteroaromatic ring electron-poor accelerates reaction with nucleophilic radicals. Thus, protonating a pyridine ring increases the rate of reaction with a primary alkyl radical by a factor of 100 to 1 000 (Table 14.2) and the rate increases with the nucleophilicity of the radical (Table 14.1).

14-2-3 Relative rate constants

Experimental determination of absolute rate constants is not always easy and requires rather elaborate physicochemical techniques. Reactivity can however be measured by a competitive method, by reacting a radical with a mixture of substituted (ArH) and unsubstituted aromatics (PhH). The ratio of the amounts of substitution products obtained gives the *overall rate factor* ($k_{X/H} = k_{ArH}/k_{PhH}$) and expresses the reactivity of the various substrates with the same radical. When the addition step (14.2) is rate determining and is irreversible or only slightly reversible, one can write:

Chapter 14: aromatic substitution

$$\text{Rad}^\bullet \begin{cases} + \text{ArH} \xrightarrow{k_{\text{ArH}}} \text{ArHRad}^\bullet \xrightarrow{[-H^\bullet]} \text{ArRad} & (14.8) \\ + \text{PhH} \xrightarrow{k_{\text{PhH}}} \text{PhHRad}^\bullet \xrightarrow{[-H^\bullet]} \text{PhRad} & (14.9) \end{cases}$$

$$k_{X/H} = \frac{k_{\text{ArH}}}{k_{\text{PhH}}} = \frac{[\text{ArRad}]}{[\text{PhRad}]} \times \frac{[\text{PhH}]}{[\text{ArH}]} \quad (14.10)$$

Benzenes

The values of $k_{X/H}$, for various radicals and substrates presented in Table 14.3 suggest the following remarks:

– a substituted benzene is almost always more reactive than benzene ($k_{X/H} > 1$) whether X is donor or attractor and Rad$^\bullet$ electrophilic or nucleophilic;

– the increase in the reactivity ($k_{X/H} \gg 1$) is always more pronounced when the polar antagonisms are well differentiated. This is the case for nucleophilic radicals ($cy\text{-}C_6H_{11}^\bullet$, Me$^\bullet$) reacting with electron-poor substrates (PhCN, PhCl) and for electrophilic radicals ($^\bullet CH_2CN$) with electron-rich substrates (PhMe, PhOMe).

Rad$^\bullet$		$cy\text{-}C_6H_{11}^\bullet$				$^\bullet CH_3$			
	Ph-X	$k_{X/H}$	F_o	F_m	F_p	$k_{X/H}$	F_o	F_m	F_p
	PhCN	27	42	3.2	72	5.1	9.0	1.68	9.5
	PhCl	3.5	5.7	3.57	2.52	1.5	3.1	0.9	0.95
	PhMe	0.75	0.72	0.96	1.12	1	1.56	0.96	0.9
	PhOMe	2.3	4.62	1.93	0.69	2.8	5.8	1	3.2
Rad$^\bullet$		$^\bullet CH_2CN$				Ph$^\bullet$			
	PhCN	0.9	1.13		0.58[a]	3.7	6.5	1.1	6.1
	PhCl	1.0	1.32	1.0	1.32	1.60	3.1	1	1.48
	PhMe	1.45	1.74	1.56	2	1.68	3.30	1	1.27
	PhOMe	2.6	3.5	1.6	5.3	1.71	3.56	0.93	1.29

Table 14.3 – *Relative reactivity ($k_{X/H}$) and regioselectivity (partial rate factors F_o, F_m, F_p (cf. para. 14-3-1) for homolytic substitution on substituted benzenes (14.1). a) Not separated.*

Another example of a marked polar effect is the increase in the rate constant for the addition of protonated aminyl radicals (14.11). Although k is not known with precision, competition with the intramolecular hydrogen transfer in the protonated aminyl radical, where R is an aliphatic chain, suggests that $k > 10^6$ M^{-1}s^{-1}, i.e. a value greater than that for many carbon radicals (Tables 14.1 and 14.2).

$$^\bullet\overset{+}{N}HR_2 + \text{C}_6\text{H}_6 \xrightarrow{k} [\text{cyclohexadienyl-}\overset{+}{N}HR_2\text{-H}] \quad (14.11)$$

Pyridines

This polar effect has been used by Minisci who reacted nucleophilic radicals with

...ated pyridines, rendered electrophilic both by protonation and by an attractor ...stituent. The addition rates for such compounds are much higher than for benzene or neutral pyridine (Tables 14.1 and 14.2) and the reaction is always regioselective in favor of the 2 position.

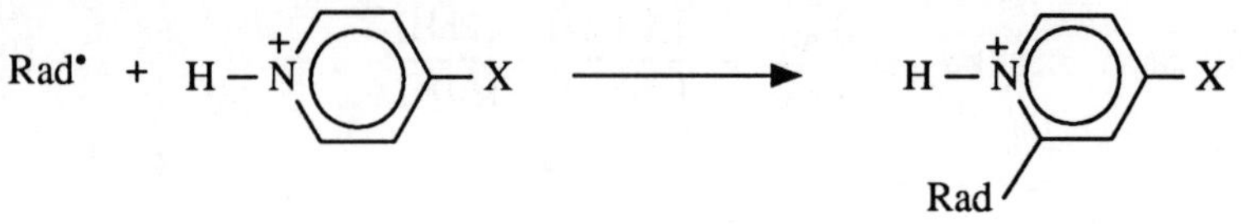

$$\text{Rad}^\bullet + \text{H}-\overset{+}{\text{N}}\!\!\bigcirc\!\!-\text{X} \longrightarrow \text{H}-\overset{+}{\text{N}}\!\!\bigcirc\!\!-\text{X} \quad (14.12)$$

The substituent effect, determined by the competitive method, is also remarkable (Table 14.4): when X is an attractor the selectivity increases with the nucleophilicity of the radical (t-Bu$^\bullet$ > n-Bu$^\bullet$) and decreases if X is a donor. The reactivity of non-protonated pyridines, however, is similar to that of benzenes.

Rad$^\bullet$ \ X	H	CN	COMe	Me	OMe
t-Bu$^\bullet$	1	1 890	144	0.15	0.006
s-Bu$^\bullet$	1	259	56	0.28	0.02
n-Bu$^\bullet$	1	20	6	0.32	0.10

Table 14.4 – *Relative rates ($k_{X/H}$) for reaction of radicals with protonated 4-X-substituted pyridines.*

14-2-4 Interpretation of results

The selectivity of radicals with regard to aromatic substrates can be interpreted in terms of the polar effect or orbital control.

Polar effect

A model of the σ^* transition state can be constructed from the orbitals involved in the reaction process:

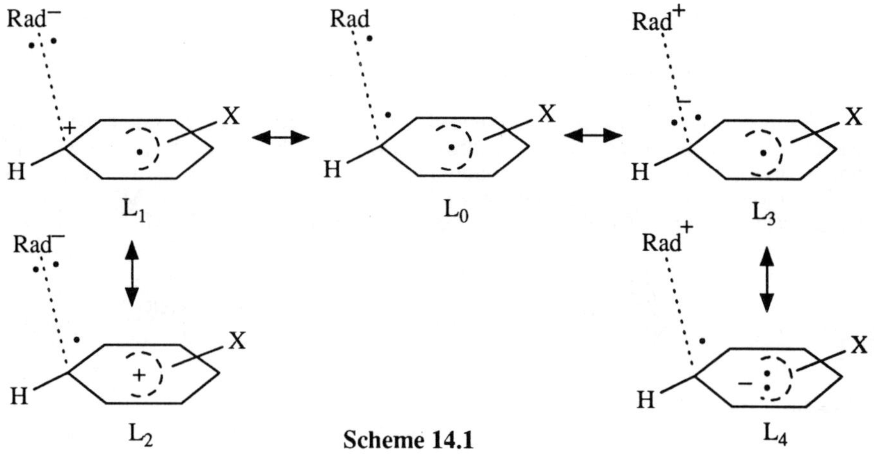

Scheme 14.1

- the SOMO of the radical, Rad•;
- the orbital localized on the aromatic carbon attacked;
- the HOMO of the remaining aromatic system.

The different canonical forms of the transition state, L_0 to L_4, are obtained by distributing three electrons between these orbitals (Scheme 14.1). If Rad• is a nucleophile and X an electron-attracting substituent or, on the contrary, Rad•, is an electrophile and X a donor, the weights of forms L_3 and L_4 or L_1 and L_2 increase. Transition state delocalization increases and its energy is reduced. This is why nucleophilic and electrophilic radicals are more selective: the first for electron-poor substrates and the second for electron-rich substrates.

In the case of protonated pyridines it is assumed that the transition state is π^* type (Figure 14.2). The second form, where the electron is transferred from the radical to the substrate, increases the reactivity. This form predominates when the substituent X makes the pyridine electron-poor and the radical is nucleophilic.

Orbital control

Orbital interactions also explain the observed selectivities. The two dominant frontier interactions are those between the radical SOMO and the substrate LUMO, a nucleophilic interaction, and between the SOMO and the HOMO, an electrophilic interaction. The effect is inversely proportional to the energy difference between the orbitals.

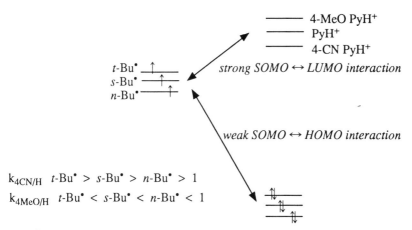

Figure 14.3 – *Orbital interactions for protonated substituted pyridines.*

When distinctly nucleophilic radicals react with electron-poor substrates, such as PhCN and PhCl (Table 14.3) and with protonated pyridines (Table 14.4), the SOMO ↔ LUMO interaction determines the reactivity (Figure 14.3):

- $k_{X/H}$ is greater than unity when the LUMO of the substituted substrate is lower than that of the unsubstituted one (X = CN). In this case, the higher the SOMO, the greater $k_{X/H}$;
- conversely, $k_{X/H}$ is lower (close to unity) when the LUMO of the substituted substrate is higher than that of the unsubstituted one (X = OMe). In this case, the higher the SOMO, the smaller $k_{X/H}$.

For the reactions of electrophilic radicals like •CH$_2$CN with substituted benzenes the electrophilic interaction, SOMO ↔ HOMO, is the more important (Figure 14.4):

– $k_{X/H}$ is greater than unity when the HOMO of the substituted substrate is higher than that of benzene (PhOMe); the lower the SOMO, the greater $k_{X/H}$;
– the selectivity is lower or of the same order of magnitude when the HOMO of the substituted substrate is lower (PhCl or PhCN).

$k_{OMe/H}$ •CH$_2$NO$_2$ > •CH$_2$CN > 1
$k_{Cl/H}$ •CH$_2$NO$_2$ < •CH$_2$CN < 1

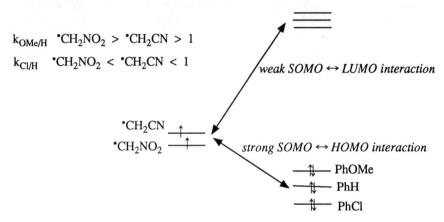

Figure 14.4 – *Orbital interactions for substituted benzenes.*

14-3 ORTHO, META, PARA REGIOSELECTIVITY

The regioselectivity depends on the electrophilicity or nucleophilicity of both the radical and the substrate.

14-3-1 Benzenes

The regioselectivity (RS_o, RS_m and RS_p), sometimes expressed in terms of *partial rate factors* (F_o, F_m and F_p), is obtained from equation (14.6) weighted for the statistical distribution of the benzene hydrogens.

$$F_o = RS_o = \frac{6}{2} \times \frac{k_o}{k_{PhH}} = \frac{6}{2} \times \frac{[ArRad_o]}{[PhRad]} \times \frac{[PhH]}{[ArH]} \quad (14.13)$$

$$F_m = RS_m = \frac{6}{2} \times \frac{k_m}{k_{PhH}} = \frac{6}{2} \times \frac{[ArRad_m]}{[PhRad]} \times \frac{[PhH]}{[ArH]} \quad (14.14)$$

$$F_p = RS_p = \frac{6}{1} \times \frac{k_p}{k_{PhH}} = \frac{6}{1} \times \frac{[ArRad_p]}{[PhRad]} \times \frac{[PhH]}{[ArH]} \quad (14.15)$$

The results in Table 14.3 show that:

– without exception, the reactivity order is *ortho > para > meta* with F_m close to unity;

- if X is attractor, the *ortho-para* regioselectivity increases with the nucleophilicity of the radical (cy-$C_6H_{11}^\bullet$ > $Me^\bullet$ > $Ph^\bullet$ > $^\bullet CH_2CN$), in parallel with the overall reactivity, $k_{X/H}$;
- if X is electron-donating the effects are small.

The *ortho-para* regioselectivity is controlled by the canonical forms depicted in Scheme 14.2, where an electron of one of the pairs of the π system is *localized* by the substituent X, the other electron being delocalized to *ortho* and *para*.

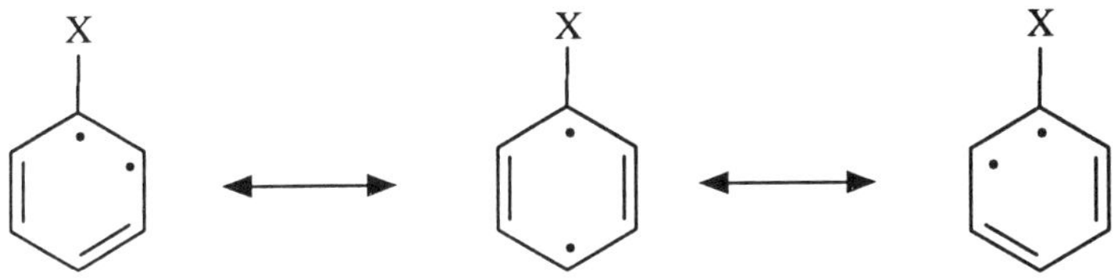

Scheme 14.2

These orientations are also given by the structure of the molecular orbitals of the π systems, which indicate that the spin density is greatest *ortho* and *para* to X. Various theoretical reactivity indices calculated from the structure of the π MOs agree with these results.

The regioselectivity control can also be explained by considering the stability of the intermediate cyclohexadienyl radical formed (σ complex). The better the unpaired electron can be localized on the carbon atom which is bonded to the substituent X, the more stable the radical. This is the situation for *ortho* and *para* attack (Scheme 14.3).

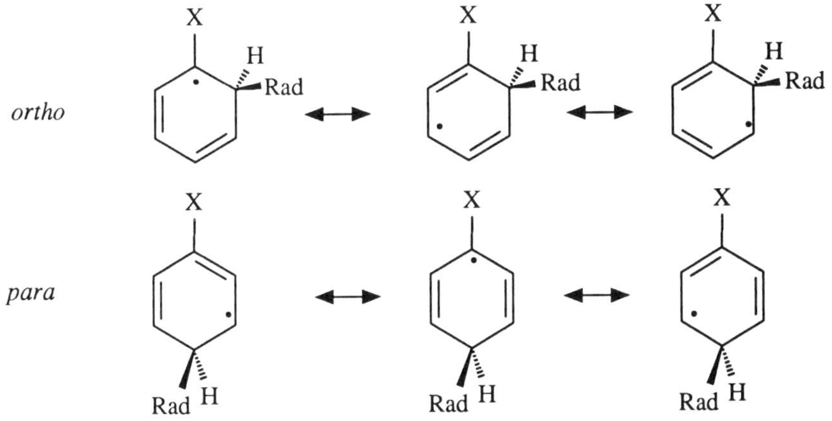

ortho

para

Scheme 14.3

If these explanations were all, the predominance of *ortho* and *para* substitution would be independent of the radical, Rad$^\bullet$. In fact, the regioselectivity depends on the radical. This is explained by polar effects. A π-donor substituent increases the electron density at the *ortho* and *para* positions, while a π-acceptor reduces the electron density at the same points (Scheme 14.4). Consequently, a nucleophilic radical will react preferentially at *ortho* and *para* if X is π-acceptor, and at *meta* if X is π-donor, and vice versa for an electrophilic radical. Indeed, the *meta/para* ratio increases with the nucleophilicity of Rad• if X is a π-donor (OMe, Cl) and decreases if X is a π-attractor (CN).

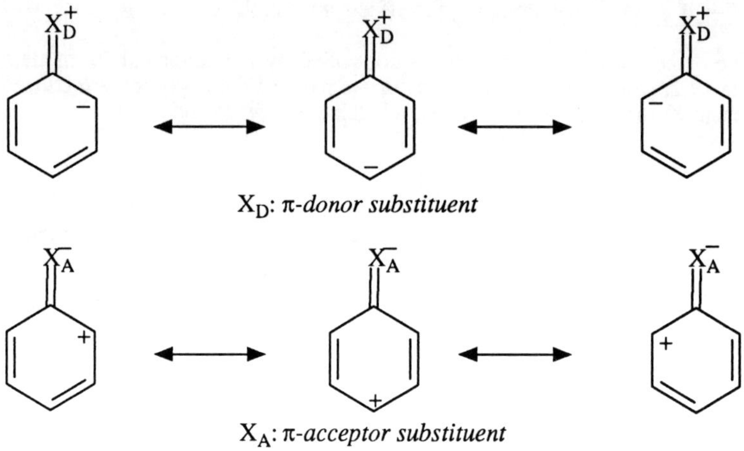

Scheme 14.4

X_D: π-*donor substituent*

X_A: π-*acceptor substituent*

Such a situation cannot arise in the *meta* position, and only a relatively weak inductive effect is observed (Scheme 14.5). Severe steric crowding (X = *t*-Bu) reduces the reactivity at *ortho* more than at *para*.

meta

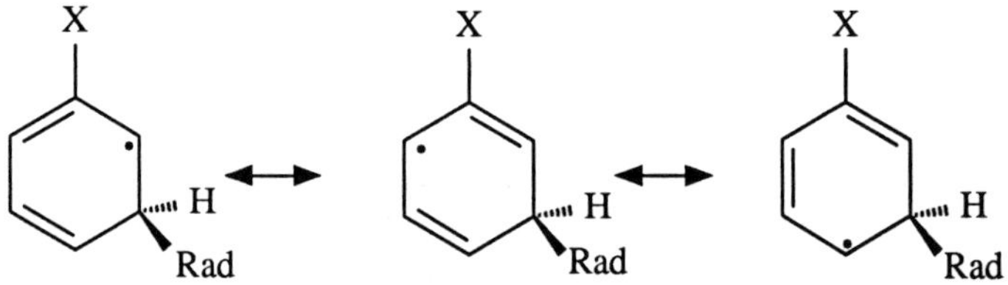

Scheme 14.5

14-3-2 Protonated pyridines

When the pyridine nucleus is not protonated the transition state is σ^* type and the regioselectivity is controlled by the same effects as in benzenes (spin density, electron delocalization, preferential stabilization of σ complexes).

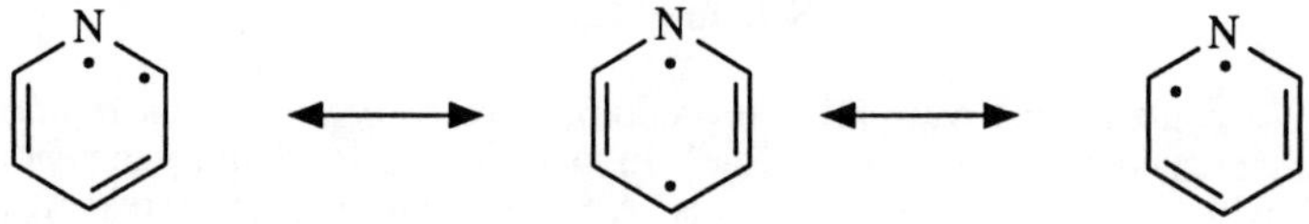

Scheme 14.6

For *protonated pyridines* there is some doubt about the exact nature of the transition state, σ* or π* (Figure 14.1). If the transition state is σ* type, the explanations given above for the regioselectivity also apply to protonated pyridines and predict greatest reactivity at the 2 and 4 positions.

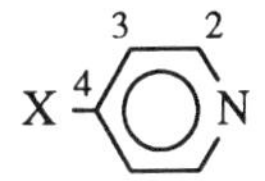		Rad• 			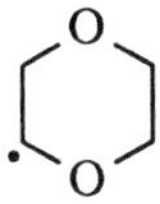						*i*-Pr•		
X		2	3	4	2	3	4	2	3	4	2	3	4
H	neutral	65	35[a]					55	31	14	65	35[a]	
	protonated	63	5	32	76[b]	ε[d]	23	64	5	31	72[c]	1	27
CN	neutral	15	85		15	85		25	75				
	protonated	100	0		100	0		62	38		100	0	
CO_2Et	neutral	40	60		40	60		30	70		40	60	
	protonated	100	0		100	0					100	0	
CH_3	neutral							48	52				
	protonated							84	16				

Table 14.5 – *Regioselectivity with respect to neutral (organic solvents) or protonated pyridines (water). a) Not separated. b) Dioxan. c) Benzene. d) Trace.*

If the transition state is π* type it resembles a charge-transfer complex (Figure 14.2) and, as is shown by the results in Table 14.5, the preference for reaction at the 2 position increases with the nucleophilicity of the entering radical and with electron attraction by the 4 substituent.

In this type of reaction the solvent can greatly influence the regioselectivity (Table 14.6); in a polar solvent such as water equilibrium (14.16) depends on the stability of the radical. Thus, the regioselectivity is about the same for Ph• (non-polar radical forming a strong bond) in water and benzene, whereas it is very different for

solvent	water[a]		benzene[a]	
Rad•	2	4	2	4
Ph•	64	31	70	24
t-Bu•	22	76	71	29

Table 14.6 – *2/4 Regioselectivity in % with respect to protonated pyridine in water or benzene. a) A few % of 3-substituted products.*

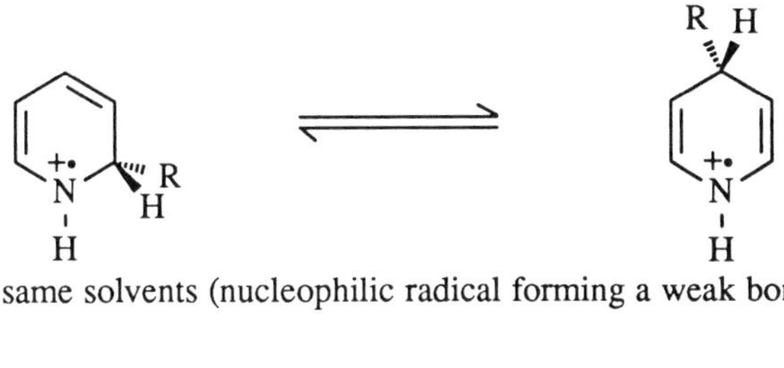

(14.16)

t-Bu• in the same solvents (nucleophilic radical forming a weak bond). If the pyridine has a very electron-attracting substituent such as CN in the 4 position the solvent has little effect on the regioselectivity of a nucleophilic radical.

14-4 ELIMINATION STEP

Unlike ionic electrophilic substitution reactions, where the aromatization step (14.3) which involves expulsion of a stable proton is very fast, the expulsion of a hydrogen atom from the cyclohexadienyl radical requires the participation of a new reactant:

— either by disproportionation with a radical of the medium (14.17);

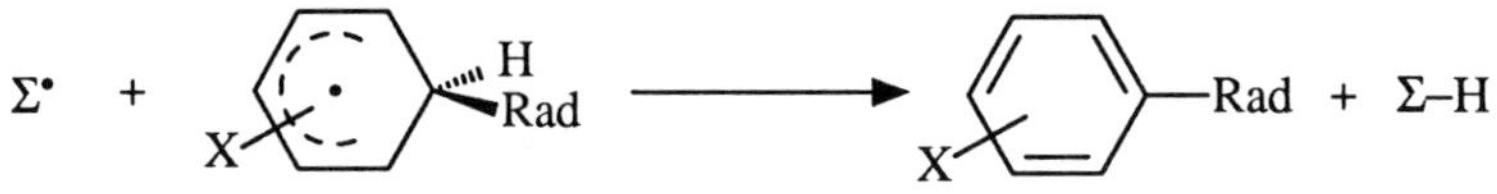

$$\Sigma^{\bullet} + \underset{X}{\text{cyclohexadienyl}}\overset{H}{\underset{\text{Rad}}{}} \longrightarrow \underset{X}{\text{arene}}-\text{Rad} + \Sigma-H \qquad (14.17)$$

— or by oxidation (14.18).

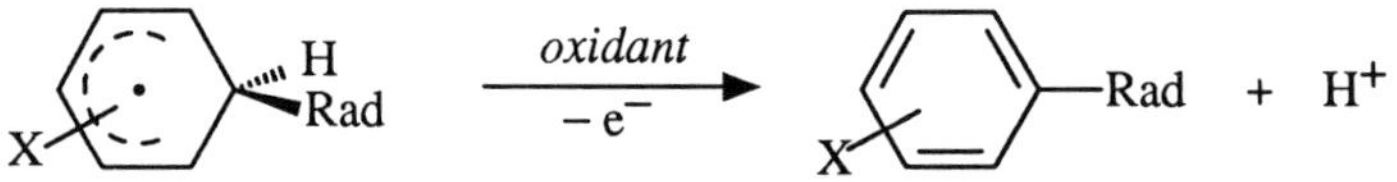

$$\underset{X}{\text{cyclohexadienyl}}\overset{H}{\underset{\text{Rad}}{}} \xrightarrow[-e^-]{oxidant} \underset{X}{\text{arene}}-\text{Rad} + H^+ \qquad (14.18)$$

The oxidant can be oxygen, a nitro derivative, a peroxide or a metal salt. A good oxidant increases the yield of substitution product. Thus, reaction of the benzoyloxyl radical, $PhCO_2^{\bullet}$, with benzene with or without oxygen gives 58 or 13% of phenyl benzoate, respectively.

14-5 REACTIONS

14-5-1 Arylation of aromatic substrates

As indicated at the beginning of this chapter, arylation usually proceeds according to the following scheme (reactions (14.19) and (14.20)). The reactivities and regioselectivities have been studied for many reactions with various combinations of $Ar^{\bullet}$ radicals, different methods of producing them, and a variety of aromatic substrates, Ar^1H; the results, which we have briefly summarized above, indicate that both are generally low, hence the limited synthetic interest of this type of reaction.

$$Ar^{\bullet} + Ar^1H \underset{k_{-add}}{\overset{k_{add}}{\rightleftharpoons}} Ar^1HAr^{\bullet} \qquad (14.19)$$

$$Ar^1HAr^{\bullet} \xrightarrow[-e^-]{oxidant} Ar^1-Ar + H^+ \qquad (14.20)$$

14-5-2 Alkylation of aromatic substrates

The alkylation of aromatic substrates follows the same scheme as arylation. This type of reaction was only studied once practical and efficient methods for producing alkyl radicals had been developed. Some examples are given in Table 14.3. As for arylations, the low reactivities and regioselectivities mean that these reactions have seen few applications in synthesis.

14-5-3 Amination of aromatic substrates

Protonated aminyl radicals, obtained by reduction of N-chloroamines, lead to aromatic amino compounds by a chain mechanism, in contrast to the previous schemes. One of the propagation steps can be a redox process.

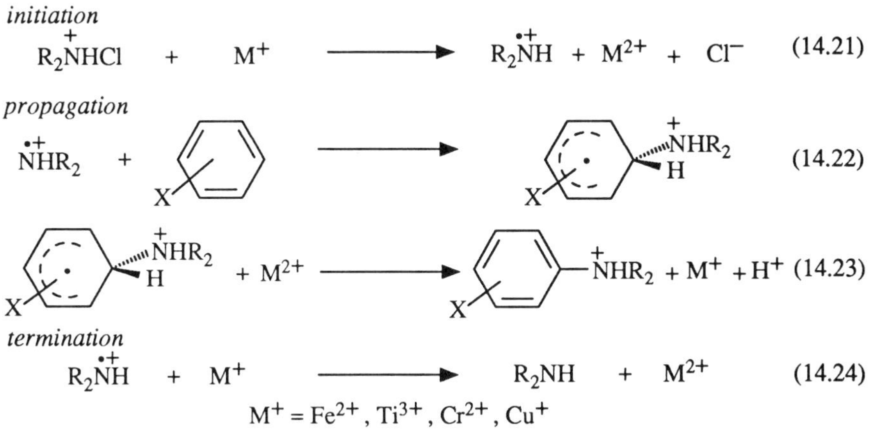

initiation

$$R_2\overset{+}{N}HCl + M^+ \longrightarrow R_2\overset{\bullet+}{N}H + M^{2+} + Cl^- \quad (14.21)$$

propagation

$$\overset{\bullet+}{N}HR_2 + \underset{X}{\bigcirc} \longrightarrow \underset{X}{\bigcirc}\overset{+}{\underset{H}{NHR_2}} \quad (14.22)$$

$$\underset{X}{\bigcirc}\overset{+}{\underset{H}{NHR_2}} + M^{2+} \longrightarrow \underset{X}{\bigcirc}-\overset{+}{N}HR_2 + M^+ + H^+ \quad (14.23)$$

termination

$$R_2\overset{\bullet+}{N}H + M^+ \longrightarrow R_2NH + M^{2+} \quad (14.24)$$

$$M^+ = Fe^{2+}, Ti^{3+}, Cr^{2+}, Cu^+$$

Scheme 14.7

Protonated aminyl radicals, which are very electrophilic, react rapidly with electron-rich aromatic substrates. The rate constant for addition (14.22) is not known but comparison with other reactions in a competitive system, in particular with intramolecular hydrogen transfer (Hofmann-Löffler-Freytag reaction), suggests it is very high. Consequently, a π^* type transition state has been proposed.

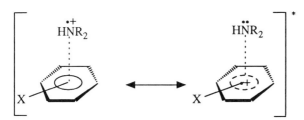

Figure 14.5 – π^* *type transition state.*

The reactivity is therefore controlled essentially by polar effects:

– π-donor substituents (X = OH, OMe, NHCOMe) increase the reactivity and, conversely, electron-attracting substituents reduce it. Thus, benzonitrile does not react with N-chlorodimethylamine. This deactivation has nevertheless an advantage: protonation makes the amino group electron-attracting and prevents a second substitution.

— substitution occurs almost exclusively in the *para* position.

14-5-4 Substitution of protonated heteroaromatic bases by nucleophilic radicals

Homolytic substitution of protonated heteroatomic bases proceeds by two elementary reactions, already described: radical addition (14.25) and rearomatization by oxidation (14.26).

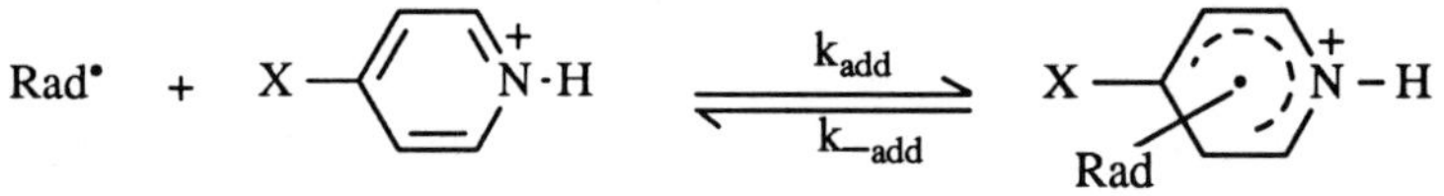

$$\text{Rad}^{\bullet} + X-\underset{}{\overset{+}{\bigcirc}}\text{N-H} \underset{k_{-add}}{\overset{k_{add}}{\rightleftarrows}} X-\underset{\text{Rad}}{\overset{+}{\bigcirc}}\text{N-H} \qquad (14.25)$$

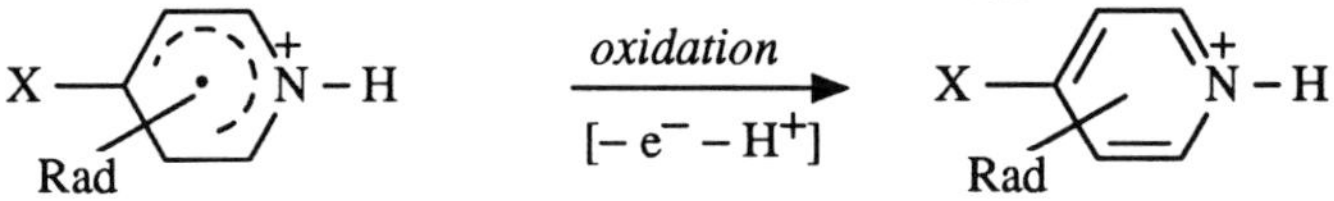

$$X-\underset{\text{Rad}}{\overset{+}{\bigcirc}}\text{N-H} \xrightarrow[{[-e^{-}-H^{+}]}]{\text{oxidation}} X-\underset{\text{Rad}}{\overset{+}{\bigcirc}}\text{N-H} \qquad (14.26)$$

Minisci's work on this type of reaction leads to the conclusions given above, which we summarize here:

— the *oxidation step* is all-important for the success of the reaction. It is very easy with the neutral pyridinyl radical (14.27);

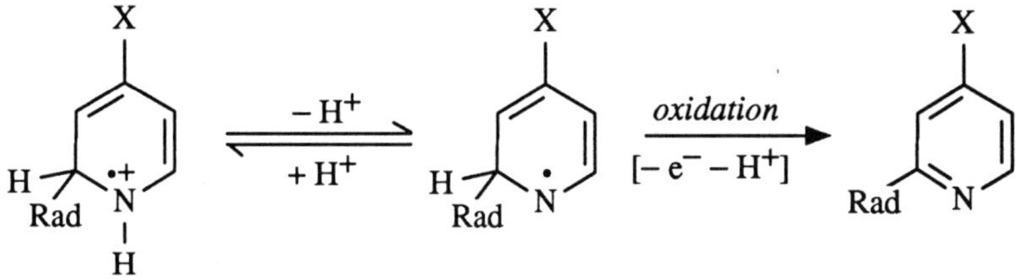

$$\underset{\substack{\text{Rad}\\ \text{H}}}{\bigcirc} \underset{+H^{+}}{\overset{-H^{+}}{\rightleftarrows}} \underset{\text{Rad}}{\bigcirc} \xrightarrow[{[-e^{-}-H^{+}]}]{\text{oxidation}} \underset{\text{Rad}}{\bigcirc} \qquad (14.27)$$

— the *rate* of the radical addition reaction (14.25) increases by a factor of 100 to 1 000 on going from a neutral pyridine to a protonated pyridine (Table 14.2). It

$(RCO_2)_2$	$\longrightarrow$	$2\ R^{\bullet} + 2\ CO_2$ (14.28)
XPy	$\underset{}{\overset{+H^{+}}{\rightleftarrows}}$	$XPyH^{+}$ (14.29)
$k_{add} \downarrow + R^{\bullet}$		$k_{addH^{+}} \downarrow + R^{\bullet}$ (14.30)
$XPyR^{\bullet}$	$\underset{}{\overset{+H^{+}}{\rightleftarrows}}$	$XPyHR^{\bullet +}$ (14.31)
$k_{elim} \downarrow + (RCO_2)_2$		$k_{elimH^{+}} \downarrow + (RCO_2)_2$ (14.32)
$XPyR +$	$\underset{}{\overset{+H^{+}}{\rightleftarrows}}$	$XPyHR^{+} +$ (14.33)
$RCO_2H + R^{\bullet} + CO_2$		$RCO_2H + R^{\bullet} + CO_2$

increases with the nucleophilicity of the radicals if the substituent X is an electron-

attractor: t-Bu• > n-Bu• (Table 14.1). It should be noted that this sequence goes against the reactivity-stability correlation. The rate decreases if X is an electron-donor (Tables 14.1, 14.2 and 14.4). Neither electrophilic radicals nor stabilized radicals such as PhCH$_2$• react;
 – the *reversibility* of reaction (14.25) depends on its enthalpy. Though irreversible for Ph• it is reversible for radicals like t-Bu•, RC•=O or –•CH-OR;
 – the *regioselectivity*: substitution α to nitrogen is very much preferred.

The role of the pH has been demonstrated in the alkylation of pyridines by a diacyl peroxide. The efficiency of this chain mechanism depends on the pH, since protonation favors addition ($k_{addH}+$ > k_{add}) but slows oxidation (14.32) which occurs preferentially with the neutral radical, XPyR• (k_{elim} > $k_{elimH}+$). Examples of the alkylation of protonated heteroatomic bases are very numerous, as are the methods of producing radicals in acid media.

14-5-5 Hydroxylation

The hydroxylation of aromatic compounds (14.34), which could be interesting for the synthesis of phenols, has proved to be relatively useless because it lacks substrate selectivity and regioselectivity. The electrophilic HO• radical adds rapidly to the aromatic nucleus ($k = 3\sim8 \times 10^9$ M^{-1}s^{-1}) but the subsequent reactions are much more complex than a simple addition-elimination process.

$$\text{Ph-X} \xrightarrow{\text{HO}^\bullet} \text{HO-C}_6\text{H}_3\text{-X} \qquad (14.34)$$

14-5-6 *Ipso* substitution

A radical can also react by an *ipso* substitution process, i.e. by replacing the substituent.

$$\text{PhX} \underset{-\text{Rad}^\bullet}{\overset{+\text{Rad}^\bullet}{\rightleftharpoons}} \text{[X, Rad cyclohexadienyl]} \xrightarrow{-\text{X}^\bullet} \text{Ph-Rad} \qquad (14.35)$$

Many examples of homolytic *ipso* substitution have been reported recently. The reaction works well provided the radical is very nucleophilic (for example, a tertiary carbon radical such as 1-adamantyl) and the substituent X is both an attractor (to make the aromatic ring electron-poor) and a good leaving radical (X = NO$_2$, RCO, PhSO$_2$, Cl, Br or I).

$$\text{NO}_2\text{-C}_6\text{H}_4\text{-NO}_2 \xrightarrow[-\text{NO}_2^\bullet]{+\text{1-Ad}^\bullet} \text{1-Ad-C}_6\text{H}_4\text{-NO}_2 \qquad (14.36)$$

14-5-7 Polynuclear aromatics

Polynuclear aromatic systems are generally more reactive.

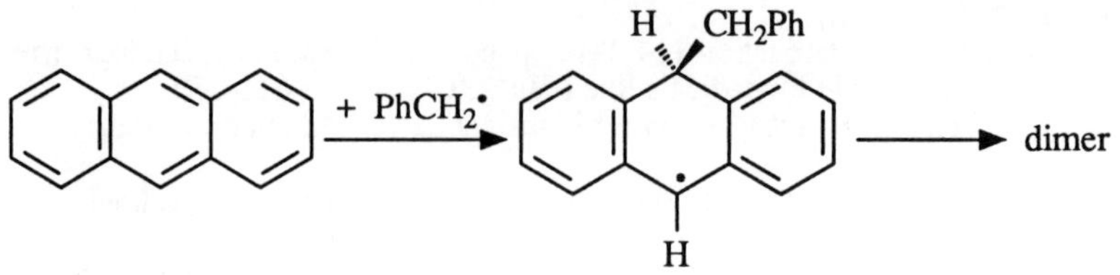

$$\text{anthracene} + PhCH_2^{\bullet} \longrightarrow \text{[9-CH}_2\text{Ph-9,10-dihydroanthracenyl radical]} \longrightarrow \text{dimer} \quad (14.37)$$

14-5-8 Intramolecular homolytic substitution

Homolytic substitution can also give rise to many intramolecular processes.

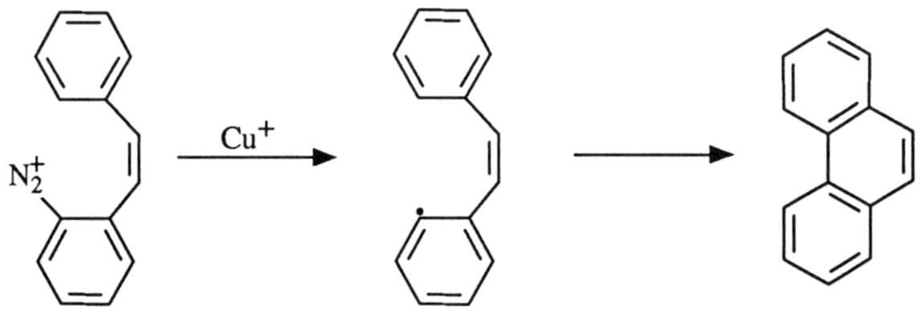

$$\text{[ArN}_2^+\text{-stilbene]} \xrightarrow{Cu^+} \text{[Ar}^{\bullet}\text{-stilbene]} \longrightarrow \text{phenanthrene} \quad (14.38)$$

14-6 CONCLUSIONS

From the very many results obtained for aromatic homolytic substitution reactions, these can be classified into two categories:

– the radical and the substrate have little or no polar antagonism (phenylation or alkylation of substituted benzenes). The *reactivity* is controlled by the enthalpy, i.e. the stability of the attacking and forming radicals. A donor or acceptor substituent increases the reactivity relative to a unsubstituted substrate. The polar effect, even when it exists, is relatively unimportant. The *regioselectivity* is governed by the localization of the orbitals of the π system and by the stability of the intermediate σ radical. The orientation is mainly *ortho* and *para*. These reactions are generally inefficient and poorly regioselective, and for this reason are synthetically uninteresting;

– the radical and the substrate have a very marked polar antagonism (amination of substituted benzenes, alkylation of protonated heteroatomic bases). The *reactivity* and the *regioselectivity* are controlled by polar effects and are very good, hence their considerable synthetic interest.

15 – REACTIONS OF CHARGED RADICALS

Charged radicals are molecular species with an unpaired electron and a positive (radical cations) or negative charge (radical anions). Charged radicals can occur either as stable, isolable species or as reaction intermediates.

15-1 RADICAL ANIONS

15-1-1 Structure

The delocalization of the unpaired electron is the chief cause of the stability of radical anions. Thus, the radical anions of naphthalene and polynuclear aromatics are more stable than that of benzene (Scheme 15.1).

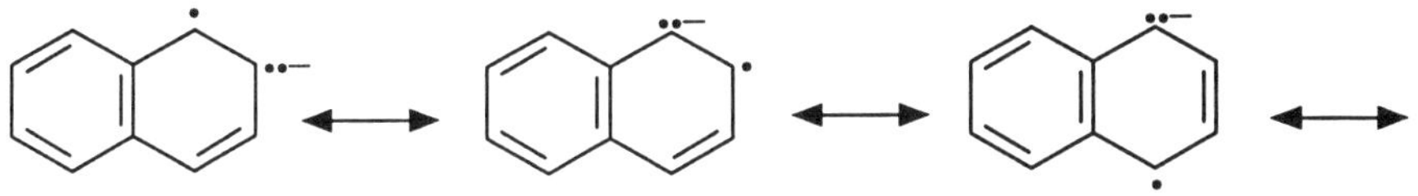

Scheme 15.1

With 7,7,8,8-tetracyanoquinodimethane (TCNQ) one obtains crystalline, isolable salts, purple or blue depending on the counter-ion (reaction 15.1).

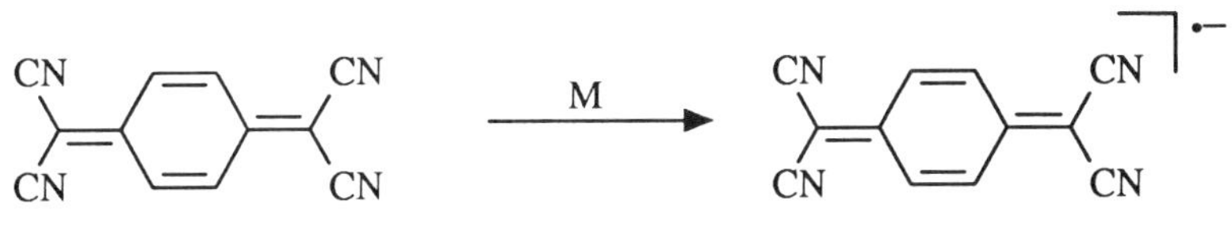

(15.1)

$M = Li^+, Na^+$ (purple), $Mn^{2+}, Fe^{2+}, Co^{2+}$ (blue)

Theoretical studies and EPR show, in agreement with the relative electronegativities, that the unpaired electron of a ketyl radical is centered on the carbon rather than the oxygen (Scheme 15.2). When the substituent R is a phenyl the electron is delocalized onto the *ortho* and *para* positions.

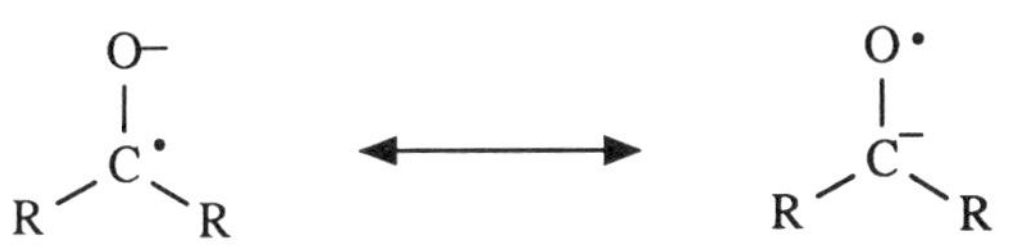

Scheme 15.2

15-1-2 Production

Radical anions are obtained either by capture of an electron by the LUMO (15.2) or by addition of a radical to an anion (15.3).

$$M \xrightarrow{+e^-} M^{\bullet -} \quad (15.2)$$

$$M^\bullet + A^- \longrightarrow MA^{\bullet -} \quad (15.3)$$

The electron can come from a metal atom, a cathode, an anion or a photochemically excited molecule. The lower the energy of the LUMO of M, the easier it is to form a radical anion. The key step in the chemistry of radical anions is often their formation. It is therefore convenient to use ideas from theoretical chemistry (low energy LUMO) or physical chemistry (UV absorption close to the visible) to get a qualitative notion of the species capable of giving this kind of reaction. The LUMO energy can be calculated or deduced from experimental measurements (spectroscopy, electrochemical reduction potential, etc.). Thus, the more the system is conjugated, the easier it is to obtain the radical anion; anthracene > naphthalene > benzene or conjugated diene > isolated double bond. In the same way, an α,β unsaturated carbonyl is better than an isolated carbonyl.

15-1-3 Reactions

Radical anions undergo the various reactions displayed in Scheme 15.3. Among the most common reactions we shall describe in more detail *nucleophilic radical substitution*, $S_{RN}1$, the dimerization of carbonyl compounds and the reactions of the superoxide anion, $O_2^{\bullet -}$.

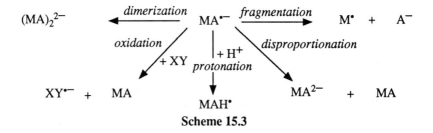

Scheme 15.3

Nucleophilic radical substitution, $S_{RN}1$

$S_{RN}1$ follows a chain mechanism which includes an initiation reaction (15.4), a

$$RX + e^- \longrightarrow RX^{\bullet -} \quad (15.4)$$

chain
$$RX^{\bullet -} \longrightarrow R^\bullet + X^- \quad (15.5)$$
$$R^\bullet + Nu^- \longrightarrow RNu^{\bullet -} \quad (15.6)$$
$$RNu^{\bullet -} + RX \longrightarrow RNu + RX^{\bullet -} \quad (15.7)$$

$$RX + Nu^- \longrightarrow RNu + X^- \quad (15.8)$$

propagation step (15.5) where R• is produced, a step where R• adds to the nucleophile (15.6) and, finally, an electron transfer step (15.7). The overall result of these reactions is the nucleophilic substitution of X by Nu (15.8). This mechanism was established for substitution at a saturated carbon by Kornblum in 1966 and for aromatic compounds by Bunnett in 1970. The latter proposed the symbol $S_{RN}1$ by analogy with S_N1, except that unimolecular bond breaking in (15.5) produces a radical and an anion, instead of a carbocation, and that it is not the rate determining step of the reaction.

Since $S_{RN}1$ is a chain reaction involving radicals and electron transfer, this process is inhibited by radical and electron traps. Competing reactions such as reduction of the radical R• followed by protonation (15.9) or hydrogen transfer from the solvent (15.10) can occur. $S_{RN}1$ reactions must therefore be performed in solvents which are non-acidic and poor hydrogen-donors.

$$R^\bullet \xrightarrow{+e^-} R^- \xrightarrow{H^+} R-H \qquad (15.9)$$

$$R^\bullet + H-\Sigma \longrightarrow R-H + \Sigma^\bullet \qquad (15.10)$$

Initiation (15.4) starts with an electron-donor (alkali metal, solvated electron, cathode, the Nu⁻ ion itself, etc.) often photochemically activated. In certain cases the reaction is thermally initiated.

The *fragmentation* step (15.5) procceds readily when X is a halogen and for nucleofugal groups of very different types (PhO⁻, ArS⁻, PhSe⁻, etc.) including groups which are poor nucleofuges in S_N1 and S_N2 reactions. Aromatic halogens react in the following order:

$$ArI > ArBr > ArCl > ArF \qquad (15.11)$$

The NO_2 group accelerates electron transfer reactions (15.4) and (15.7) at a saturated carbon, but blocks the overall $S_{RN}1$ reaction on an aromatic substrate because the intermediate radical anion, ArX•⁻, is too stable.

Radical addition to the anion (15.6) is usually the rate determining step. Unlike S_N2, it is relatively insensitive to steric hindrance. This characteristic, which makes it possible to obtain very branched molecules, distinguishes $S_{RN}1$ from S_N2. Thus, the ambident anion from 2-nitropropane leads to C-alkylation by $S_{RN}1$ and O-alkylation by S_N2.

$$(15.12)$$

In the case of *p*-nitrocumyl chloride only the C–alkylation product is obtained, the S_N2 reaction being very slow.

$$\text{NO}_2\text{-C}_6\text{H}_4\text{-C(CH}_3)_2\text{Cl} \ \xrightarrow{+ (CH_3)_2\bar{C}\text{-NO}_2 \atop -Cl^-} \ \text{NO}_2\text{-C}_6\text{H}_4\text{-C(CH}_3)_2\text{-C(CH}_3)_2\text{NO}_2 \quad (15.13)$$

90%

With aromatics, since the Ar• radical formed is σ type, the reaction is very *regioselective*, the nucleophile taking the place of the nucleofuge (reaction (15.14)). The presence of *ortho* substituents does not hinder the reaction. $S_{RN}1$ works just as well with charged anions (enolate, nitroalkanes) as with heteroatomic anions (nitrite, azide, thiolate, aryloxide) but certain nucleophilic compounds which react with aliphatics prove to be unreactive with aromatics.

$$\text{Py-CH}_2^- + \text{Br-Mes} \ \xrightarrow{-Br^-} \ \text{Py-CH}_2\text{-Mes} \quad (15.14)$$

87%

The *electron transfer* reaction (15.7) takes place between an easily oxidized radical anion and an easily reduced substrate. For this reason, in most cases the substrate, RX or ArX, must carry a strong electron-attracting group, which with aromatics is generally NO_2 (Kornblum).

The $S_{RN}1$ reaction is therefore very important. It can be used to perform nucleophilic substitution on aromatic, vinylic and aliphatic compounds in situations where steric and electronic factors or ring strain prevent S_N2 and S_N1 reactions (perfluoroalkyl iodides, bridgehead, neopentyl or cyclopropyl halides, etc.).

Dimerization

A carbonyl group readily accepts an electron from a metallic reducing agent and then, under aprotic and anhydrous conditions, the radical anion dimerizes to give an α-diol after acidification.

$$R_2C=O \ \xrightarrow{+e^-} \ R_2\dot{C}\text{-}O^- \quad (15.15)$$

$$2 \ R_2\dot{C}\text{-}O^- \ \longrightarrow \ R_2C(O^-)\text{-}C(O^-)R_2 \quad (15.16)$$

This type of reaction when applied to diesters leads to medium- and large-ring acyloins (reaction (15.17)).

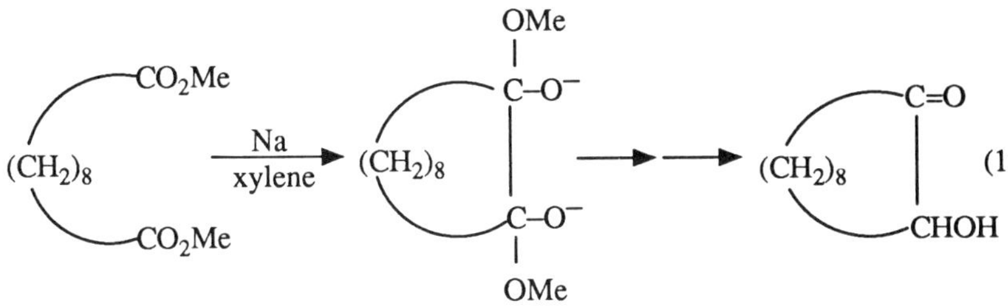

$$\underset{(CH_2)_8}{\overset{CO_2Me}{\diagdown}} \xrightarrow{\text{Na, xylene}} \underset{(CH_2)_8}{\overset{C-O^-}{\diagdown}}\overset{|}{\underset{|}{\text{OMe}}}\overset{}{\underset{\text{OMe}}{C-O^-}} \longrightarrow \longrightarrow \underset{(CH_2)_8}{\overset{C=O}{\diagdown}}_{CHOH} \quad (15.17)$$

Radical anion, $O_2^{\bullet-}$

Dioxygen can be reduced (–0.33 V, normal hydrogen electrode in water at pH 7) to a radical anion $O_2^{\bullet-}$, called the *superoxide anion*. This has the twofold character of a radical and an anion, and has the following chemical properties:

- powerful nucleophile under aprotic conditions;
- strong Brønsted base;
- one-electron reducing agent;
- oxidizing agent, due probably to O_2 and H_2O_2 produced by proton-assisted disproportionation. In water, $O_2^{\bullet-}$ is a better oxidant than O_2.

$$2\ O_2^{\bullet-} + 2\ H^+ \longrightarrow O_2 + H_2O_2 \quad (15.18)$$

Despite the presence of an unpaired electron (delocalized over two centers) few radical reactions are observed.

15-2 RADICAL CATIONS

15-2-1 Structure

In radical cations, obtained by loss of one of the electrons of a lone pair, the charge and the unpaired electron are on the same atom (Scheme 15.4).

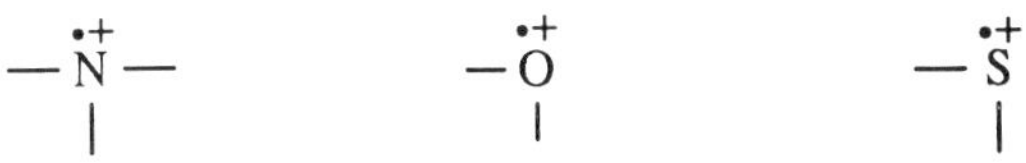

$$-\overset{\bullet+}{\underset{|}{N}}- \qquad -\overset{\bullet+}{\underset{|}{O}} \qquad -\overset{\bullet+}{\underset{|}{S}}$$

Scheme 15.4

In the case of *distonic* radicals, often observed in mass spectroscopy, the charge and the unpaired electron are on different atoms (Scheme 15.5). These are generally more stable than the previous species.

$$^{\bullet}CH_2\text{-}(CH_2)_n CH_2\overset{+}{O}H_2 \qquad ^{\bullet}CH_2\text{-}(CH_2)_n CH_2\overset{+}{N}H_3$$

Scheme 15.5

A similar situation arises for radical cations obtained from an unsaturated compound (15.19).

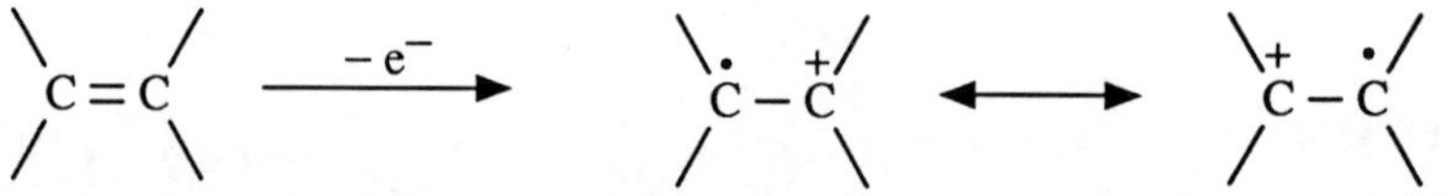

$$\diagup_{/}^{\backslash}C=C_{\backslash}^{\diagup} \xrightarrow{-e^-} \diagup_{/}^{\backslash}\overset{\bullet}{C}-\overset{+}{C}_{\backslash}^{\diagup} \longleftrightarrow \diagup_{/}^{\backslash +}C-\overset{\bullet}{C}_{\backslash}^{\diagup} \qquad (15.19)$$

The delocalization of the unpaired electron in conjugated systems leads to very stable radical cations such as the salts of Wurster **1** (violenes), of Weitz **2** (viologens) or the anthracenic compounds **3**. These isolable compounds, deep blue or red depending on the substituents, have very characterisitic UV and EPR spectra. Their stability depends on R and the pH.

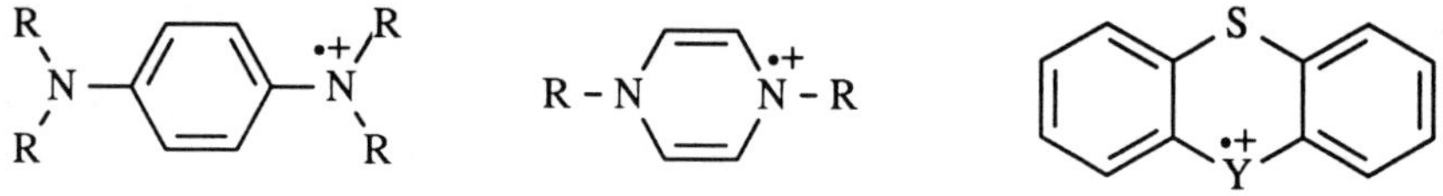

Scheme 15.6

15-2-2 Production

A radical cation is obtained:

— either by abstracting an electron from a neutral molecule (15.20). This oxidation can be achieved by chemical oxidizing agents $((NH_4)_2S_2O_8$ or conc. $H_2SO_4)$, by metal ions (Mn^{3+}, Co^{3+}, Ce^{4+}, Ag^{2+}, etc.), by Lewis acids ($AlCl_3$, $SbCl_3$, etc.), by photoionization, pulse radiolysis, electron impact (mass spectroscopy) or by anodic oxidation;

$$M \xrightarrow{-e^-} M^{\bullet +} \qquad (15.20)$$

— or by the combination of a neutral radical and a positively charged species, such as a protonated heteroatomic base (15.21).

$$R^{\bullet} + M^+ \longrightarrow RM^{\bullet +} \qquad (15.21)$$

The aptitude of an organic molecule to release an electron is controlled by the energy of the highest occupied molecular orbital (HOMO) which is equal to the negative of the ionization potential (IP) and is related to the oxidation potential in solution ($E_{1/2}$). The smaller these values, the easier it is to oxidize the molecule. Inspection of Table 15.1 shows that the easiest molecules to oxidize are those which contain π electrons or heteroatoms with lone pairs of electrons.

15-2-3 Reactions

Radical cations can be very reactive. They behave like *radicals* in dimerization, disproportionation, fragmentation, addition and hydrogen transfer and like *cations* in their reactions with various nucleophiles, whether charged (halide ions, carboxylates, etc.) or not (water, alcohols, amines, etc.). They can also behave as oxidizing agents or strong acids. The following scheme summarizes the possible reactions.

compound	IP(eV)	$E_{1/2}(V)^a$
ethylene	10.51	2.90
benzene	9.24	2.04
cyclohexene	8.95	1.98
thiophene	8.86	1.70
phenol	8.50	1.04
naphthalene	8.12	1.34
anthracene	7.23	0.84
N,N-dimethylaniline	7.14	0.45

Table 15.1 – *Ionization and oxidation potentials. a) Measured in MeCN.*

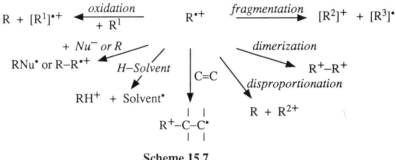

Scheme 15.7

Radical cations are therefore involved in many processes of organic chemistry. A very classical reaction of these radical cations is stabilization by fragmentation to a radical and a cation. The cation is often the most stable one, i.e. the proton ($[R^2]^+$ = H^+) as we have already seen or will see in the following examples. The reactions of protonated aminyl radicals are mentioned in Chapter 8 (addition and hydrogen transfer) and in Chapter 14 (aromatic substitution). Here we shall simply give some examples of the oxidation of aromatic and ethylenic compounds by metal complexes, where the formation of the products is explained by the reaction of radical cation intermediates.

Aromatic compounds

Oxidative substitution of these compounds by Co^{3+} salts, for example, proceeds via an intermediate radical cation reacting with a nucleophile such as AcO^- from the solvent acetic acid. The electron transfer reaction (15.22) is the rate determining step of the process.

$$\langle \overset{\cdot\;-}{\bigcirc} \rangle \overset{X}{\underset{H}{\diagdown}} + Co^{III}X_3 \xrightarrow{[-H^+ + Nu]}_{X = CH_3CO_2} \langle \bigcirc \rangle - X + Co^{II}X_2 \quad (15.24)$$

For alkylaromatics the overall reaction can lead to products of substitution in the chain, because of the deprotonation reaction (15.26).

$$ArCH_3 + Mn^{3+} \rightleftharpoons [ArCH_3]^{\bullet+} + Mn^{2+} \quad (15.25)$$

$$[ArCH_3]^{\bullet+} \xrightarrow{+ Nu} Ar\overset{\bullet}{C}H_2 + NuH^+ \quad (15.26)$$

$$Ar\overset{\bullet}{C}H_2 + Mn^{3+} \longrightarrow Ar\overset{+}{C}H_2 + Mn^{2+} \quad (15.27)$$

$$Ar\overset{+}{C}H_2 + CH_3CO_2H \xrightarrow[\text{fast}]{+ Nu} ArCH_2OCOCH_3 + NuH^+ \quad (15.28)$$

Reaction (15.29) is determined by stereoelectronic control and not by the thermodynamic stability of the radical. In the preferred conformation of the molecule the C–H bond of the isopropyl group is in the nodal plane of the π system of the aromatic nucleus. In the transition state the stabilizing *C–H bond-π system* interaction is therefore minimized and makes C–H bond breaking unfavorable. This is not the case for the methyl group, where one of the C–H bonds is out of the nodal plane of the π system.

$$\text{(15.29)}$$

The fate of the radical cation depends on the structure of the aromatic compound, on the oxidizing metal salt and on the reaction medium. Thus, Mn(OAc)$_3$ oxidizes *p*-methoxytoluene but not toluene, whereas Co(OAc)$_3$ oxidizes both. The ease of oxidation is not solely related to the $M^{n+}/M^{(n-1)+}$ oxido-reduction potentials determined in aqueous solution. Other factors, such as the nature of the ligands, the solvent, etc., are involved.

Phenols can be oxidized by a great number of one-electron oxidants, including enzymes (15.30). The radical cation is deprotonated very rapidly. This reaction, predominant in acid media or in apolar solvents, is very common in the biosynthesis of natural phenolic products by coupling of ArO$^\bullet$ radicals. It explains the antioxidant role of phenols.

$$ArOH \xrightarrow{-e^-} Ar\overset{\bullet+}{O}H \xrightarrow{+ Nu} ArO^\bullet + NuH^+ \quad (15.30)$$

Alkenes

Olefins have IPs close to those of aromatics. As previously, the evolution of the radical cation obtained by oxidation by metal complexes depends on the nature of the complex and the reaction medium. By reaction with oxygen-centered nucleophiles a C–O bond is formed (15.31).

$$\text{cyclohexene} \xrightarrow{M^{n+}} \text{[cyclohexene]}^{+\bullet} \xrightarrow[X = H, R, RCO]{XOH} \text{[C}_6\text{H}_{10}\text{-OX]}^{\bullet} + H^+ + M^{(n-1)+} \quad (15.31)$$

The radical obtained can abstract a hydrogen from the solvent, H–Σ, or be oxidized to an unsaturated compound.

$$\text{[C}_6\text{H}_{10}\text{-OX]}^{\bullet} \xrightarrow{H\Sigma} \text{C}_6\text{H}_{11}\text{-OX} + \Sigma^{\bullet} \quad (15.32)$$

$$\text{[C}_6\text{H}_{10}\text{-OX]}^{\bullet} \xrightarrow{M^{n+}} M^{(n-1)+} + \text{[C}_6\text{H}_{10}\text{-OX]}^{+} \longrightarrow \text{C}_6\text{H}_9\text{-OX} + H^+ \quad (15.33)$$

The interest in radical cations is relatively recent. It will certainly grow, because of the importance of these intermediates in many oxidation and reduction reactions observed in biological phenomena.

16 – FREE RADICALS IN BIOCHEMISTRY

Radical reactions are important in living systems since they are involved in normal biological processes and energy production. They are also induced by external factors such as radiation, medicines, foodstuffs, etc.

16-1 BIOLOGICAL OXIDATION: A SOURCE OF ENERGY

The energy required by the organism is provided by the oxidation of foods in the mitochondria and is stored in the form of ATP (adenosine triphosphate). For example, glucose is transformed into CO_2 and H_2O (16.1) with the release of 673 kcal/mol.

$$C_6H_{12}O_6 \; + \; 6 \; O_2 \; \xrightarrow{\Delta H = -673 \text{ kcal/mol}} \; 6 \; CO_2 \; + \; 6 \; H_2O \quad (16.1)$$

This reaction is not simple: glucose does not react with molecular oxygen at ambient temperature. The biological reaction proceeds by a series of oxido-reduction steps, some of which are performed by enzyme systems, in which the C–H bonds are oxidized one-by-one and the oxygen is reduced to H_2O. This is referred to as a *respiratory chain*. These oxido-reduction reactions occur by one- or two-electron transfers. It is the first category which interests us, insofar as they imply neutral or charged free radical intermediates.

Four important classes of oxido-reduction enzymes are involved in the transfer of an electron from a substrate to molecular oxygen. They are, in order of their redox potentials:

- *pyridine nucleotide dehydrogenases*, which use NAD or NADP as coenzymes;
- *flavin dehydrogenases*;
- *semiquinones* or coenzymes Q;
- *cytochromes* which contain an iron porphyrin system.

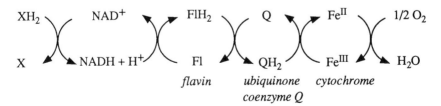

Scheme 16.1

16-1-1 NAD$^+$ → NADH Oxidation

NAD$^+$ is a two-electron oxidant (hydride captor) which oxidizes, for example, alcohol functions (16.2).

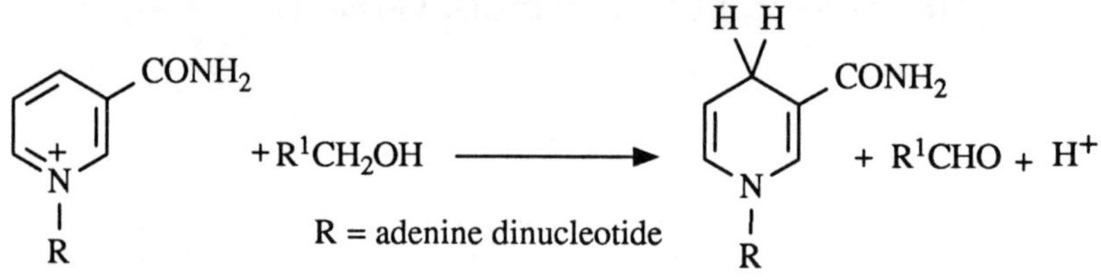

$$\text{NAD}^+ \text{ (oxidized form)} + R^1CH_2OH \longrightarrow \text{NADH (reduced form)} + R^1CHO + H^+ \quad (16.2)$$

R = adenine dinucleotide

The one-electron reduction of NAD$^+$ gives a NAD$^\bullet$ radical which is capable of reducing molecular oxygen.

$$\text{NAD}^+ \xrightarrow{-e^-} \text{NAD}^\bullet \xrightarrow{+O_2} \text{NAD}^+ + O_2^{\bullet -} \quad (16.3)$$

16-1-2 Oxidation by flavoproteins

NADH formed in a step such as (reaction (16.2)) is reoxidized to NAD$^+$ by flavoproteins. This oxidation can consist of a two-electron transfer or two successive one-electron transfers.

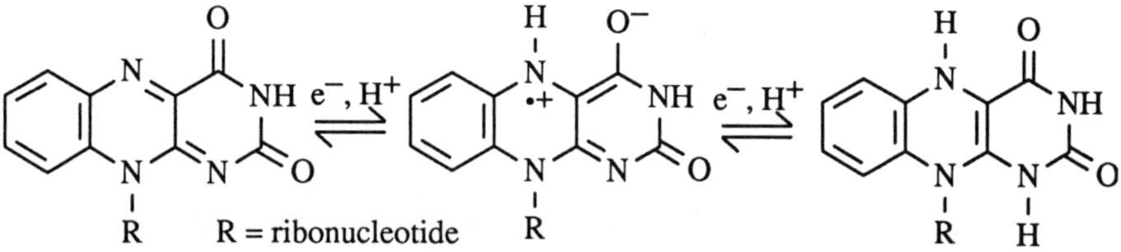

Fl *(flavoquinone)* ⇌ FlH$^\bullet$ *(flavosemiquinone)* ⇌ FlH$_2$ *(flavohydroquinone)*

R = ribonucleotide

Scheme 16.2

16-1-3 Ubiquinone

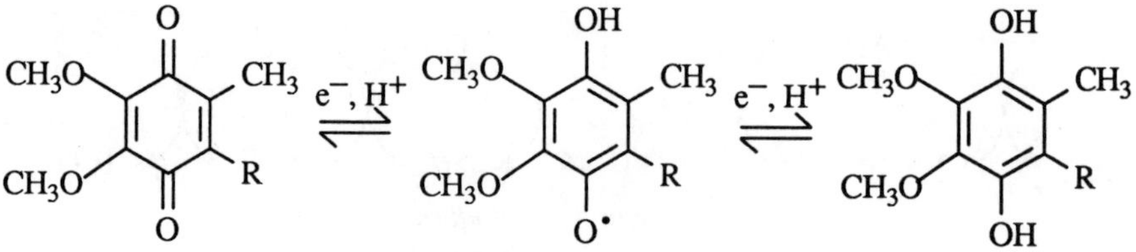

Scheme 16.3

Quinones, which are good one-electron oxidants, are converted into hydroquinone in two steps (Scheme 16.3). The coenzymes Q (and vitamin K) in which the isoprene chain, R, varies from C_6 to C_{10} depending on the organism, are molecules with the quinone moiety, whence their oxidizing ability.

16-1-4 Cytochromes

Cytochromes are proteins containing a porphyrin moiety with an iron atom which is responsible for one-electron transfer. Of the five cytochromes in the mitochondrial respiratory chain, only one, cytochrome a, is able to reduce oxygen to a radical anion which then leads to the formation of water.

$$Fe^{II} + O_2 \rightleftharpoons Fe^{III} + O_2^{\bullet -} \qquad (16.4)$$

16-2 DIOXYGEN EFFECT

Oxygen is a source of energy and of life. Paradoxically, oxygen-centered radicals derived from it have potential for causing widespread damage in organisms.

16-2-1 Production of oxygen-centered radicals

Molecular oxygen, a biradical triplet state, is very unreactive but can be reduced to *superoxide anion* (Chap. 15).

$$O_2 + e^- \xrightarrow{-330 \text{ mV}} O_2^{\bullet -} \qquad (16.5)$$

Many electron-donors, present in biological systems, can perform this reaction. This is the case of metalloenzymes, e.g. xanthine oxidase (a flavoprotein containing molybdenum and iron), or non-enzymatic compounds (hydroquinone, thiols, leukotrienes, etc.).

The superoxide anion evolves in various ways, notably by disproportionation (16.6). This reaction which is very fast in the presence of SOD (superoxide dismutase, an enzyme containing Fe or Mn or the Zn-Cu couple) is important in the regulation of the reactions of oxygen.

$$O_2^{\bullet -} + O_2^{\bullet -} + 2 H^+ \longrightarrow O_2 + H_2O_2 \qquad (16.6)$$

The resulting hydrogen peroxide is consumed in various reactions.

$$H_2O_2 + H_2O_2 \xrightarrow{catalase} 2 H_2O + O_2 \qquad (16.7)$$

$$H_2O_2 + RH_2 \xrightarrow{peroxidase} 2 H_2O + R \qquad (16.8)$$

$$H_2O_2 + Fe^{2+} \longrightarrow HO^- + HO^{\bullet} + Fe^{3+} \qquad (16.9)$$

The conjugate acid of the superoxide radical anion is the peroxyl radical (16.10). At pH 6, frequently encountered in biological systems, equilibrium (16.10) corresponds to 4% of HOO•, which is by no means negligible.

$$O_2^{•-} + H^+ \underset{}{\overset{pK_a = 4.7}{\rightleftharpoons}} HOO^{•} \qquad (16.10)$$

16-2-2 The toxicity of oxygen-centered radicals. Protection of the organism

The reduction of oxygen gives three radical species: $O_2^{•-}$, HOO• and HO•. $O_2^{•-}$ is unreactive; HOO• is moderately reactive (since BDE(HOO–H) = 90 kcal/mol) but it can induce autoxidation reactions by hydrogen abstraction. On the other hand, HO• is very reactive and can attack any biological molecule, in particular:

– the molecules which make up DNA, causing structural modifications and breaking the strands, either by abstracting hydrogen α to the intracyclic oxygen of *deoxyribose* (which provokes both dephosphorylation and ring opening (Scheme 16.4))

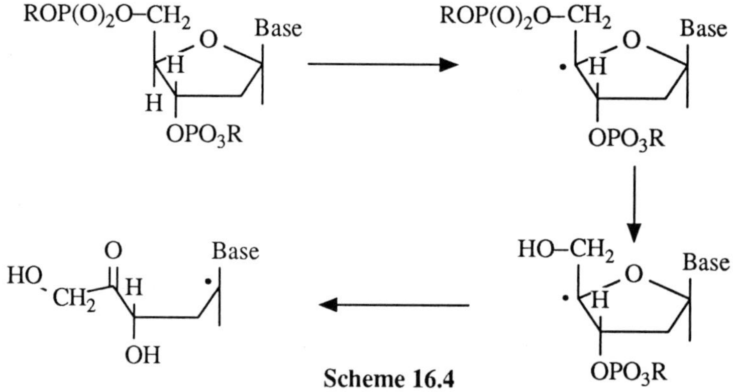

Scheme 16.4

or by addition to the unsaturated systems of the *pyrimidine* (Scheme 16.5) or *purine* bases (16.11);

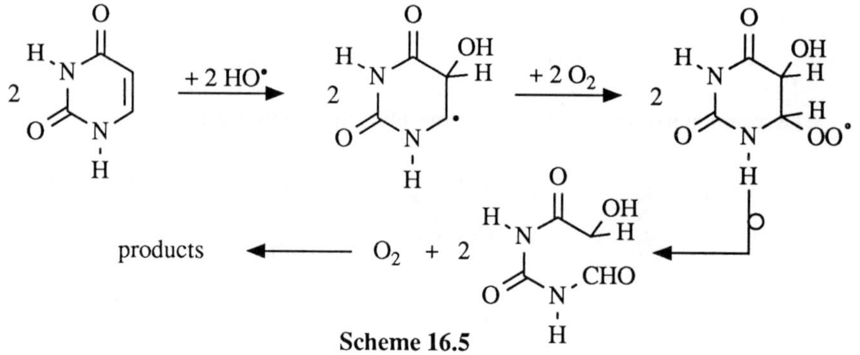

Scheme 16.5

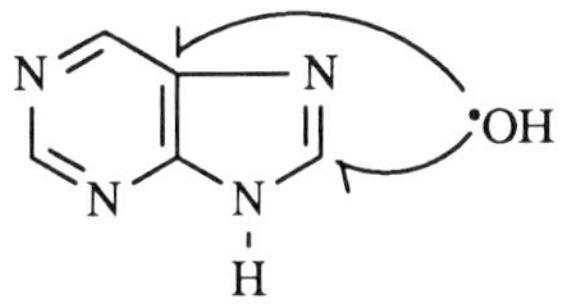

$$\text{[pyrrolopyrimidine]-OH} \xrightarrow{+O_2} \text{degradation products} \quad (16.11)$$

– the *aminoacids* of proteins, by abstracting hydrogen α to nitrogen, addition to the unsaturated systems of tryptophan or histidine, or oxidation of thioether functions (e.g. methionine) to sulfoxide and sulfone;
– *unsaturated fatty acids* of phospholipids by abstraction of allylic hydrogen (para. 16.3).

In general the HO• radical induces autoxidation reactions.

$$HO^\bullet + H-\Sigma \longrightarrow H_2O + \Sigma^\bullet \quad (16.12)$$

$$\Sigma^\bullet + O_2 \longrightarrow \Sigma OO^\bullet \quad (16.13)$$

Organisms have several means of defense against these potentially harmful radical reactions. These include:

– enzyme systems which regulate concentrations: SOD for $O_2^{\bullet-}$, catalase and peroxidase for H_2O_2. It is believed that glutathione peroxidase, GS–H, removes hydroperoxides by the following reaction:

$$ROOH + 2\ GS-H \xrightarrow{SOD} ROH + GS-SG + H_2O \quad (16.14)$$

– non-enzymic biochemical antioxidants such as α-tocopherol (vitamin E) and vitamin C. Apart from their ability to trap radicals, the synergy of these two vitamins is no doubt due to the fact that the first is liposoluble and the second hydrosoluble.

$$R^\bullet + vit.E \longrightarrow RH + vit.E^\bullet \quad (16.15)$$

$$vit.E^\bullet + vit.C \longrightarrow vit.E + vit.C^\bullet \quad (16.16)$$

16-3 AUTOXIDATION OF LIPIDS

The autoxidation of unsaturated lipids deserves special attention. It is known that this phenomenon is behind the taste and the smell of rancid foodstuffs. In biological systems phospholipids are the basic materials of cell membranes. They contain, amongst other things, certain polyunsaturated fatty acids. The reaction of these with oxygen is particularly important and their autoxidation has two-fold and contradictory consequences:

– harmful, because of the radicals and the products to which they give rise;
– beneficial, because of the biosynthesis of certain compounds, such as prostaglandins and leukotrienes, whose biological activity is important.

Unsaturated fatty acids contain one or more non-conjugated *cis* double bonds. Oxygen reacts with the allylic CH_2 groups by the following radical chain mechanism:

initiation

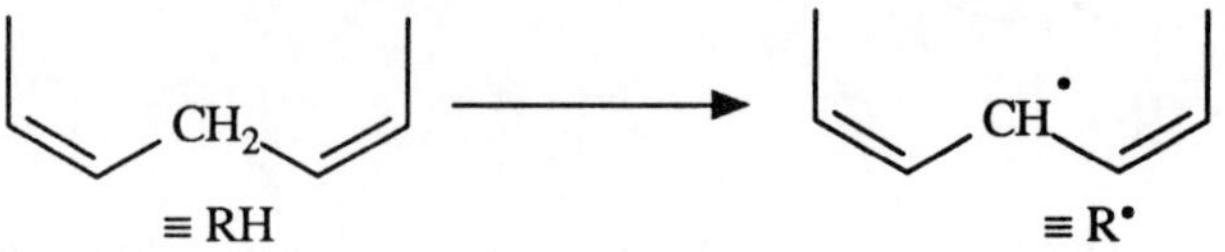

$$\equiv RH \longrightarrow \equiv R^\bullet \qquad (16.17)$$

propagation

$$R^\bullet + O_2 \longrightarrow ROO^\bullet \qquad (16.18)$$

$$ROO^\bullet + RH \longrightarrow ROOH + R^\bullet \qquad (16.19)$$

termination

$$2\ ROO^\bullet \longrightarrow products \qquad (16.20)$$

In fact, for several reasons, the mechanism is more complex than it appears:

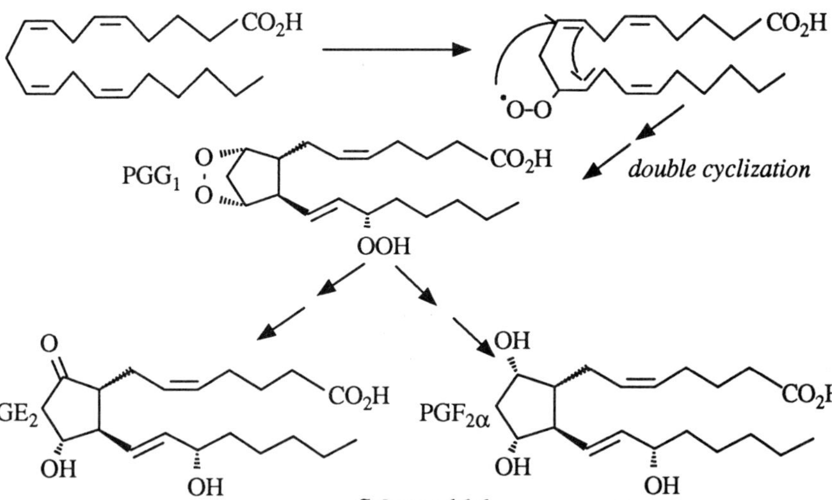

Scheme 16.6

- *allylic isomerization* leads to several peroxyl radicals and, consequently, to several hydroperoxides. This explains, in particular, the biosynthesis of prostaglandin endoperoxides (PGG_1) from arachidonic acid (C_{20} acid with 4 double bonds);
- the *evolution of the oxyl radicals*, $RO^\bullet$, arising from termination reactions of the peroxyl radicals or from hydroperoxide decomposition, often catalyzed by Fe^{2+} (Fenton reaction), leads to peroxides (16.21) - (16.22)

$$ROOH + Fe^{2+} \longrightarrow RO^\bullet + HO^- + Fe^{3+} \qquad (16.21)$$

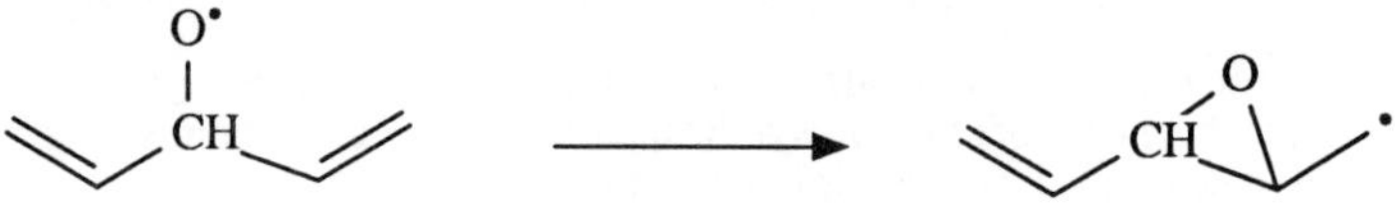

or to products of the β-fragmentation of aliphatic chains (16.23).

$$\underset{RCH_2}{}\overset{\overset{O^\bullet}{|}}{\underset{}{CH}}R^1 \longrightarrow RCH_2^\bullet \; + \; R^1-\overset{\overset{O}{\nearrow}}{\underset{H}{C}} \qquad (16.23)$$

Finally, among the products of the peroxidation of unsaturated lipids, one finds ethane, propane, pentane and malonaldehyde, this latter coming from the fragmentation of endoperoxides (16.24).

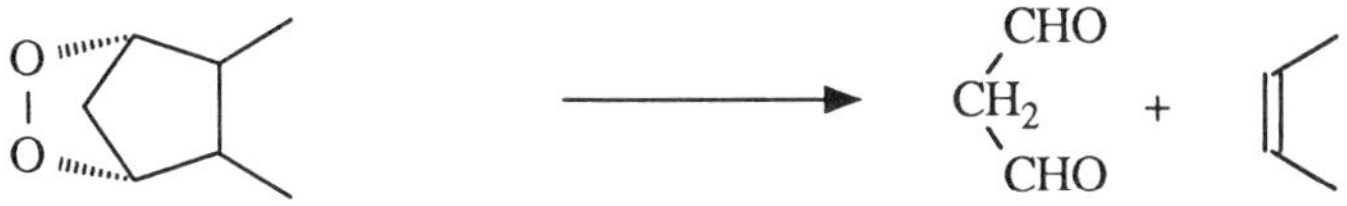

$$\text{(endoperoxide)} \longrightarrow \underset{CHO}{\overset{CHO}{CH_2}} \; + \; \text{(diene)} \qquad (16.24)$$

Of all the reactions of the autoxidation chain, the most difficult is the initiation step (16.17). The direct action of oxygen (16.25) is a very endothermic bimolecular reaction (BDE(H–OO$^\bullet$) = 47 kcal/mol), even when a doubly allylic hydrogen is involved, as in the case of polyunsaturated acids (ΔH = 24 kcal/mol). In biological systems this reaction is performed by lipoxygenase (an enzyme containing iron). On the other hand, reaction (16.19), which has been extensively studied both *in vitro* and *in vivo*, is very fast (10^8 to 10^9 M^{-1}s^{-1}).

$$RH \; + \; O_2 \longrightarrow R^\bullet \; + \; HOO^\bullet \qquad (16.25)$$

To summarize, the autoxidation of unsaturated lipids has two aspects:

– *harmful* as regards the formation of toxic products (hydroperoxides, epoxides), the breaking of fatty acid chains, and attack on purine bases by HO$^\bullet$ leading to modification of genetic material. Lipid peroxidation is believed to cause ageing of tissues and various illnesses or ailments (cancer, atherosclorosis, inflammation, etc.). Oxidation is inhibited by substances such as vitamins C and E (tocopherol) which trap free radicals;
– *beneficial* as regards the biosynthesis of prostaglandins and leukotrienes whose physiological activity is complex (regulation of cell metabolism, hormone transmission, blood circulation).

16-4 ISOMERIZATION REACTIONS

Certain 1,2 isomerizations (16.26), which do not occur in *in vitro* radical reactions, are observed in biological systems because of coenzyme B_{12} (adenosyl cobalamine Co^{III}). Radical intermediates have been detected by EPR, in the conversion of methylmalonyl coenzyme A to succinyl coenzyme A (16.27) and the transformation of 1,2-propanediol into propanal (16.28), for example.

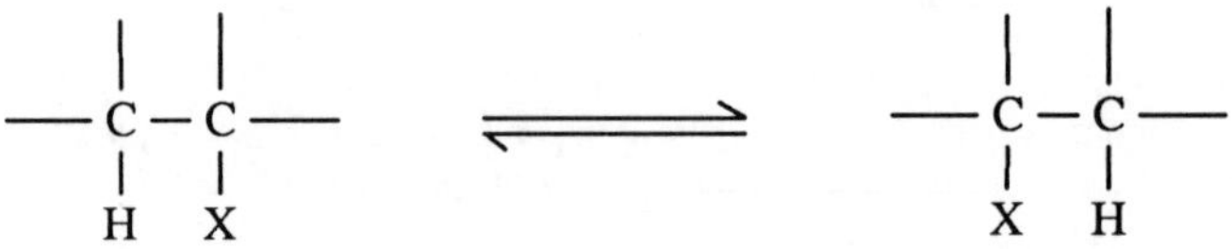

$$-\overset{|}{\underset{H}{C}}-\overset{|}{\underset{X}{C}}- \quad \rightleftharpoons \quad -\overset{|}{\underset{X}{C}}-\overset{|}{\underset{H}{C}}- \quad (16.26)$$

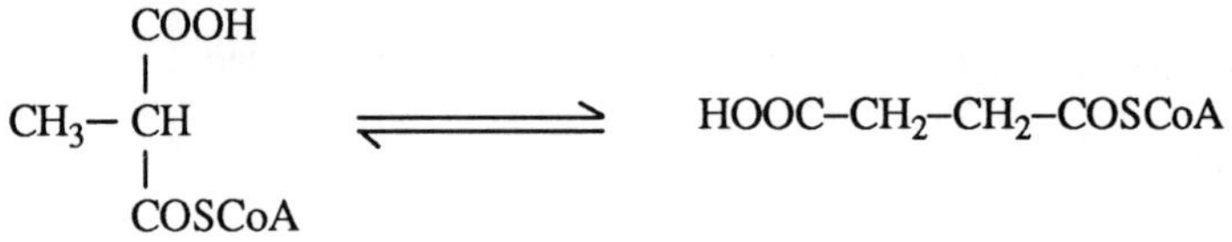

$$\underset{COSCoA}{\overset{COOH}{CH_3-CH}} \quad \rightleftharpoons \quad HOOC-CH_2-CH_2-COSCoA \quad (16.27)$$

$$\underset{OH}{CH_3-CH-CH_2OH} \longrightarrow CH_3-CH_2-\underset{OH}{\overset{OH}{CH}} \xrightarrow{-H_2O} CH_3-CH_2-CHO \quad (16.28)$$

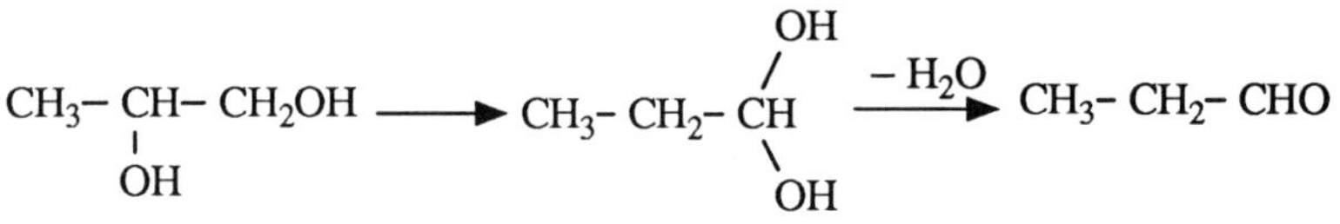

16-5 HYDROXYLATION BY CYTOCHROME P450

In a mono-oxygenating enzyme system cytochrome P450 ($HpFe^V=O$, iron porphyrin) transforms a C–H bond into C–OH by transfer of an oxygen atom from molecular oxygen (Scheme 16.7).

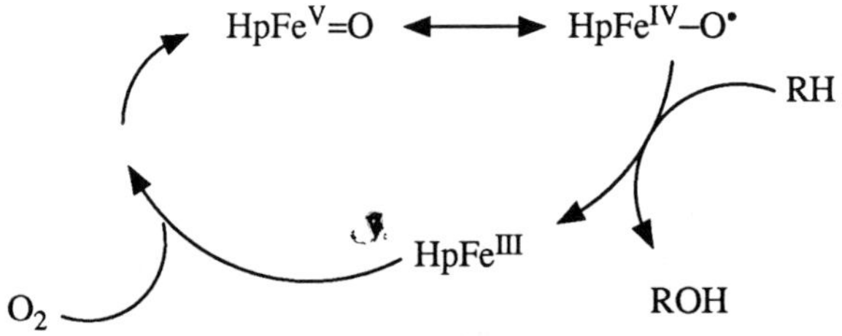

$$HpFe^V=O \longleftrightarrow HpFe^{IV}-O^\bullet$$

Scheme 16.7

16-6 REACTIONS INDUCED BY EXOGENOUS AGENTS

A good number of extracellular agents, both chemicals and radiations, may induce radical reactions in organisms.

16-6-1 Radiation

Both X-rays and γ-radiation, which are highly energetic, can cause severe lesions. Attention has been focused on their effects particularly on account of the dangers for personnel working with X-ray apparatus or in nuclear power stations. The principal reaction is the radiolysis of water (16.29).

$$H_2O \xrightarrow{X \text{ or } \gamma} HO^\bullet, e^-_{aq}, H^\bullet, H^+, H_2O_2, H_2 \quad (16.29)$$

Of all the radicals produced, the HO$^\bullet$ radical, which is particularly reactive, is the major cause of damage to biological molecules.

UV radiation is less aggressive but, nevertheless, it can cause skin cancers either directly or via sensitizers. These can act by contact (tars, psoralens from plants or perfumes) or be systemic in the form of medicines (antibacterial sulfamides, diuretics, thiazides, sulfonylurea antidiabetics, phenothiazines, tetracycline antibiotics). It is well known that after consuming tetracyclines exposure to the sun should be avoided.

16-6-2 Halogen derivatives

The toxicity of carbon tetrachloride in the liver is due to $^\bullet CCl_3$ radicals, which are generated by cytochrome P450-catalyzed enzyme reduction. These radicals can then induce peroxidation of lipids, either directly (16.30)

$$^\bullet CCl_3 + RH \longrightarrow HCCl_3 + R^\bullet \xrightarrow{+ O_2} ROO^\bullet \quad (16.30)$$

or indirectly after reaction with oxygen (16.31) - (16.34).

$$^\bullet CCl_3 + O_2 \longrightarrow CCl_3OO^\bullet \quad (16.31)$$
$$CCl_3OO^\bullet + RH \longrightarrow CCl_3OOH + R^\bullet \quad (16.32)$$
$$R^\bullet + O_2 \longrightarrow ROO^\bullet \quad (16.33)$$
$$CCl_3OOH \longrightarrow COCl_2 + HOCl \quad (16.34)$$

16-6-3 Nitrogen oxides

Nitrogen oxides, NO$^\bullet$ and NO$_2^\bullet$, are free radicals. As such, they are toxic since, by addition to double bonds or by abstraction of allylic hydrogen, they indirectly initiate peroxidation. NO$^\bullet$, which is relatively inert, is oxidized to the more active NO$_2^\bullet$ (16.35). These oxides are therefore partly responsible for the toxicity of the products of combustion (vehicle exhaust gases) and even of tobacco smoke. Their lifetime in the gas phase can exceed 5 mn, which gives them ample time to reach the mucous membranes of the throat and the lungs. In this respect, they are perhaps more toxic than tars.

$$NO^\bullet \xrightarrow[\text{slow}]{+ 1/2\, O_2} NO_2^\bullet \xrightarrow{\text{olefins}} R^\bullet \xrightarrow[\text{fast}]{+ O_2} ROO^\bullet \quad (16.35)$$

16-6-4 Ozone

The concentration of ozone in the atmosphere which we breathe (0.1-0.2 ppm) is enough to initiate lipid autoxidation. It has been shown that ozone reacts with various types of organic molecules both by ionic and radical processes. The biradical intermediate **1** of reaction (16.37) disappears by decomposition or reaction with a substrate.

$$RH + O_3 \longrightarrow [R^\bullet ---O_2---{}^\bullet OH] \longrightarrow ROH + O_2 \quad (16.36)$$

$$\underset{/}{\overset{\backslash}{C}}=\underset{\backslash}{\overset{/}{C}} \xrightarrow{+O_3} \underset{\overline{/}}{\overset{O\overset{O}{\diagdown}O}{\underset{C-C}{|\quad|}}}\underset{\backslash}{\overset{}{}} \longrightarrow \underset{\underset{\mathbf{1}}{O^\bullet}}{\overset{\backslash}{\underset{|}{C}}-O} + O=\underset{\backslash}{\overset{/}{C}} \quad (16.37)$$

16-7 CONCLUSIONS

Work on the involvement of free radicals in biological systems is currently in a state of rapid development and interests many researchers, chemists, biologists and doctors. At the moment it is difficult to establish clearly the nature of the processes in which radicals are implicated, and in this chapter we have simply presented some problems. Some biochemical reactions are clearly explained by the formation of free radicals: lipid peroxidation, isomerization by coenzyme B_{12}, etc. EPR and the spin-trapping technique contribute greatly to the detection and identification of the free radical intermediates in this field.

On the other hand, the fundamental and essential relationship of cause and effect between the presence of free radicals and the perturbation of biological processes, i.e. the appearance of certain illnesses and ailments, is more difficult to establish with certainty. These difficulties are due to the complexity of the cellular and molecular structures, and the multiplicity of possible reactions, due as much to the nature of the free radicals as to the variety of targets.

Nonetheless, it would appear certain that free radicals are involved in ageing. They seem also to be involved in the appearance of certain illnesses and cell transformations (cancers, mutations, atherosclorosis, cardiovascular disorders, inflammations). They explain the toxicity of exogenous products (medicines, halo compounds, etc.).

The radical reactions which we have described can have dramatic consequences. Fortunately, organisms have remarkable endogenous defense systems (for example, SOD and vitamins C and E) and means of repairing damaged molecules.

17 – HOW TO PROVE THAT A REACTION IS A RADICAL PROCESS

The question often arises as to whether a reaction or a step in a complex process involves radicals or not. The ways of answering this question depend on whether the radical intermediate is persistent, stable or short-lived. In the first two cases direct observation by EPR is the most rapid and the most conclusive. For short-lived radicals physical methods (EPR either direct or after trapping, CIDNP (Chap. 2)) can also be used. In this chapter we present the chemical methods for detecting and identifying a radical; these are based on the reaction principles discussed in the previous chapters. A short-lived radical ($< 10^{-3}$s) must be trapped, in a reaction which is both fast and characteristic, either by a reactive molecule in a transfer reaction or in an addition or reaction, or by a stable radical which by combination leads to a non-radical product. These compounds are called *radical traps* or scavengers. The kinetic order of the reaction also gives useful information (Chap. 7).

17-1 TRANSFER REACTION

The principle is summarized by reactions (17.1) and (17.2).

$$\text{Rad}^\bullet + \text{X}-\Sigma \xrightarrow[X = H, I, Br, Cl]{k_X} \text{Rad}-X + \Sigma^\bullet \quad (17.1)$$

$$2\ \Sigma^\bullet \longrightarrow \Sigma-\Sigma \quad (17.2)$$

A radical, Rad$^\bullet$, is trapped by a fast transfer reaction (17.1). The radical formed, $\Sigma^\bullet$, is stable or relatively unreactive. It disappears mainly by dimerization (17.2). *In this case finding both Rad–X and Σ–Σ among the reaction products is unambiguous proof of the existence of Rad$^\bullet$*.

17-1-1 Hydrogen transfer

When X is hydrogen, reaction (17.1) is fast if it is exothermic, therefore, if BDE(Rad–H) > BDE(Σ–H). This implies a weak Σ–H bond. The most commonly used hydrogen-donors are listed in Table 17.1.

Metal hydrides
Metal hydrides provide a useful range of BDE values (17.3). Tin compounds are toxic and can be replaced by $(Me_3Si)_3Si$–H, whose BDE(Si–H) is about the same as that of Bu_3Sn–H, or by diphenylsilane (para. 9-2-3).

$$R_3Sn\text{–}H > R_3Ge\text{–}H \geq R_3Si\text{–}H \quad (17.3)$$

Phenols, thiols and thiophenols
The last are the most efficient since their BDEs are the lowest. These compounds are frequently used as antioxidants.

Σ–H	BDE[a] kcal/mol	RCH$_2^\bullet$	PhCH$_2^\bullet$	Ph$^\bullet$
			k_H (M^{-1}s^{-1})	
PhS–H	80	9.2 x 10^7	3.0 x 10^5	1.9 x 10^9
n-Bu$_3$Sn–H	74	2.7 x 10^6	3.6 x 10^4	5.9 x 10^8
R$_2$P–H	74	1 x 10^6	2.5 x 10^3	
1,4-cyclohexadiene	73	4.8 x 10^5	1 x 10^2	
n-Bu$_3$Ge–H	84	1 x 10^5		2.6 x 10^8
Et$_3$Si–H	90	7.0 x 10^3		
Σ–X			k_X (M^{-1}s^{-1})	
Cl$_3$C–Br	56	7.0 x 10^6		1.4 x 10^9
t-Bu–I	54	3 x 10^6		2.0 x 10^9
Cl$_3$C–Cl	71	7.2 x 10^3		6.0 x 10^6
t-Bu–Br	70	4.6 x 10^3		1 x 10^6
t-Bu–Cl	81	6 x 10^2		

Table 17.1 – *Rate constants for hydrogen transfer (k_H) and halogen transfer (k_X) for various radicals. a) BDE = Σ–H or Σ–X bond energy.*

1,4-Cyclohexadiene
The BDE(C–H) of 1,4-cyclohexadiene is low because the radical is doubly allylic. It reacts by hydrogen transfer (17.4) rather than by addition to a double bond. The cyclohexadienyl radical is converted into benzene by reaction (17.5) with a second radical.

$$\text{Rad}^\bullet + \text{C}_6\text{H}_8 \longrightarrow \text{Rad–H} + \text{C}_6\text{H}_7^\bullet \quad (17.4)$$

$$\text{Rad}^\bullet + \text{C}_6\text{H}_7^\bullet \longrightarrow \text{Rad–H} + \text{C}_6\text{H}_6 \quad (17.5)$$

Cumene
Because of its low BDE(C–H), 79 kcal/mol, cumene reacts readily with non-stabilized carbon radicals. The stable tertiary benzyl radical obtained leads to an easily identified dimer.

Dicyclohexylphosphine
Dicyclohexylphosphine is a good trap. However, its pK of about 38 makes it a good hydride-donor for simple carbocations. These are also trapped by *tert*-butylamine, which does not react with free radicals.

17-1-2 Halogen transfer

It is necessary to choose a halogen derivative such that reaction (17.1) is fast. The rate constant, k_X, depends on the BDE(Σ–X) of the halogen in the order:

$$k_I > k_{Br} > k_{Cl} \gg k_F \quad (17.6)$$

Iodine and iodo compounds are efficient radical traps. Fluoro compounds have too strong a bond and cannot be used. Polyhalomethanes such as $CBrCl_3$ and CCl_4 are much used since the $^•CCl_3$ radical dimerizes easily. Finding Rad–Br or Rad–Cl and C_2Cl_6 together is very good proof of the presence of Rad$^•$. *tert*-Butyl hypochlorite is also a very good chlorine-donor (7.69). This is a radical trap about 100 times better than Ph_3SnH is as a hydrogen-donor.

17-1-3 Substitution at a divalent atom

Diphenylsulfides and diphenylselenides react rapidly with carbon radicals to give sulfides and selenides (17.7).

$$Rad^• + Ph-X-X-Ph \xrightarrow{X = S \text{ or } Se} Rad-X-Ph + Ph-X^• \quad (17.7)$$

17-2 COUPLING WITH A STABLE RADICAL

The coupling reaction (17.8) is very fast (k ≈ 10^9 M^{-1} s^{-1}) even when the radicals are stable. Certain radicals ($\Sigma^•$) are sufficiently stable for them to be handled in the crystalline state and to be introduced in known amounts into a reaction medium to trap short-lived radicals (Rad$^•$). Their rate of disappearance and, therefore, that of the appearance of Rad$^•$ is followed by UV, EPR, etc. The most used radicals are DPPH, galvinoxyl and various nitroxide radicals.

$$Rad^• + \Sigma^• \longrightarrow Rad-\Sigma \quad (17.8)$$

Diphenylpicrylhydrazyl
DPPH **1** is a violet crystalline compound which can be stored for several months in the dark. Different radicals attack either at N or on Ar. DPPH cannot be used in the presence of peroxides.

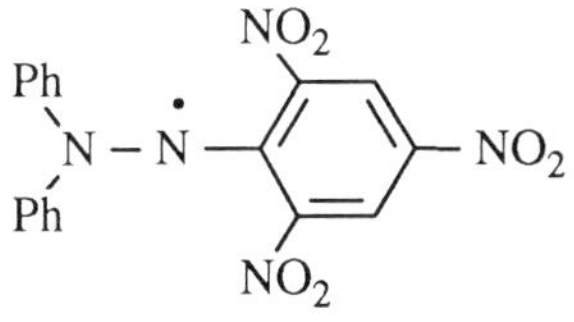

Scheme 17.1

Galvinoxyl
Galvinoxyl **2**, like DPPH, owes its stability to the delocalization of the unpaired

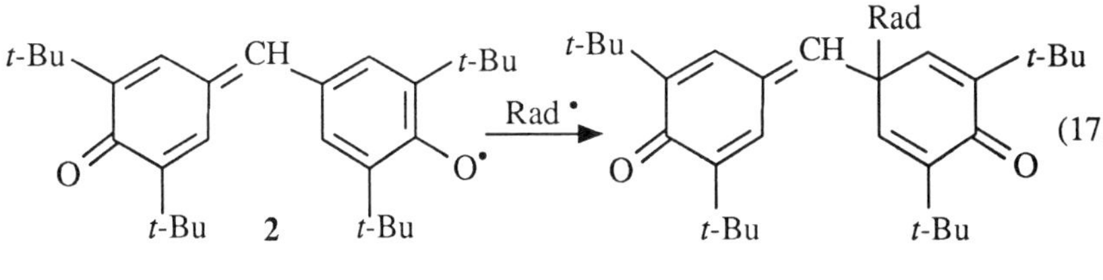

(17.9)

electron. It is an excellent trap for alkyl, alkoxyl or alkylperoxyl radicals. Coupling generally occurs at the carbon atom (17.9).

Nitroxide (Aminoxyl) radicals

Nitroxide (Aminoxyl) radicals can be used as radical traps. Note that with 2,2,6,6-tetramethyl-1-piperidinyloxy (TEMPO) and 1,1,3,3-tetramethyl-2-isoindolinyloxy (T) free radicals, the alkoxylamines formed can be isolated and identified.

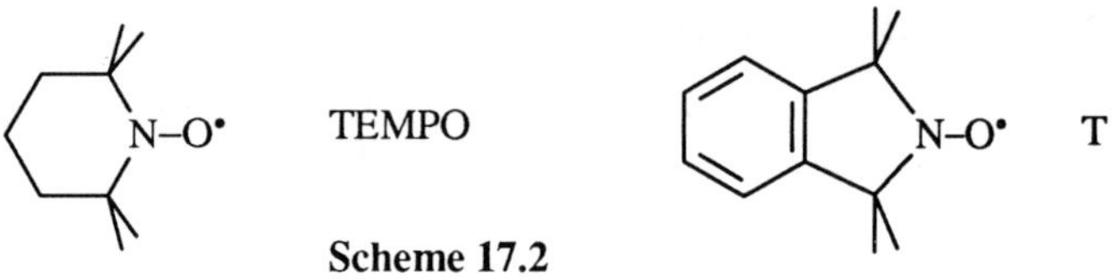

Scheme 17.2

For alkyl or primary benzyl radicals k (17.10) is about $10^9 \text{ M}^{-1}\text{s}^{-1}$.

$$\text{>N-O}^\bullet + \text{Rad}^\bullet \xrightarrow{k} \text{>N-O-Rad} \qquad (17.10)$$

17-3 ISOMERIZATION OR REARRANGEMENT

Many free radicals undergo characteristic isomerization reactions. The isomerized or rearranged radical is trapped by a transfer reaction. The choice of X–Σ and its concentration depends on k_i and k_X, since Rad_i–X will not be detected if $k_X[\text{X–Σ}] \gg k_i$. Fortunately, a fairly extensive range of isomerization and transfer reactions is available for this purpose.

$$\text{Rad}^\bullet \xrightarrow{\text{isomerization}, k_i} \text{Rad}_i{}^\bullet \xrightarrow{\text{X–Σ}} \text{Rad}_i\text{–X} + \Sigma^\bullet \qquad (17.11)$$

$$\text{Rad}^\bullet \xrightarrow{\text{X–Σ}, k_X} \text{Rad–X} + \Sigma^\bullet \qquad (17.12)$$

17-3-1 Intramolecular hydrogen transfer

Intramolecular hydrogen transfers (17.13) are very regioselective, 1,5 (n = 3) and 1,6 (n = 4) being much faster than any other. The $k_{1,5}/k_{1,6}$ ratio is 3.3 and 10 for A = CH_2 and O, respectively. This contrasts with carbocation chemistry where hydride transfers are common but unregioselective.

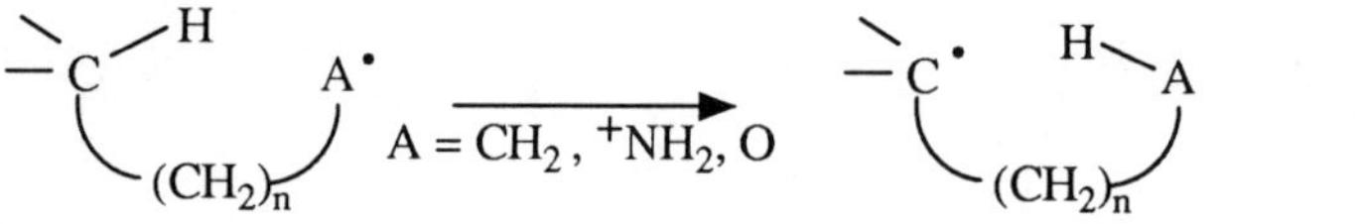

$$\begin{array}{c} \text{–C–H} \cdots \text{A}^\bullet \\ | \\ (CH_2)_n \end{array} \xrightarrow{A = CH_2, {}^+NH_2, O} \begin{array}{c} \text{–C}^\bullet \cdots \text{H–A} \\ | \\ (CH_2)_n \end{array} \qquad (17.13)$$

17-3-2 Ring opening and cyclization

Ring opening and cyclization are very characteristic and common reactions of free radicals.

$$\text{(17.14)}$$

Because of differences in ring strain the rates of opening of cycloalkoxyl vary in the order given in (13.37), p. 161. Ring opening is also observed for aminyl radicals (17.15).

$$2.5 \times 10^7 \text{ s}^{-1} \quad \text{(17.15)}$$

The 5-*exo* cyclization of the 5-hexenyl radical (Table 13.3, p. 154) is often used to establish the existence of a radical. Potential ionic intermediates, carbocations and carbanions, give 6- and 5-membered rings, respectively. Confusion with the latter case is therefore possible. However, carbanion cyclization differs significantly from radical cyclization:

 - it is 10^8 to 10^9 times slower;
 - a 1,4 proton transfer causes double bond isomerization;
 - the stereochemistry observed for a 1-methyl-5-hexenyl moiety favors the *trans* compound for a carbanion and the *cis* isomer for a radical ((13.8), p. 154);
 - a carbanion can be trapped by *tert*-butylamine, which is a poor hydrogen-donor.

$$\text{(17.16)}$$

The cyclization-opening combination of the 2,2-dimethyl-3-butenyl radical is particularly well adapted to strongly basic media ((13.7), p. 153). The 5-*exo* cyclization ((13.3), p. 151) is not an absolute rule. If the initial radical is stabilized the reaction becomes reversible and the thermodynamically preferred 6-*endo* product ((13.11), p. 155) is obtained. Another case is that of a carbon radical α to silicon, where 6-*endo* cyclization is preferred (17.17).

$$66\% \quad + \quad 34\% \quad \text{(17.17)}$$

The rate constants for a certain number of radical isomerizations, cyclization or ring opening (Table 13.3, p.154), have been accurately determined. They are used as references in competitive systems, hence the name *radical clocks* or clock reactions.

17-4 ADDITION TO UNSATURATED COMPOUNDS

A radical can be trapped by addition to an unsaturated compound, M (17.18). M must be very reactive and the radical adduct, Rad–M•, characteristic. Various systems meet these criteria. Obviously, it is best to work in a solvent which is a poor hydrogen donor, unless the intention is to identify the radical species as Rad–H or Rad–M–H.

$$\text{Rad}^\bullet \ + \ M \longrightarrow \text{Rad–M}^\bullet \qquad (17.18)$$

17-4-1 Olefin and transfer agent

Anti-Markovnikov regioselectivity in the addition of Rad• to the less substituted carbon of a terminal olefin (17.19) is in itself an indication of a radical process. If the radical is nucleophilic k_{add} can be increased by using electron-poor olefins. The rate of reaction (17.20) must be adapted to the system in order to avoid trapping the initial radical, Rad•.

$$\text{Rad}^\bullet \ + \ CH_2=CH-R \ \xrightarrow{k_{add}} \ \text{Rad–}CH_2\text{–}\overset{\bullet}{C}H\text{–}R \qquad (17.19)$$

$$\text{Rad–}CH_2\text{–}\overset{\bullet}{C}H\text{–}R \ + \ X\text{–}\Sigma \ \xrightarrow{k_{trans}} \ \text{Rad–}CH_2\text{–}CHX\text{–}R \ + \ \Sigma^\bullet \qquad (17.20)$$

Because of ring strain *norbornene* is very reactive. Moreover, the intermediate 2-norbornyl radical undergoes no skeletal rearrangement (17.21), unlike the corresponding carbocation.

To trap nucleophilic carbon radicals, heteroatomic bases such as quinoxaline or 4-cyanopyridine (Tables 14.1 and 14.2) are useful, the reactions being run under acid conditions (para. 14-5-4). As for electrophilic radicals, such as •CH_2CO_2Me, they react readily with styrene ($k \approx 10^6$ M^{-1} s^{-1}).

17-4-2 Olefin and oxidation

The oxidation of the radical adduct, Rad–M•, to olefin by metal salts is a means of terminating the addition reaction when the radical itself is difficult to oxidize; this is the case, for example, with oxygen-centered radicals (17.22) and (17.23).

$$PhCO_2^\bullet \ + \ \diagup\!\!\!\diagdown\!R \longrightarrow PhCO_2\diagup\!\!\!\diagdown\overset{\bullet}{}\diagup\!\!\!R \qquad (17.22)$$

$$PhCO_2\diagup\!\!\!\diagdown\overset{\bullet}{}\diagup\!\!\!R \ + \ Cu^{2+} \ \xrightarrow[-\ NuH^+]{+\ Nu} \ PhCO_2\diagup\!\!\!\diagdown\!=\!\diagup\!\!\!R \ + \ Cu^+ \qquad (17.23)$$

Addition to unsaturated compounds 207

The system becomes catalytic when radical production itself is the result of a redox reaction (17.24).

$$(PhCO_2)_2 + Cu^+ \longrightarrow PhCO_2^{\bullet} + PhCO_2^- + Cu^{2+} \quad (17.24)$$

17-4-3 Olefin and fragmentation

Isonitriles can be used to trap radicals and to obtain a nitrile by an addition-fragmentation process:

$$R^{\bullet} + {}^-C\equiv\overset{+}{N}-t\text{-Bu} \longrightarrow R-\overset{\bullet}{C}=N-t\text{-Bu} \xrightarrow{\beta\text{-scission}} R-C\equiv N + t\text{-Bu}^{\bullet} \quad (17.25)$$

17-4-4 Olefin and dimerization

Addition to an olefin can lead to a stable radical which for this reason dimerizes (17.26).

$$R^{\bullet} + CH_2=CMePh \longrightarrow R-CH_2-\overset{\bullet}{C}MePh \longrightarrow (R-CH_2-\underset{Me}{\overset{Ph}{\underset{|}{\overset{|}{C}}}}-)_2 \quad (17.26)$$

A captodative olefin is particularly suitable for this sort of reaction (17.27).

$$R^{\bullet} + CH_2=CSMeCN \longrightarrow R-CH_2-\underset{CN}{\overset{SMe}{\underset{|}{\overset{|}{C^{\bullet}}}}} \longrightarrow (R-CH_2-\underset{CN}{\overset{SMe}{\underset{|}{\overset{|}{C}}}}-)_2 \quad (17.27)$$

17-4-5 Olefin and stable radical

Nitrones and nitroso compounds are much used to trap radicals since they lead to nitroxide radicals which are sufficiently stable to be studied by EPR or even to be isolated:

$$R^{\bullet} + Ph-CH=\underset{\text{nitrone}}{\overset{O^-}{\underset{+}{\overset{|}{N}}-t\text{-Bu}}} \longrightarrow \underset{Ph}{\overset{R}{\diagdown}}CH-\overset{\overset{\bullet}{O}}{\underset{|}{N}}-t\text{-Bu} \quad (17.28)$$

$$R^{\bullet} + \underset{\text{nitroso}}{R^1-N=O} \longrightarrow \underset{R}{\overset{R^1}{\diagdown}}N-O^{\bullet} \quad (17.29)$$

N-Dicyanomethylene aniline, proposed as an analogue for the nitroso compound, ArNO, reacts poorly with carbon or alkoxyl radicals but is, however, an excellent trap for silyl, germyl and stannyl radicals, with which it gives persistent radicals:

$$\text{Ar-N=C(CN)}_2 \; + \; \text{R}_3\text{M}^\bullet \longrightarrow \text{Ar-N(MR}_3)-\overset{\bullet}{\text{C}}(\text{CN})_2 \quad (17.30)$$

A thiocarbonyl compound such as Ph$_3$SiHC=S is a very good trap for t-BuO$^\bullet$ and t-BuC$^\bullet$=O radicals, with which it gives persistent radicals having half-lives of several minutes and suitable for EPR study:

$$\text{Ph}_3\text{Si-C}\!\!\begin{array}{c}\nearrow\!\!\text{S}\\ \searrow\!\!\text{Ph}\end{array} \; + \; t\text{-BuO}^\bullet \longrightarrow \text{Ph}_3\text{Si-}\overset{\bullet}{\text{C}}\!\!\begin{array}{c}\nearrow\!\!\text{S-O}t\text{-Bu}\\ \searrow\!\!\text{Ph}\end{array} \quad (17.31)$$

17-5 INHIBITORS AND KINETICS

The addition of inhibitors often proves or confirms a radical chain mechanism. The inhibitor reacts rapidly with a radical species and leads to another, more stable, radical which does not function as a chain carrier. The reaction rate is considerably reduced and the product yield falls dramatically.

Good hydrogen- or halogen-donors or stable radicals are used as inhibitors. Oxygen is a good inhibitor of carbon radicals, with which it reacts very rapidly ($\approx 10^9$ M^{-1}s^{-1}).

One should be careful, however, not to commit the error of discarding a radical mechanism simply because a trace of hydroquinone, the most common inhibitor, does not modify the products. It is only if the rate decreases markedly that one can conclude that a radical reaction is involved.

A fractional kinetic order may also be taken as evidence for a radical chain mechanism.

PART III

APPLICATIONS IN SYNTHESIS

18 – FUNCTIONALIZATION OF THE C–H BOND

The ability to create a radical by substitution at a C–H bond, even when it is not activated, is a characteristic of radical reactions, which makes it possible to functionalize saturated hydrocarbons. However, the replacement of hydrogen by a functional group, X, is limited both by the nature of X and by the regioselectivity. This type of reaction can be performed either *inter-* or *intra*molecularly. Halogenation and hydroxylation are important examples.

18-1 HALOGENATION

18-1-1 Intermolecular reactions

The halogenation of C–H bonds follows a classical chain mechanism:

initiation $\quad$ X–X $\quad \xrightarrow{h\nu} \quad$ 2 X• $\qquad$ (18.1)

propagation $\quad$ X• + R–H $\quad \longrightarrow \quad$ X–H + R• $\qquad$ (18.2)

$\qquad\qquad\quad$ R• + X–X $\quad \longrightarrow \quad$ R–X + X• $\qquad$ (18.3)

One can understand by looking at the thermochemical data in Table 8.4 why only chlorination, where both steps (18.2) and (18.3) are reasonably exothermic, has any practical utility. With fluorine the reactions are so exothermic that they become explosive and one must either dilute the F_2/RH mixture or work at low temperature. With Br• and I• reaction (18.2) is endothermic and for this reason the chain stops quickly.

Chlorination by molecular chlorine; solvent effect
Reaction (18.1) is performed photochemically (sunlight is adequate) and the chain length can be very high (> 1 000). A certain selectivity can only be achieved by introducing a polar effect by means of a substituent, X (Table 18.1).

X	CH_3 —	CH_2 —	CH_2 —	CH_2 — X
H	1	3.6	3.6	1
Cl	1	3.7	2.1	0.8
C(O)Cl	1	3.9	2.1	0.2
CO_2H	1	4.1	1.5	0

Table 18.1 – *Relative reactivity for chlorination (Cl_2 in gas phase) of n-butyl derivatives.*

The electrophilic chlorine radical, Cl•, is less reactive with hydrogens α to an electron-attracting substituent, which reduces the electron density of the C_α–H bond. The marked electron-attracting character of Cl$^\bullet$ means that it can be complexed with solvents such as benzene (Scheme 18.1).

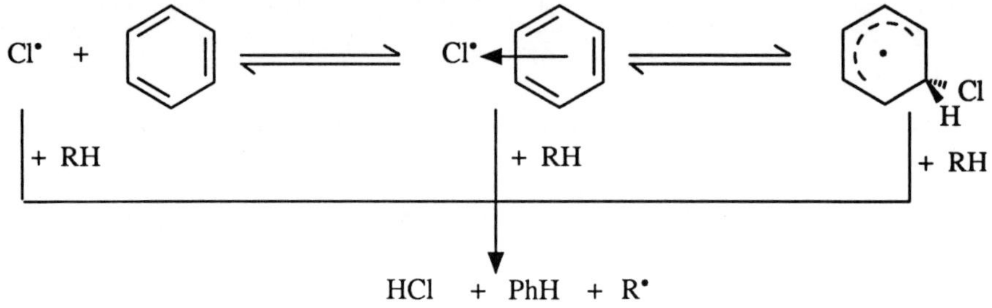

Scheme 18.1

The complex turns out to be less reactive but more selective. Thus, for 2,3-dimethylbutane chlorination the selectivity H_{tert}/H_{prim} ($RS_{t/p}$, eqn 11.8) changes with the solvents and the concentration: 4.2 in the pure hydrocarbon at 25°C, from 11 to 49 and 15 to 109, respectively in benzene and CS_2, when the concentration in hydrocarbon varies from 2 M to 8 M.

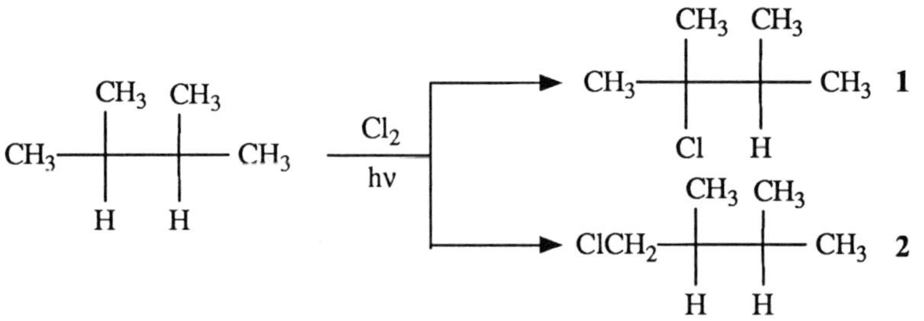

$$(18.4)$$
$$(18.5)$$

Other halogenation agents

Tert-butyl hypochlorite is a good chlorinating agent, since the energy of the breaking O–Cl bond is less than that of Cl–Cl (BDE(t-BuO–Cl) = 48 kcal/mol < BDE(Cl–Cl) = 58 kcal/mol) and the energies of the bonds formed are the same (BDE(t-BuO–H) ≈ BDE(Cl–H)). The intermediate t-BuO$^\bullet$ radical, which is also electrophilic, is a little more selective than Cl$^\bullet$. Moreover, unlike Cl$^\bullet$, t-BuO$^\bullet$ has an interesting feature in that it prefers to abstract allylic hydrogen rather than add to a C=C double bond (k_{all}/k_{add} ≈ 30). With toluene, benzyl chloride is obtained selectively.

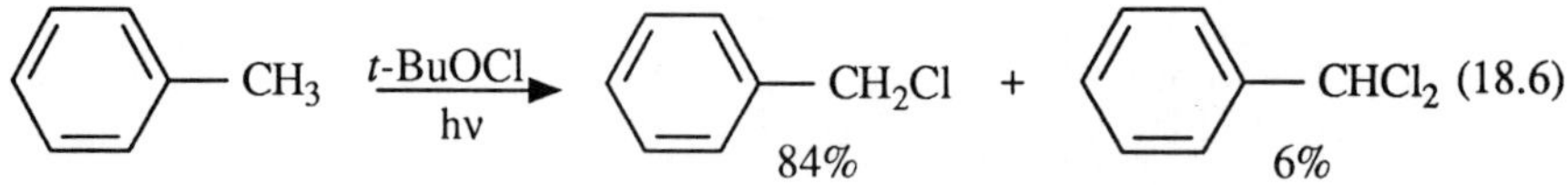

$$(18.6)$$

Other compounds such as *sulfuryl chloride* (18.7) or *trichloromethane sulfonyl chloride* (18.8) are also used, with no particular advantage, except that they are easier to handle and slightly more selective than chlorine.

$$SO_2Cl_2 \xrightarrow{h\nu} {}^{\bullet}SO_2Cl + Cl^{\bullet} \quad (18.7)$$

$$Cl_3CSO_2Cl \xrightarrow{h\nu} Cl_3CSO_2^{\bullet} + Cl^{\bullet} \quad (18.8)$$

On the other hand, the halogenation of aliphatic compounds with an electron-attracting substituent by *N-chloroamines* in acid is much more regioselective (Table 18.2). The reaction is usually initiated by a redox system (18.9):

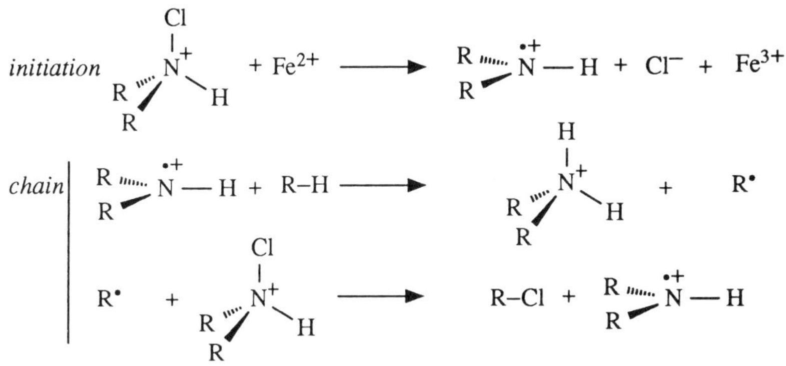

initiation: $R_2N^+(Cl)H + Fe^{2+} \longrightarrow R_2\overset{\bullet+}{N}-H + Cl^- + Fe^{3+}$ (18.9)

chain:
$R_2\overset{\bullet+}{N}-H + R-H \longrightarrow R_2N^+H_2 + R^{\bullet}$ (18.10)

$R^{\bullet} + R_2N^+(Cl)H \longrightarrow R-Cl + R_2\overset{\bullet+}{N}-H$ (18.11)

N-chloroamine	aliphatic compounds							
Me$_2$N–Cl	CH$_3$– 2	CH$_2$– 72	CH$_2$– 19	CH$_2$– 7	CH$_2$– ε^a	CH$_2$–Cl -		
Me$_2$N–Cl *i*-Bu$_2$N–Cl	CH$_3$– 3 3	CH$_2$– 72 83	CH$_2$– 20 13	CH$_2$– 5 1	CH$_2$– 1 -	CH$_2$–CO$_2$CH$_3$ ε^a -		
i-Pr$_2$N–Cl	CH$_3$– 1	CH$_2$– 92	CH$_2$– 3	CH$_2$– 1	CH$_2$– 2	CH$_2$– 2	CH$_2$– -	CH$_2$–OH^b -
Me$_2$N–Cl	CH$_3$– 1	CH$_2$– 65	CH$_2$– 23	CH$_2$– 11	CH$_2$– 	CH$_2$– 	CH$_3$ 	

Table 18.2 – *Isomers obtained (%) in chlorination of saturated aliphatic compounds. by N-chloroamines in acid (70% H_2SO_4), with 60-80% conversion.*
a) Trace. b) OH protonation makes this group electron-attracting.

The high selectivity can be attributed to the following factors:

– inductive effect of the substituent transmitted over a considerable distance, well beyond 3 carbons;
– difference between the BDEs of secondary and primary C–H;
– ω–1 effect (substitution at the last but one carbon) which is explained by a steric effect. This ω–1 selectivity increases with the steric size of the alkyl substituents of the N-chloroamine.

This reaction is easy to perform and, since chlorination deactivates the C–H bonds with respect to the electrophilic radical, $R_2NH^{\bullet+}$, no disubstituted products are obtained. The yields are generally satisfactory since reactions (18.10) and (18.11) are both exothermic ($k_{18.11} = 7 \times 10^2$ to 1×10^4 $M^{-1}s^{-1}$) and the termination reaction involving two aminyl radical cations is slow. Bromination can be performed in the same way and shows similar selectivity, the radical intermediate of reaction (18.10) being the same as in chlorination.

However, the concentrated acid medium (70-80% H_2SO_4) is incompatible with the presence of other functions, such as a double bond, an aromatic or ester grouping, etc., which are acid-sensitive.

Allylic substitution

Allylic halogenation, interesting from the synthetic point of view, encounters difficulties due to the competition between addition (18.12) and allylic hydrogen abstraction (18.13). At low temperature $k_{add} > k_{all} > k_{elim}$; when the temperature rises k_{all} and k_{elim} increase faster than k_{add} since the activation energies are in the order: $E_{elim} > E_{all} > E_{add}$. Allylic halogenation (18.13) is therefore favored by a high temperature and a low X_2 concentration.

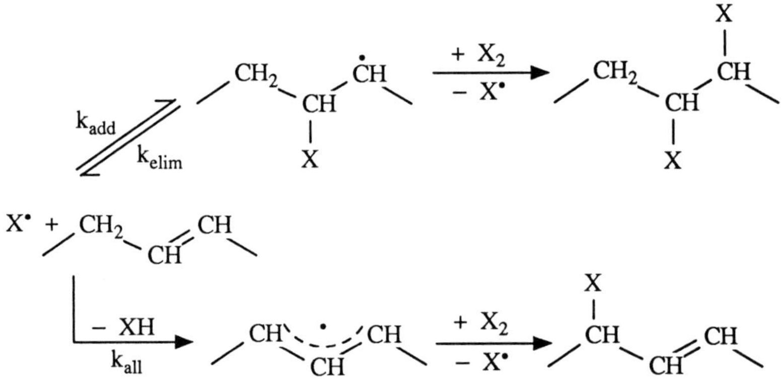

Scheme 18.2

To displace this competition in favor of allylic halogenation one can use:

— either *t*-BuOCl (chlorination), since the *t*-BuO$^{\bullet}$ radical adds to the double bond 30 times slower than it abstracts allylic hydrogen;
— or N-bromosuccinimide (bromination) which serves as a source of a low concentration of Br_2 by reacting with HBr. The mechanism is given by reactions (18.14) to (18.16);

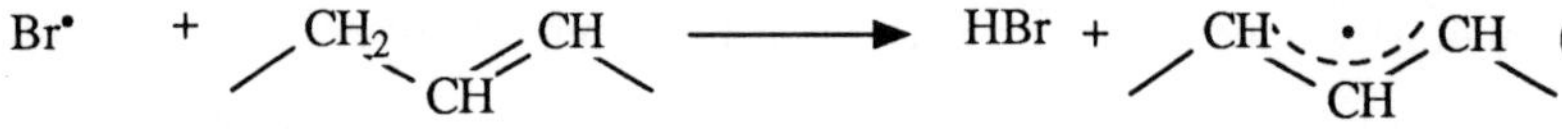

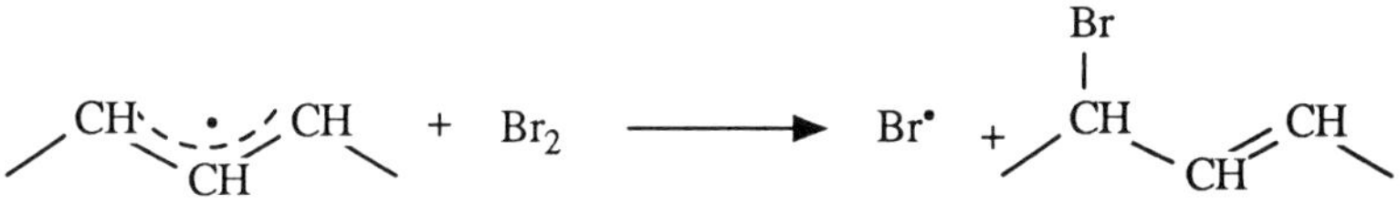

$$\underset{CH}{CH \cdots CH} + Br_2 \longrightarrow Br^\bullet + \underset{CH}{\overset{Br}{\underset{|}{CH}}} {=} CH \quad (18.16)$$

It should be noted that it is not the succinimidyl radical which abstracts hydrogen. These reagents can also be used to perform selective benzylic halogenation.

18-1-2 Intramolecular reactions

The starting materials for intramolecular chlorinations are either hypochlorites (Barton type reaction) or N-chloroamines (Hofmann-Löffler-Freytag reaction). These reactions are very regioselective.

Hypochlorites

With aliphatics the reaction is initiated photochemically and leads to a 10/1 mixture of δ-chloro and ε-chloroalcohols with an overall yield of 80% (Scheme 18.3). Treatment of these chloroalcohols with base gives tetrahydrofuran and tetrahydropyran derivatives with more than 90% of the former.

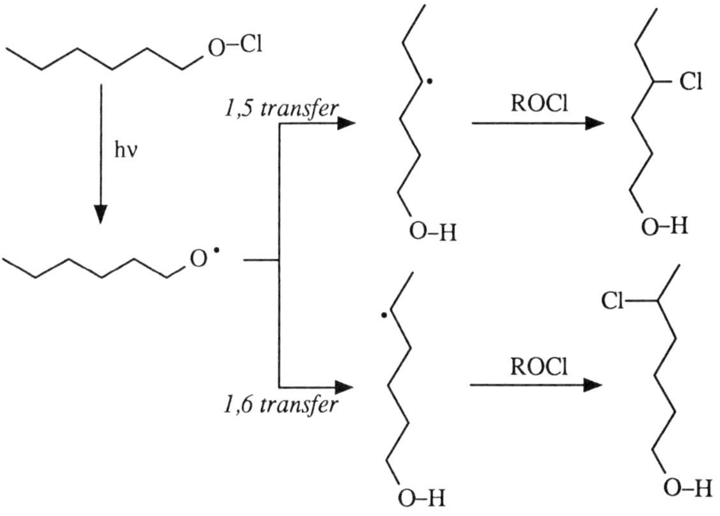

Scheme 18.3

(18.17)

(18.18)

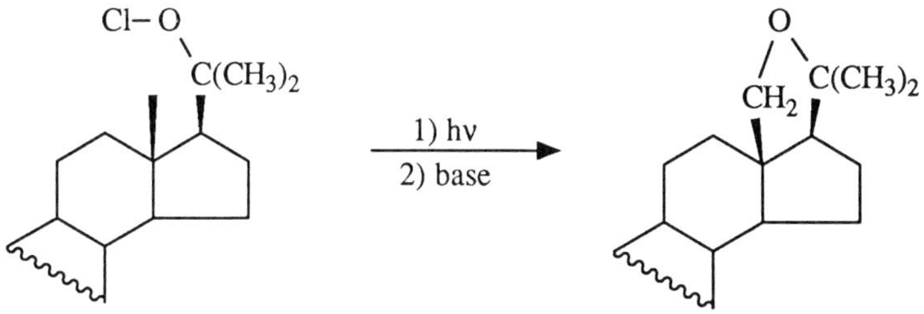

(18.19)

When the structure of the molecule is appropriate the reaction is very regioselective; for example, in steroids (18.19).

The β-fragmentation of the intermediate alkoxyl radical (18.20), which depends on the structure and the solvent, can compete with cyclization.

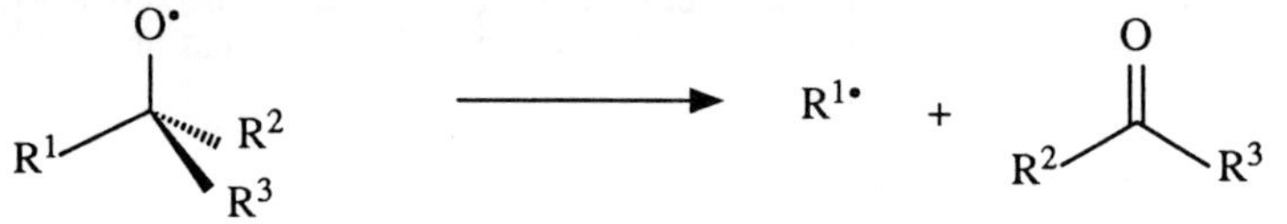

$$R^1-C(O^\bullet)(R^2)(R^3) \longrightarrow R^{1\bullet} + R^2-C(=O)-R^3 \quad (18.20)$$

N-Chloroamines

By irradiation or a redox reaction in strong acid a N-chloroamine gives a protonated amidyl radical which undergoes intramolecular hydrogen transfer; the δ-chloro and ε-chloroamines obtained are then transformed into pyrrolidine and piperidine derivatives, respectively (Scheme 18.4).

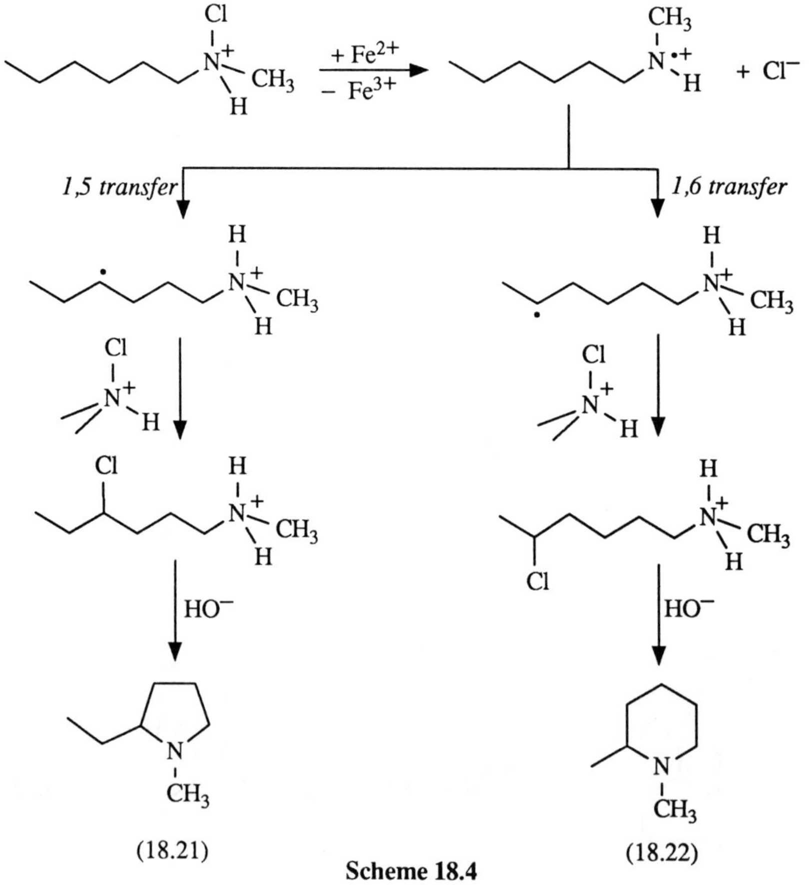

Scheme 18.4

(18.21)   (18.22)

Hydroxylation

The rate constant for intramolecular 1,5 hydrogen transfer is very high ($\approx 10^6$ s^{-1}). The overall yield is 80% in 5N H_2SO_4 and only 40% for 1N H_2SO_4. The ratio of 1,5 to 1,6 hydrogen transfer (and therefore of the pyrrolidine and piperidine derivatives) is about 9. This type of reaction can be used for the synthesis of bicyclic compounds (18.23)

$$\text{(18.23)}$$

or alkaloids such as conanine (18.24).

$$\text{(18.24)}$$

However, as in the case of intermolecular reactions, the need for a strongly acidic medium means that there must be no acid-sensitive functions in the molecule.

18-2 HYDROXYLATION

Aromatic peracids, provided they are used in boiling benzene in order to eliminate oxygen, replace hydrogen by hydroxyl via a chain mechanism, (18.25) to (18.27). The selectivity, $H_{tert} > H_{sec} > H_{prim}$, results from the hydrogen abstraction reaction (18.26). Furthermore, the tertiary radical reacts more rapidly at the electrophilic oxygen of the peracid O–O bond than a secondary or primary radical, because the substitution (18.27) depends on the nucleophilicity of the attacking radical.

PhCO$_3$H		$\longrightarrow$	PhCO$_2$$^\bullet$ + $^\bullet$OH		(18.25)
PhCO$_2$$^\bullet$ + R–H		$\longrightarrow$	PhCO$_2$H + R$^\bullet$		(18.26)
R$^\bullet$ + PhCO$_3$H		$\longrightarrow$	ROH + PhCO$_2$$^\bullet$		(18.27)

With methylcyclohexane, adamantane or *trans*-decalin the tertiary alcohol is obtained in 65-80% yield, the remainder being a mixture of secondary alcohols. Moreover, with *trans*-decalin the reaction occurs with 95% retention of configuration.

An aliphatic peracid, RCO$_3$H, is unsuitable for this hydroxylation reaction since the RCO$_2$$^\bullet$ radical decarboxylates very rapidly (k $\approx 10^{10}$ s^{-1} as against 10^6 s^{-1} for

$PhCO_2^\bullet$) and the nucleophilic radical, $R^\bullet$, obtained induces the decomposition of the initial peracid.

$$RCO_2^\bullet \longrightarrow R^\bullet + CO_2 \quad (18.28)$$
$$R^\bullet + RCO_3H \longrightarrow ROH + RCO_2^\bullet \quad (18.29)$$

18-3 AUTOXIDATION

Under mild conditions (below 100°C) very many organic compounds react with molecular oxygen to give hydroperoxides and other oxygen-containing species such as peroxides, alcohols, ketones, aldehydes and acids. In previous chapters reactions with O_2 (8.76) and (9.23), between peroxyl radicals (Scheme 10.3) and saturated fatty acids with O_2 (Section 16.3) have already been mentioned. Here we take another look at the general scheme for autoxidation. There is often a complex series of reactions, the simplest and most fundamental form of which is a radical chain process.

18-3-1 Chain reaction

The overall reaction, exothermic by 18 ~ 24 kcal/mol, for hydroperoxide formation:

$$R-H + O_2 \longrightarrow R-OO-H \quad (18.30)$$

is broken down as follows:

initiation $R-H \longrightarrow R^\bullet$ (18.31)
propagation $R^\bullet + O_2 \longrightarrow R-OO^\bullet$ (18.32)
   $R-OO^\bullet + R-H \longrightarrow R-OO-H + R^\bullet$ (18.33)
termination $2\ R-OO^\bullet \longrightarrow$ non-radical products (18.34)

Initiation

This is without doubt the step which is the least well understood. Direct hydrogen abstraction

$$R-H + O_2 \longrightarrow R^\bullet + H-OO^\bullet \quad (18.35)$$

is endothermic by 40 ~ 50 kcal/mol and at room temperature it is too slow to generate radicals at a significant rate. Often the role of initiator is attributed to "impurities", without these being clearly identified. It is true that autoxidation reactions go better when a peroxide initiator is added. Certain authors have proposed an ionic mechanism:

$$R_2CH_2 + O_2 \rightleftharpoons [R_2\overset{+}{C}H \cdots \overset{-}{O}_2H] \longrightarrow R_2CHOOH \quad (18.36)$$

Whatever the case, observations and synthetic operations show that autoxidations are naturally initiated in the presence of atmospheric oxygen.

Propagation
Reaction (18.32) is very fast, k being about 10^9 $M^{-1}s^{-1}$. An R• radical is therefore trapped by oxygen as soon as it is formed.

Given the dissociation energy of the ROO–H bond (88 kcal/mol), reaction (18.33) only works with hydrogens whose bond energies are relatively low (allylic, tertiary, α to oxygen, etc.); some examples are given below.

Termination and reactions of ROO•
Termination occurs by the equilibrated recombination of two ROO• radicals (18.37). The tetroxide decomposes in various ways but, in particular, it can give either a peroxide or two RO• radicals (see para. 10-1-3).

$$\text{ROO}^• + \text{ROO}^• \rightleftharpoons \text{ROO–OOR} \qquad (18.37)$$

ROO• radicals can also react by addition to double bonds to form epoxides and RO• radicals:

$$\text{ROO}^• + \underset{/}{\overset{\backslash}{C}}=\underset{\backslash}{\overset{/}{C}} \longrightarrow \overset{\text{RO–O}}{\underset{/}{\overset{\backslash}{C}}-\underset{\backslash}{\overset{/}{\dot{C}}}} \longrightarrow \text{RO}^• + \underset{/}{\overset{\backslash}{C}}\overset{O}{-}\underset{\backslash}{\overset{/}{C}} \qquad (18.38)$$

It should be noted that RO• radicals are perfectly capable of propagating the chain by means of a reaction of type (18.33) and that this reaction is more exothermic than with ROO• radicals.

18-3-2 Examples of autoxidation reactions

Phenol is prepared industrially by the autoxidation of cumene (see para. 24-4-2). Hydroperoxide precursors of prostaglandins are obtained by autoxidation of polyunsaturated fatty acids. Unsaturated cyclic compounds and tetrahydrofuran give hydroperoxides readily. For examples:

tetralin + O_2 $\xrightarrow{70°C / 48 h}$ tetralin-OOH ≈ 50%   (18.39)

cyclohexene + O_2 $\xrightarrow{55°C / 8 h}$ cyclohexenyl-OOH 95%   (18.40)

THF + O_2 $\xrightarrow{25 \sim 40°C / 8 h}$ 2-OOH-THF   (18.41)

This hydroperoxide synthesis, which is a chain mechanism, works better in the presence of an initiator such as benzoyl peroxide or AIBN.

Aldehydes are autoxidized to the corresponding acids by a radical chain mechanism which has been known for a long time:

$$R-\underset{H}{\overset{O}{\overset{\|}{C}}} \longrightarrow \longrightarrow R-\overset{O}{\overset{\|}{C^{\bullet}}} \quad (18.42)$$

chain

$$R-\overset{O}{\overset{\|}{C^{\bullet}}} + O_2 \longrightarrow R-\underset{O-O^{\bullet}}{\overset{O}{\overset{\|}{C}}} \quad (18.43)$$

$$R-\underset{O-O^{\bullet}}{\overset{O}{\overset{\|}{C}}} + R-\underset{H}{\overset{O}{\overset{\|}{C}}} \longrightarrow R-\underset{O-O}{\overset{O}{\overset{\|}{C}}}\overset{H}{|} + R-\overset{O}{\overset{\|}{C^{\bullet}}} \quad (18.44)$$

$$R-\underset{O-O}{\overset{O}{\overset{\|}{C}}}\overset{H}{|} + R-\underset{H}{\overset{O}{\overset{\|}{C}}} \longrightarrow 2\ R-\underset{O-H}{\overset{O}{\overset{\|}{C}}} \quad (18.45)$$

It should be remembered that hydroperoxides at high concentration can cause explosions when heated. It is therefore important to destroy them by means of $FeSO_4$ when, for example, recovered tetrahydrofuran is to be distilled.

19 – TRANSFORMATION OF FUNCTIONAL GROUPS

Radical processes can be used to replace a functional group by a hydrogen or by another group, most often a halogen. The advantages of these reactions are that they are selective, they do not affect other functions in the molecule and in most cases they work in neutral media.

19-1 HALOGEN REDUCTION

Halides can be reduced (19.1) by means of tin, germanium or silicon hydrides, via a chain mechanism described in Chapter 8-1.

$$R-X \longrightarrow R-H \qquad (19.1)$$

The reaction is initiated (19.2) by means of AIBN or a peroxide. Fluoro compounds do not react. Reaction (19.3) is exothermic. The rate, which depends on the halogen, increases as the bond strengths decrease (Table 19.1).

$$\text{initiation} \quad R_3M-H \xrightarrow{\text{AIBN or peroxide}} R_3M^\bullet \qquad (19.2)$$

$$\text{chain} \begin{cases} R^1-X + R_3M^\bullet \longrightarrow R^{1\bullet} + R_3M-X & (19.3) \\ R^{1\bullet} + R_3M-H \longrightarrow R^1-H + R_3M^\bullet & (19.4) \end{cases}$$

$$\text{termination} \quad \text{two radicals} \longrightarrow \text{non-radical products} \qquad (19.5)$$

The data presented in this table show that $Bu_3Ge^\bullet$ and $Bu_3Sn^\bullet$ radicals are about equally reactive and that $Et_3Si^\bullet$ abstracts halogen atoms the fastest.

halide	n-Bu$_3$Sn$^\bullet$	n-Bu$_3$Ge$^\bullet$	Et$_3$Si$^\bullet$
CH$_2$=CH(CH$_2$)$_4$Cl	6.6 x 10^3	1.2 x 10^4	2.6 x 10^5
t-BuCl	2.7 x 10^4	<5 x 10^4	2.5 x 10^6
PhCH$_2$Cl	1.1 x 10^6	1.3 x 10^6	1.4 x 10^7
CH$_2$=CH(CH$_2$)$_4$Br	5.0 x 10^7	4.6 x 10^7	5.5 x 10^8
t-BuBr	1.4 x 10^8	8.6 x 10^7	1.1 x 10^9
PhCH$_2$Br	1.7 x 10^9	8.6 x 10^7	1.1 x 10^9
CH$_3$I	4.3 x 10^9		8.1 x 10^9

Table 19.1 – *Rate constants for halogen abstraction by $R_3M^\bullet$ radicals (19.3) (k in $M^{-1}s^{-1}$ at 25°C).*

The relative rates for the hydrogen transfer reaction (19.4) to a primary radical vary in the order:

$$R_3Sn-H \quad > \quad R_3Ge-H \quad > \quad R_3Si-H \quad (19.6)$$

The rate also depends on the strength of the C–H bond and, therefore, on the type of radical, $R^{1\bullet}$ (Table 19.2).

$R^{1\bullet}$	R_3Sn-H	R_3Ge-H	R_3Si-H
$Ph^\bullet$	5.9×10^8	2.6×10^8	
$cy\text{-}C_3H_5^\bullet$	8.5×10^7	1.3×10^7	
$^\bullet CH_3$	1.2×10^7	5.0×10^7	
$RCH_2CH_2^\bullet$	2.7×10^6	1.0×10^5	7.0×10^3

Table 19.2 – *Rate constants for H abstraction by hydrides, R_3M-H (19.4) (k in $M^{-1}s^{-1}$ at 25°C).*

The data in the two Tables show that on the whole the chain propagation steps, (19.3) and (19.4), are fairly fast and that the reduction is therefore very efficient. Bu_3Sn-H is the most commonly employed reagent but, because of its toxicity and the problem of eliminating organotin compounds at the end of the reaction, it can be replaced by $(Me_3Si)_3Si-H$, where the Si–H bond strength is about the same as that of Sn–H (para. 9-2-3).

Polyhalogenated compounds can be reduced selectively (19.7) by adapting the conditions of temperature and concentration (see also examples (8.7) and (8.8)).

$$\text{[structure with Br and X]} \xrightarrow[< 40°C]{n\text{-}Bu_3SnH,\ 1\ eq\ /\ AIBN} \text{[structure with H and X]} \quad (19.7)$$

X = Br, 62 ~ 87%
X = Cl, 97%

Advantage has been taken of this Br/Cl chemoselectivity to perform the cyclization in (19.8).

$$\text{[acyclic structure with CO}_2\text{Et, Br, Cl]} \xrightarrow[\text{in } C_6H_6 \text{ at } 80°C]{n\text{-}Bu_3SnH,\ 1\ eq\ /\ AIBN} \text{[cyclopentane with CO}_2\text{Et and Cl]} \quad (19.8)$$

60%

19-2 REPLACEMENT OF –COOH BY ANOTHER GROUP

The general scheme (19.9) implies the decarboxylation of $RCO_2^\bullet$. There are three main methods for bringing about this transformation: the decomposition of peracids, the Hunsdiecker reaction and that of thiohydroxamic esters.

$$RCO_2H \longrightarrow RCO_2^\bullet \xrightarrow{-CO_2} R^\bullet \xrightarrow[X = H \text{ or halogen}]{X-\Sigma} RX + \Sigma^\bullet \quad (19.9)$$

19-2-1 Peracid decomposition

Aliphatic peracids, RCO_3H, are easily obtained from acids. Their radical decomposition leads to the corresponding alcohol, ROH, or, by hydrogen transfer from a suitable donor, to the hydrocarbon, RH.

$$RCO_2H \xrightarrow{H_2O_2 / H^+} RCO_3H \quad (19.10)$$
$$RCO_3H \xrightarrow{\Delta} RCO_2^\bullet + HO^\bullet \quad (19.11)$$
$$RCO_2^\bullet \longrightarrow R^\bullet + CO_2 \quad (19.12)$$
$$R^\bullet + RCO_3H \longrightarrow ROH + RCO_2^\bullet \quad (19.13)$$
$$R^\bullet + H-\Sigma \longrightarrow R-H + \Sigma^\bullet \quad (19.14)$$

This reaction is useful for obtaining the bridgehead alcohol from bicyclo[2.2.1]-heptane (norbornane) which is difficult to synthesize in any other way (19.15).

(19.15)

19-2-2 Hunsdiecker reaction

The reaction of bromine with the silver salt of a carboxylic acid leads to a bromide by a chain mechanism (19.16) - (19.19). The bromide yields are not always good, in particular because of traces of water. Moreover, this reaction, of historical interest (1939), is limited to bromide synthesis. Nowadays the much more general method using hydroxamic esters is preferred.

$$RCO_2Ag + Br_2 \longrightarrow RCO_2Br + AgBr \quad (19.16)$$
$$RCO_2Br \longrightarrow RCO_2^\bullet + Br^\bullet \quad (19.17)$$
$$RCO_2^\bullet \longrightarrow R^\bullet + CO_2 \quad (19.18)$$
$$R^\bullet + RCO_2Br \longrightarrow R-Br + RCO_2^\bullet \quad (19.19)$$

Other oxidizing salts of heavy metals can be used; for example, $Pb(OAc)_4$ with carboxylic acids, in the presence of the appropriate lithium halides, gives halides.

$$RCO_2H \xrightarrow{Pb(OAc)_4} R^\bullet + XLi \xrightarrow{-Li} R-X \qquad (19.20)$$

19-2-3 Reaction of thiohydroxamic esters

The photochemical or thermal reaction of hydroxamic esters, in particular those obtained from N-hydroxy-2-thiopyridone, has been developed by Barton. It is at present the most flexible and best adapted method for the transformation of a –COOH group into many other functions. The intermediate radical, $R^\bullet$, resulting from decarboxylation, can be oriented in various ways (substitution, addition, etc.). The overall addition-fragmentation-rearomatization process, which we have described above (Chap. 9), is generally energetically favorable. It can be applied to molecules where R has other functions (unsaturation, ether, ester, etc.) without these being modified. Scheme 19.1 gives the general principle of the reaction where the thiohydroxamic ester is the source of $R^\bullet$ radicals.

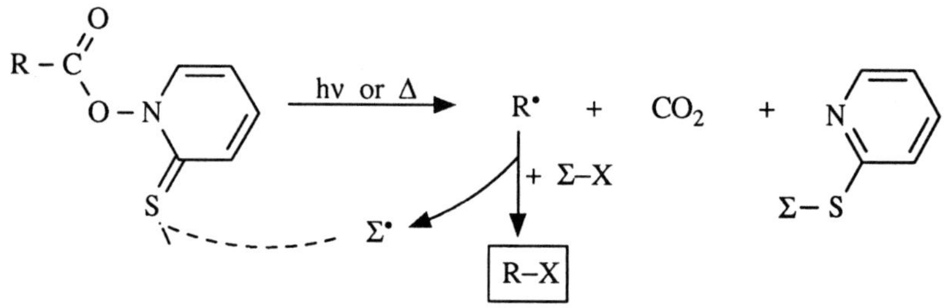

Scheme 19.1

The possible structures of R are very varied: unsaturated aliphatic (oleic or linoleic acid derivatives, for example), cyclohexyl, adamantyl, etc. By substitution, the radical $R^\bullet$ leads to various products, depending on the nature of $\Sigma-X$, in very good yields (Table 19.3).

$\Sigma-X$	R–X	yield %
n-Bu$_3$Sn–H or PhS–H	R–H	70 ~ 80
CHI$_2$–I	R–I	60 ~ 75
Cl$_3$C–Br	R–Br	72 ~ 98
Cl$_3$C–Cl	R–Cl	70 ~ 95
p-MeC$_6$H$_4$SO$_2$–CN or MeSO$_2$–CN	R–CN	65 ~ 95
p-MeC$_6$H$_4$SO$_2$–NCS or MeSO$_2$–NCS	R–SCN	60 ~ 90
PhSe–SePh	R–SePh	98

Table 19.3 – *Products and yields obtained by the Barton reaction (Scheme 19.1)*

The radical, R•, can also be trapped by oxygen. In the presence of a thiol as hydrogen-donor it is then transformed into hydroperoxide, ROOH (19.21). This reaction is particularly interesting, since it is difficult to prepare primary and secondary hydroperoxides in any other way.

$$R• \xrightarrow{O_2} ROO• \xrightarrow{t\text{-BuSH}} ROOH + t\text{-BuS}• \quad (19.21)$$

The transformation of –COOH into halide, nitrile, thiocyanate or selenide is not always easy by other methods than this. However, one limitation is the rate of decarboxylation of $RCO_2•$. This is very high when R is aliphatic ($\approx 10^9$ s^{-1}) but lower when R is aryl ($\approx 10^6$ s^{-1}). In the latter case either one has to increase the temperature (130°C instead of 80°C) or modify the conditions. For example, in the case where Σ–X is $CBrCl_3$, the addition of an external initiator such as AIBN makes it possible to increase the instantaneous concentration of •CCl_3 and therefore that of $RCO_2•$ and, consequently, of R•.

19-3 ALCOHOL DEOXYGENATION

Many natural products contain hydroxyl groups. In the synthesis of molecules by the lengthening of carbon chains the OH group is often introduced at an intermediate stage (Grignard reaction). It is therefore frequently necessary to replace the C–OH moiety by another group, commonly the transformation into C–H. The classical procedure is to reduce a tosylate by a hydride. This is, however, an expensive method which requires that other reducible groups be protected and which is difficult to apply to tertiary alcohols. It is easy to appreciate therefore why chemists have become very interested in the reactions described below which, moreover, constitute an excellent method for generating radicals.

The deoxygenation of alcohols (19.22) by the reaction of $Bu_3Sn•$ radicals with

$$ROH \longrightarrow RH \quad (19.22)$$

benzoate esters is an addition-fragmentation process (Scheme 19.2). Since the fragmentation rate depends on the stability of R• the yields are better for R = benzyl, triphenyl or cinnamyl than for R = n-Bu, Et, etc. In the case of xanthates, thioesters or

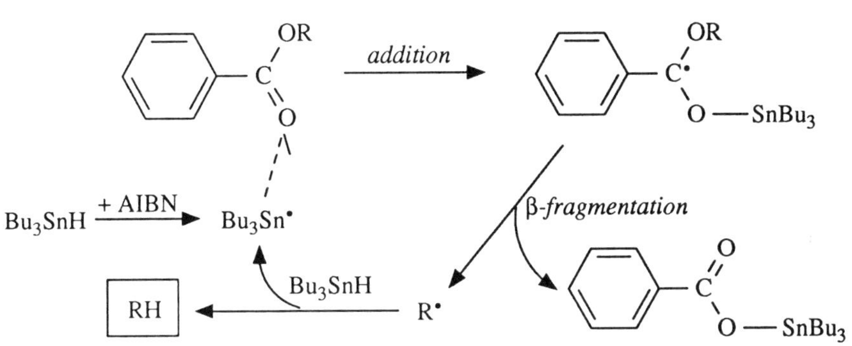

Scheme 19.2

hydroxamic esters addition takes place at the thiocarbonyl bond, C=S, instead of at C=O and, for this reason, gives better results. The starting material, which includes the RO fragment, is both the radical source and the reactant. As previously, the key step of the reaction is the addition-fragmentation process.

Primary alcohols

With primary alcohols one must either work at higher temperature in xylene or use the imidazole thiocarbonyl **1** or the thiobenzoate **2**.

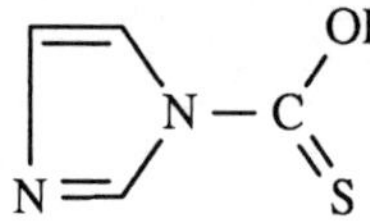

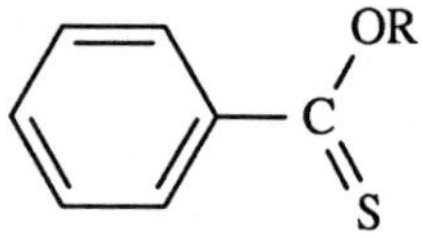

Scheme 19.3

The deoxygenation of primary and secondary alcohols can be improved by increasing the reactivity of the C=S double bond with radicals by using thiocarbonates **3** with substituted aryl groups such as 2,4,6-trichlorophenyl, pentafluorophenyl or 4-fluorophenyl.

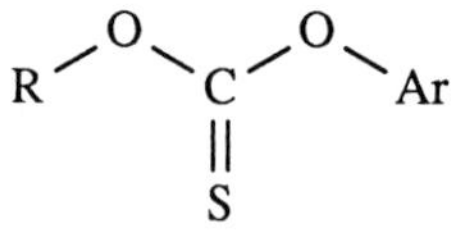

Scheme 19.4

Secondary alcohols

Secondary alcohol dithiocarbonates (xanthates) are frequently used in solution at 80°C (Barton-McCombie reaction) (Scheme 19.5).

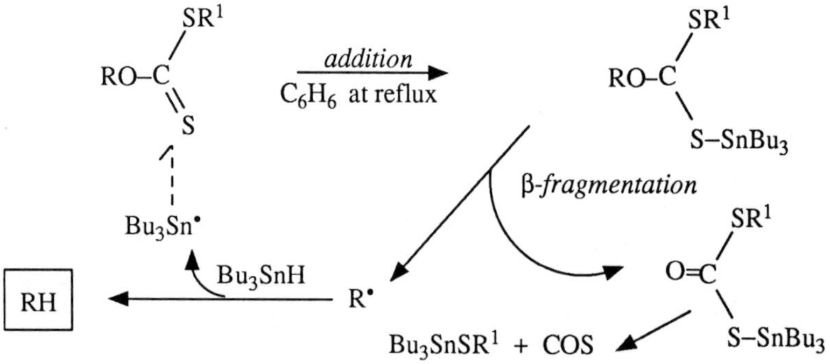

Scheme 19.5

Tertiary alcohols

Since tertiary alcohol dithiocarbonates are often thermally unstable, mixed oxalic-thiohydroxamic esters are used in the same way as for the decarboxylation of acids (Scheme 19.1). The reaction is run at 80°C in benzene, in the presence of t-BuSH or Bu_3SnH as hydrogen-donor (Scheme 19.6).

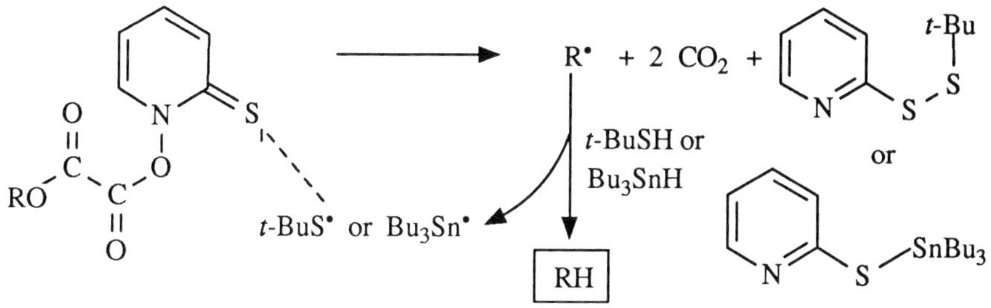

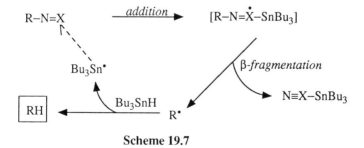

Scheme 19.6

In general, the yields for the conversion of alcohol to RH are good (60-95%). As previously, the method can be applied to molecules with other functional groups, in particular to glucosides and aminoglucosides.

Because of its toxicity, Bu_3SnH, as mentioned above in section 19-1, can be replaced by Ph_2SiH_2 or $(Me_3Si)_3SiH$, or even by dimethyl or diethyl phosphite in the presence of benzoyl peroxide as initiator. The reaction temperature is lowered by using Et_3B in the presence of controlled amounts of oxygen as the initiator system.

$$Et_3B \xrightarrow{O_2} Et^\bullet \qquad (19.23)$$

19-4 DEAMINATION

$$RNH_2 \longrightarrow RH \qquad (19.24)$$

This reaction is not always easy in ionic chemistry, particularly with aliphatics. To do it by a radical process requires that the amine be first converted into an isonitrile, thiocyanate or selenocyanate.

$$RNH_2 \longrightarrow \underset{X = C\overset{..}{:},\ C=S,\ C=Se}{R-N=X} \qquad (19.25)$$

Treatment of these R–N=X derivatives with Bu_3SnH in the presence of AIBN leads to the product, RH, in good yields (Scheme 19.7).

Scheme 19.7

Aromatic isonitriles do not react. Other functions such as OH, acetoxy, acetate, etc. are inert under the reaction conditions. This methodology has been applied to the chemistry of aminoacids, peptides and β-lactams (preparation of 6-alkylpenicillinates) with, moreover, very good stereoselectivity.

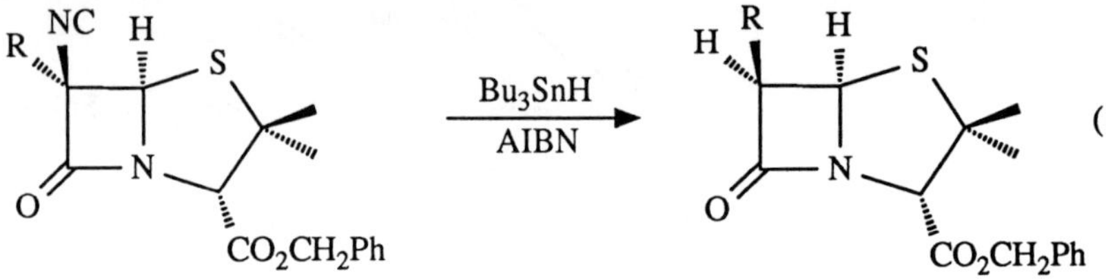

(19.26)

19-5 DENITRATION

$$RNO_2 \longrightarrow RH \quad (19.27)$$

Nitro derivatives are reduced, as above, by Bu_3SnH in the presence of AIBN in refluxing benzene or toluene, via the mechanism shown in Scheme 19.8.

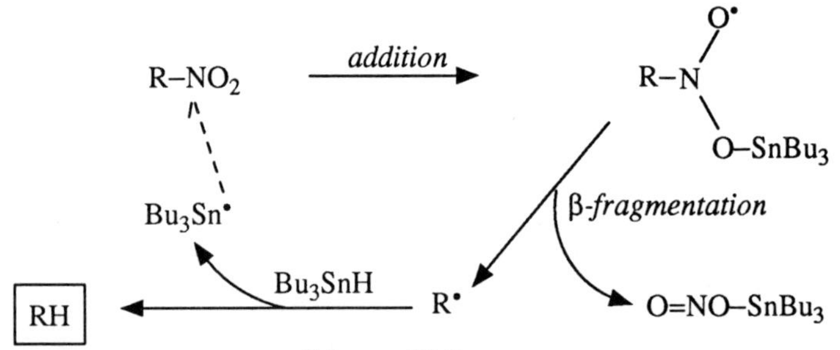

Scheme 19.8

Only secondary or tertiary nitro derivatives, or those with an activating substituent on the same carbon, such as benzylic, allylic, ketone or ester groups, give good yields of RH. The reaction is very selective and other functions such as keto, ester, cyano, chloro or sulfide are not affected.

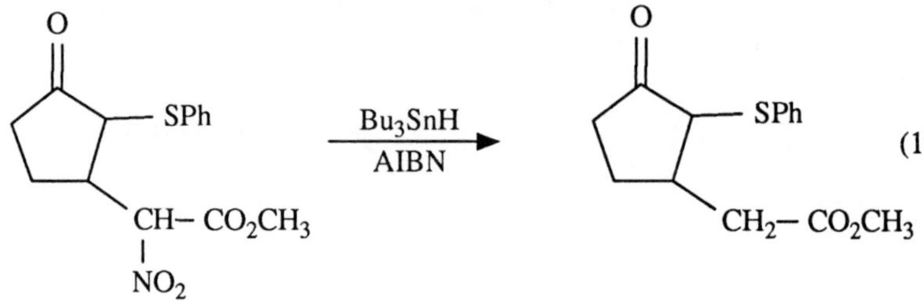

(19.28)

20 – FUNCTIONALIZATION OF MULTIPLE BONDS

The functionalization of double or triple bonds by radical addition (Chap. 12) is a very general method, which can be applied to many compounds and which allows the synthesis of organic molecules with C–C, C–N, C–S, C–P or C–M (M = Si, Ge or Sn) bonds.

$$A–X \; + \; \overset{\displaystyle \diagdown}{\underset{\displaystyle \diagup}{C}} = \overset{\displaystyle \diagup}{\underset{\displaystyle \diagdown}{C}} \; \longrightarrow \; \overset{A}{\underset{\diagup\;\;\diagdown}{C}} - \overset{\diagup\diagup}{\underset{X}{C}} \qquad (20.1)$$

$$A–X \; + \; —C\equiv C— \; \longrightarrow \; \overset{A}{\underset{\diagup}{C}} = \overset{}{\underset{X}{C}} \qquad (20.2)$$

20-1 KINETIC ASPECTS

Certain kinetic conditions must be satisfied in order to obtain a good yield of the addition product (1:1 adduct), ACCX, by a chain mechanism (20.1).

$$\text{propagation} \begin{cases} A^\bullet + C{=}C \xrightarrow{k_{add}} ACC^\bullet & (20.3) \\ ACC^\bullet + AX \xrightarrow{k_{trans}} ACCX + A^\bullet & (20.4) \end{cases}$$

$$\text{termination} \quad A^\bullet + A^\bullet \xrightarrow{k_{term}} \text{non-radical products} \qquad (20.5)$$

The rate of formation of ACCX is equal to the rate of propagation, v_{pro}, of the chain and, therefore, to the rates of addition (20.3) and transfer (20.4) (Chap. 7).

$$v_{pro} = k_{add} [A^\bullet][C{=}C] = k_{trans}[ACC^\bullet][AX] \qquad (20.6)$$
$$v_{term} = 2\, k_{term} [A^\bullet]^2 \qquad (20.7)$$

In order to have a good yield of ACCX, one must have a deficiency of the unsaturated reactant (then the termination reaction takes place between two $A^\bullet$ radicals) and have a rate of propagation larger than the rate of termination.

$$v_{pro} > v_{term} \;\Rightarrow\; \frac{v_{pro}}{v_{term}} = \frac{k_{add}\,[C{=}C]}{2\, k_{term}\,[A^\bullet]} > 1 \qquad (20.8)$$

In general, k_{term} is of the order of 10^9 to 10^{10} M^{-1}s^{-1} for radical recombination in the liquid phase. The concentrations of C=C and AX are determined by the experimental conditions; the ratio [C=C]/[AX] is usually about 1/10. The radical concentrations in chain reactions depend on the initiation and termination steps (Chap. 7) and are about 10^{-7} to 10^{-8} M. Since [C=C] is generally about 1 M, to get $v_{pro} > v_{term}$, k_{add} must be greater than 10^2 M^{-1}s^{-1}. Only systems satisfying these criteria can be used in synthesis.

20-2 ADDITION OF CARBON RADICALS: C–C BOND FORMATION

$$\begin{array}{c}\diagdown \\ C-X \\ \diagup\end{array} + \begin{array}{c}\diagdown \quad \diagup \\ C=C \\ \diagup \quad \diagdown\end{array} \xrightarrow{X = H \text{ or halogen}} \begin{array}{c} -C \\ \diagdown \\ C-C \\ \diagup \quad \diagdown \\ \quad\quad X \end{array} \qquad (20.9)$$

The synthetic procedure is very important; it depends on the nature of the reactants and the target. The structure of the products obtained and the yields can vary greatly depending on the method of initiation and the reaction conditions.

Two very general methods can be distinguished, depending on whether the addition step is coupled with:

— *direct transfer* of hydrogen or halogen;
— the use of a *mediator*, a direct or indirect radical source.

20-2-1 Direct atom transfer

Many radical additions involve *hydrogen atom transfer* (Scheme 20.1, X = H).

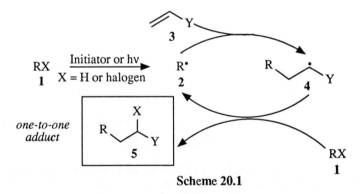

Scheme 20.1

The reaction is initiated either photochemically or by the addition of small amounts of an initiator, AIBN or peroxide (Chap. 9). Optimization of the yield depends on several factors: the reactants, molar ratios, temperature, addition mode, solvent, etc. The reaction works best if each step of the chain is exothermic and polar effects are favorable. The many investigations of the mechanisms and the kinetics of radical reactions provide detailed information about:

- the role and the efficiency of chemical initiators, the choice of initiator depending on the reactants and the conditions;
- the bond dissociation energies, which can be used to calculate the enthalpy of reaction;
- the absolute rate constants for addition and transfer reactions.

The first step (addition of a radical to olefin) is generally problem-free. On the other hand, for the transfer reaction to be efficient the C–X bond broken must not be stronger than the new bond. Thus, when X = H the C–H bond must be weakened by a substituent effect (Chap. 4). This is the case with aldehyde C–H bonds, or C–H bonds β to ketone, ester, amide, amine or alcohol functions. Some examples of the addition of such compounds to terminal olefins are given in Table 20.1 and to 1,2-disubstituted or cyclic olefins in Table 20.2.

R•	–Y in 3	initiator[a]	yield[b]
$CH_3\overset{\bullet}{C}=O$	$-(CH_2)_5CH_3$	hν (40°C)	64
cyclohexanone radical (=O)	$-(CH_2)_7CH_3$	TBP (130°C)	75
cyclopentanone radical (=O)	$-(CH_2)_7CH_3$	TBP (130°C)	80
$CH_3CH_2\overset{\bullet}{C}HOH$	$-(CH_2)_5CH_3$	TBP (115°C)	46
$CH_3\overset{\bullet}{C}HOH$	$-(CH_2)_5CH_3$	TBP (130°C)	28
$\overset{\bullet}{C}H_2CO_2H$	$-(CH_2)_5CH_3$	TBP (120°C)	60 ~ 70
$Cl\overset{\bullet}{C}HCO_2H$	$-(CH_2)_8CO_2Me$	TBH (120°C)	30 ~ 60
$\overset{\bullet}{C}H(CO_2Et)_2$	$-(CH_2)_3CH_3$	TBH (120°C)	56
$CH_3\overset{\bullet}{C}HNHC(O)CH_3$	$-CO_2Me$	TBP (155°C)	70
pyrrolidinyl radical (N)	$-(CH_2)_5CH_3$	TBP (120°C)	70

Table 20.1 – Addition of carbon radicals R•, obtained from R–H, to terminal olefins 3 (Scheme 20.1). a) TPB: di-tert-butyl peroxide, TBH: tert-butyl hydroperoxide. b) Yield of 5 in % with respect to olefin 3.

R•	olefin	initiator[a]	yield[b]
$CH_3(CH_2)_2\overset{\bullet}{C}=O$	dimethyl maleate	BP (82°C)	84
$CH_3CH_2\overset{\bullet}{C}HOH$	maleic acid	TBP (115°C)	96
$\overset{\bullet}{C}H(CO_2Et)_2$	cyclohexene	TBP (105°C)	13

Table 20.2 – Addition of carbon radicals, R•, obtained from RH, to 1,2-disubstituted and cyclic olefins. a) BP: benzoyl peroxide, TBP: di-tert-butyl peroxide. b) Yield in % with respect to olefin.

Polyhalomethanes generally give good yields of addition products (Table 20.3), in the order:

$$CBr_4 \approx BrCCl_3 > CCl_4 \gg HCCl_3 \qquad (20.10)$$

The C–H bond strength of chloroform, which is greater than that of the C–Br or C–Cl bonds of the other compounds, explains the very low yield, the transfer reaction being more difficult.

X–R	–Y in 3	initiator[a]	yield[b]
Br–CH$_2$CO$_2$CH$_3$	–(CH$_2$)$_8$CO$_2$Me	BP (90°C)	88
Cl–CCl$_3$	–(CH$_2$)$_4$CH$_3$	BP (80°C)	80
Br–CBr$_3$	–Ph	hν (80°C)	96
Br–CCl$_3$	–CH$_3$	hν or AP (80°C)	95
H–CCl$_3$	–(CH$_2$)$_5$CH$_3$	BP (80°C)	22

Table 20.3 – *Addition of carbon radicals, R°, to terminal olefin 3 (Scheme 20.1, X=H). a) BP: benzoyl peroxide, AP: acetyl peroxide. b) Yield in % with respect to olefin 3.*

The bond energies are also responsible for the different chemoselectivities for H–C–Br (Br transfer, cf. line 1 of Table 20.3) and for H–C–Cl (hydrogen transfer, cf. last line of Table 20.3). In the second case the chemoselectivity changes on going from a peroxide (hydrogen transfer) to a catalytic redox system (Cl transfer).

$$HCCl_3 + CH_2=CH-R \xrightarrow{\text{peroxide}} Cl_3C-CH_2-CH_2-R \qquad (20.11)$$

$$\xrightarrow[M = Cu^{2+}, Fe^{3+}, Ru^{2+}]{MCl_n} Cl_2CH-CH_2-CHCl-R \qquad (20.12)$$

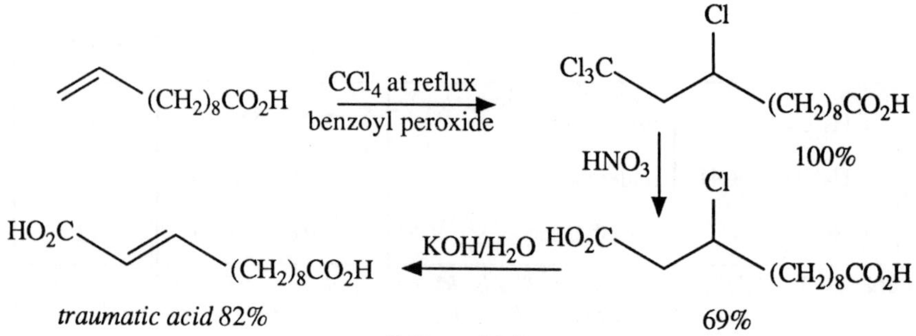

Scheme 20.2

Polyhalo compounds are interesting starting materials for the synthesis of other more complex molecules. Thus, the plant hormone, traumatic acid, is prepared from undecylenic acid, an industrial product (Scheme 20.2).

20-2-2 Mediators

Two mediators are frequently employed, tri-*n*-butyltin hydride and mercury salts. The mechanisms involved and the applications are somewhat different.

Tri-n-butyltin hydride

The reaction is initiated photochemically or by AIBN at 80°C. The chain mechanism develops as shown in Scheme 20.3. RX can be a halide, a selenide or a sulfide.

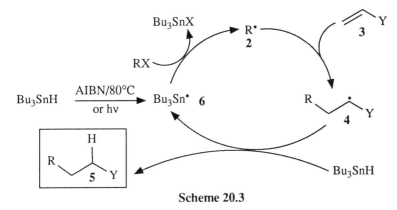

Scheme 20.3

Three radicals, **2**, **4** and **6**, are necessary for the formation of adduct **5**. To obtain a good yield of **5** each radical must have a well defined chemoselectivity: **2** must add to **3**, while **4** and **6** must react with Bu$_3$SnH and RX, respectively. The rate constants and the polar effects in these different reactions are known. This makes it possible to choose the reactants and the experimental conditions in order to control the competitive systems as required.

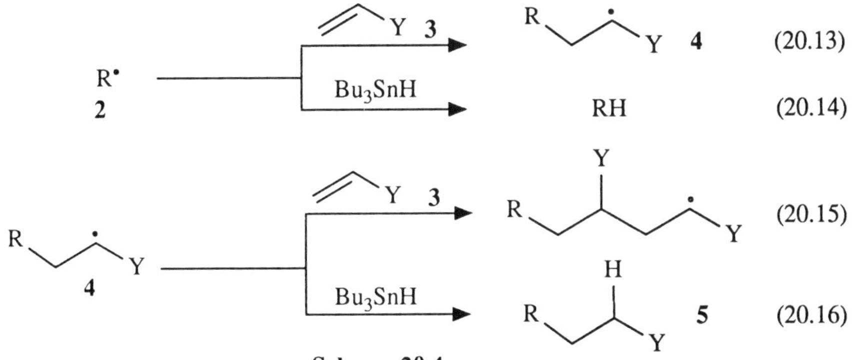

Scheme 20.4

Radicals **2** and **4** can undergo the competing reactions which are shown in

R–X	alkene	initiator	product	yield[a]
$n\text{-}C_6H_{13}I$	$\diagup\!\!\diagdown\text{CN}$	hν	$n\text{-}C_6H_{13}\diagdown\!\!\diagup\!\!\diagdown\text{CN}$	80
$t\text{-BuBr}$	$\diagup\!\!\diagdown\text{CN}$	AIBN	$t\text{-Bu}\diagdown\!\!\diagup\!\!\diagdown\text{CN}$	98
cyclohexyl–I	$\diagup\!\!\diagdown\text{CHO}$	AIBN	cyclohexyl–CH_2CH_2CHO	90
cyclohexyl–I	$\diagdown\!\!\diagup\text{CN}$	hν	cyclohexyl–CH(CH$_3$)CN	86
tetra-O-acetyl glucopyranosyl bromide	$\diagup\!\!\diagdown\text{CN}$	AIBN	tetra-O-acetyl glucopyranosyl–CH_2CH_2CN	70
tetra-O-acetyl glucopyranosyl bromide	$\diagup\!\!\diagdown\text{CN}$	AIBN	tetra-O-acetyl glucopyranosyl–CH_2CH_2CN	68
$(EtO_2C)_2CHCl$	$\diagup\!\!\diagdown Ot\text{-Bu}$	hν	$(EtO_2C)_2CH\diagdown\!\!\diagup\!\!\diagdown Ot\text{-Bu}$	60
$NH_2\text{-}C_6H_4\text{-I}$	$\diagup\!\!\diagdown\text{CN}$	hν	$NH_2\text{-}C_6H_4\text{-}CH_2CH_2\text{CN}$	50

Table 20.4 – *Mediated addition of Bu_3SnH (Scheme 20.3).*
a) Yield in % with respect to olefin.

Scheme 20.4. The substituent, Y, determines if the polar effects are compatible with addition of R• (20.13) and incompatible with that of **4** (20.15). Excess of olefin **3** favors reaction (20.13) but, unfortunately, also reaction (20.15). The instantaneous concentration of Bu_3SnH can be kept low, either by slow addition during the reaction or by *in situ* regeneration, an equimolar amount of $NaBH_4$ being added to a tri-*n*-butyltin halide.

$$Bu_3Sn\text{–}X \xrightarrow{NaBH_4} Bu_3Sn\text{–}H \qquad (20.17)$$

The $Bu_3Sn•$ radical has two possibilities, (20.18) and (20.19). It is essential to enhance reaction (20.19) since it regenerates R•. For this purpose, the iodide is preferred because it is 10-100 times as reactive as the bromide. With chlorides reaction (20.19) is generally too slow.

$$Bu_3Sn• \begin{array}{c} \xrightarrow{\diagup\!\!\diagdown Y\ \mathbf{3}} Bu_3Sn\diagdown\!\!\diagup\!\!\diagdown Y\quad (20.18) \\ \xrightarrow{R\text{–}X} R• + Bu_3SnX \quad (20.19) \\ \mathbf{2} \end{array}$$

Some examples of the very many radical addition reactions performed by this procedure are given in Table 20.4. The experimental conditions are particularly mild and this method can be applied to polyfunctional compounds such as glucosides (Table 20.4, lines 5 and 6). Aryl (line 8) and vinyl radicals, which are not accessible by the *direct transfer* method, can also be generated.

Since tin compounds are toxic and difficult to eliminate, Bu_3SnH can be replaced by $(Me_3Si)_3SiH$ in which the Si–H bond strength is about the same as Sn–H.

Instead of halo compounds one can use phenylselenides or phenylsulfides (Chap. 9),

$$Bu_3Sn^\bullet + \text{R–SePh or R–SPh} \longrightarrow R^\bullet + Bu_3SnSePh \text{ or } Bu_3SnSPh \quad (20.20)$$

the reactivity order being:

$$I > Br > PhSe > Cl > PhS \quad (20.21)$$

Organomercury mediators

An alkylmercury halide or acetate, reduced by a hydride such as $NaBH_4$ or Bu_3SnH, gives an organomercury hydride which is an excellent hydrogen-donor and a source of radicals (Scheme 20.5).

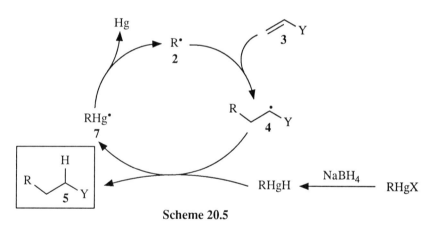

Scheme 20.5

This process is formally analogous to *direct hydrogen transfer* (Scheme 20.1) in which an extra fragmentation step is included. The main differences from the previous methods are as follows:

- since RHgH is a very good hydrogen-donor it will only work for olefins (acetonitrile, vinyl ketones, acrylates, etc.) which are very reactive towards nucleophilic radicals;
- the reaction is very fast and takes only a few minutes under very mild conditions: normal temperature, in the dark.

The organomercury hydride can be obtained by hydroboration of a double bond (20.22)

$$\ce{>C=CH2} \xrightarrow[\text{2) Hg(OAc)}_2]{\text{1) B}_2\text{H}_6} \ce{>C(H)-CH2HgOAc} \xrightarrow{\text{3) NaBH}_4} \ce{>C(H)-CH2HgH} \quad (20.22)$$

or by solvomercuration (20.23).

$$\ce{>C=CH2} \xrightarrow[\text{in AH}]{\text{Hg(OAc)}_2} \ce{>C(A)-CH2HgOAc} \xrightarrow{\text{NaBH}_4} \ce{>C(A)-CH2HgH} \quad (20.23)$$

A = OH, OR, OAc, NHAc

By this means it is possible to reductively couple an electron-rich terminal double bond with an electron-poor double bond in one step.

$$\ce{>C1=C2} + \ce{C3=C4-Y} \xrightarrow[\text{3) NaBH}_4]{\text{1) B}_2\text{H}_6 \text{ 2) Hg(OAc)}_2} \ce{C1-C2-C3-C4-Y} \quad (20.24)$$

The intermediate radical, R•, bears either a hydrogen or a functional group, A, depending on whether the mercury derivative is obtained by hydroboration (20.22) or solvomercuration (20.23). There is therefore a wide range of synthetic possibilities with interesting regioselectivity (Table 20.5).

olefin[a]	olefin	product	yield
$\ce{CH3-C(C2H5)=CH2}$	$\ce{CH2=CH-CN}$	$\ce{CH3-CH(C2H5)(CH2)3CN}$	65
cy-C_6H_{11}–CH=CH_2	$\ce{CH2=CH-CN}$	cy-$C_6H_{11}(CH_2)_4$CN	54
cyclopentene	$\ce{CH2=CH-CO2Et}$	cyclopentyl-OCH$_3$, (CH$_2$)$_2$CO$_2$Et	60[b]

Table 20.5 – *Organomercury-mediated addition (20.24). a) Olefin precursor of the organometal hydride. That of the last line is obtained by solvomercuration in methanol (20.23). b) trans/cis = 88/12.*

20-3 ADDITION OF AMINYL GROUPS: C–N BOND FORMATION

Protonated aminyl radicals, obtained from N-chloroamines in acid, in the dark at 30°C add very easily to double bonds to form a C–N bond by a chain process (Scheme 20.6). Neutral aminyl radicals give little or no radical addition.

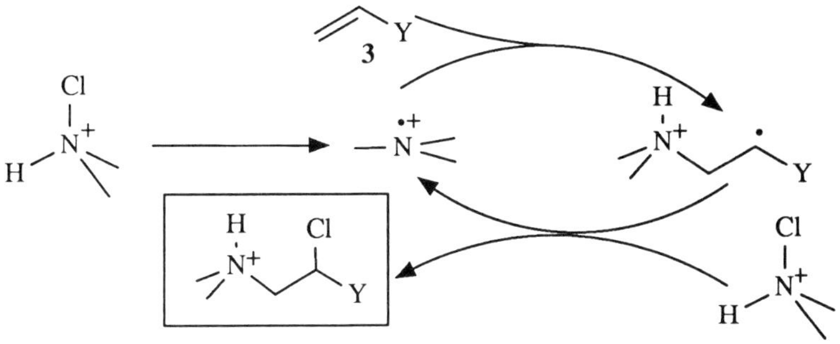

Addition of aminyl groups: C–N bond formation

Scheme 20.6

Radical aminochlorination works well with conjugated and slightly deactivated olefins (Table 20.6). If the substituent, Y, is strongly electron-attracting (CN, COOR, COR) there is no reaction, because the protonated aminyl radical is electrophilic.

N-chloroamine	olefin	product	yield
$(n\text{-}C_4H_9)_2NCl$	butadiene	$(n\text{-}C_4H_9)_2NCH_2CH=CHCH_2Cl$	60
$(C_2H_5)_2NCl$	 cyclooctadiene	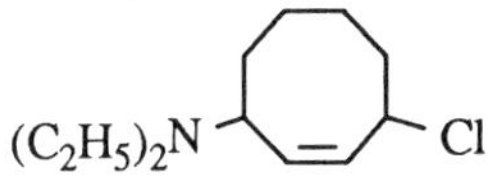 $(C_2H_5)_2N\text{—}\text{—}Cl$	68
$(C_2H_5)_2NCl$	$CH_2=CHCl$	$(C_2H_5)_2NCH_2CHCl_2$	82
 piperidinyl N–Cl	$CH_2=C(Cl)CH_3$	 piperidinyl N–$CH_2CCl_2CH_3$	92
 piperidinyl N–Cl	$CH_2=CHCH_2OEt$	 piperidinyl N–$CH_2CHClCH_2OEt$	88

Table 20.6 – *Addition of N-chloroamines to olefins (4M H_2SO_4 in acetic acid at 30°C under nitrogen).*

A serious limitation of this process is competition with electrophilic chlorination. To avoid this one uses a redox system, the reaction with a ferrous salt, for example, being very rapid (Scheme 20.7).

Under such conditions the addition of N-chloroamines to cyclohexene in neutral media gives mainly the *cis* isomer. This stereoselectivity is attributed to the coordination complex of the *non*-protonated amino group with the ferric salt responsible for chlorine atom transfer (Scheme 20.8). The protonated amino radical cannot form such a complex. In acid, therefore, addition leads to a mixture of *cis* and *trans* stereoisomers.

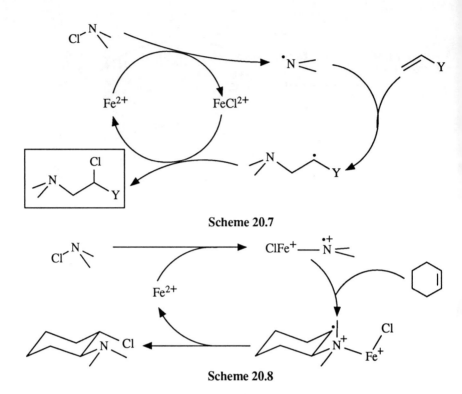

Scheme 20.7

Scheme 20.8

20-4 ADDITION OF SULFUR- AND PHOSPHORUS-CENTERED RADICALS

Compounds with an S–H bond such as thiols or thioacids, with low dissociation energies, are particularly suitable for the direct transfer process. The addition of RS• is often reversible; as transfer is very efficient the system proceeds easily to the one-to-one adduct (Scheme 20.9). In this way compounds with a C–S bond are obtained in very good yields (Table 20.7, first 5 lines).

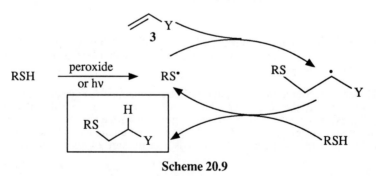

Scheme 20.9

The addition of phosphorus compounds, various phosphines, phosphorus trichloride, dialkyl phosphite, etc. also gives addition products with a C–P bond. These reactions are much used (Table 20.7, last 3 lines).

reactant	olefin	initiator[a]	product	yield[b]
C_2H_5SH	1-octene	BP + hν	$C_2H_5S-C_8H_{17}$	75
C_2H_5SH	$HC\equiv C-CO_2H$	BP + hν	$C_2H_5S-CH=CHCO_2H$	74
$CH_3C(O)SH$	Ph\C=CH₂ / CH₃	BP	Ph\CHCH₂-SC(O)CH₃ / CH₃	90
$Cl_3CC(O)SH$	cyclohexene	BP	cyclohexyl-S-C(=O)-CCl₃	97
PhSH	1-butene	BP	$PhS-C_4H_9$	73
PH_3	1-octene	BP	$H_2P-C_8H_{17}$	50
PCl_3	1-octene	BP	$Cl_2P-CH_2CHClC_6H_{13}$	60
$(MeO)_2P(O)H$	1-hexene	γ	$(MeO)_2P(O)-C_6H_{13}$	85

Table 20.7 – *Addition of sulfur and phosphorus compounds to olefins. Formation of C–S and C–P bonds. a) BP: benzoyl peroxide. b) Yield in % with respect to olefin.*

olefin	hydride	initiator[a]	product	yield[b]
$CH_2=CH(CH_2)_2CH_3$	Cl_3SiH	TBP	$Cl_3Si(CH_2)_4CH_3$	82
$CH_2=CHCH_2OEt$	Cl_3SiH	TBP	$Cl_3Si(CH_2)_3OEt$	84
$CH_2=CH(CH_2)_5CH_3$	$PhSiH_3$	TBPB	$PhSiH_2(CH_2)_7CH_3$	70
$CH_2=CH(CH_2)_8CHO$	Ph_3GeH	AIBN	$Ph_3Ge(CH_2)_{10}CHO$	80
$CH_2=CHCN$	Ph_3GeH	hν	$Ph_3Ge(CH_2)_2CN$	83
$CH_2=CH(CH_2)_2CH_3$	Ph_3GeH	hν	$Ph_3Ge(CH_2)_4CH_3$	91
$CH_2=CHCN$	Ph_3SnH	AIBN	$Ph_3Sn(CH_2)_2CN$	98
$CH_2=CHCH_2CH(CO_2Et)_2$	Ph_3SnH	AIBN	$Ph_3Sn(CH_2)_3CH(CO_2Et)_2$	90

Table 20.8 – *Hydrometallation of ethylenic compounds. a) TBP: di-tert-butyl peroxide; TBPB: tert-butyl perbenzoate. b) Yield in % with respect to olefin.*

20-5 ADDITION OF GROUP 14 RADICALS: C–M BOND FORMATION

In the presence of AIBN, a peroxide or light, the hydrides of Si, Ge and Sn add to olefins (20.25) by a chain process involving direct hydrogen transfer; good yields of products with a C–M bond are obtained (Table 20.8).

$$R_3MH \; + \; \diagup\!\!\!\diagdown Y \xrightarrow[\text{AIBN at 80°C or hv}]{\text{peroxide or}} R_3M\diagdown\!\!\diagup\!\!\!\diagdown Y \qquad (20.25)$$

The enthalpy of reaction (20.25) is –18, 3 and 5 kcal/mol for M = Si, Ge and Sn, respectively. As in the case of thiols, the addition of stannyl radicals is reversible and hydrogen transfer very easy; the yield of addition product is therefore very good. For silyl radicals there may be competition with addition to the oxygen of a carbonyl when the unsaturated compound also contains this function.

Radical chemistry is doubtless one of the easiest means of obtaining organometallic compounds.

R–X	conditions	product	yield
PhI	AIBN 80°C toluene	Ph-allyl	60
cyclohexyl-I with acetal	hv 20°C benzene	cyclohexyl-allyl with acetal	81
sugar-Br (RO, OH, OBz, CH$_3$O)	AIBN 80°C toluene	sugar-allyl	92
cyclopentyl-Br with OH	hv 20°C benzene	cyclopentyl-allyl with OH	67
		trans/cis = 10/1	
β-lactam-Br with S, Me, Me, CO$_2$Me	AIBN 80°C benzene	β-lactam-allyl	90

Table 20.9 – *Allylation (20.26). Yield in %.*

20-6 ADDITION-FRAGMENTATION: ALLYLATION REACTION

Grafting an *allyl* group is particularly useful in organic synthesis. It can be achieved by reacting a halide, R–Hal, a sulfide, R–SPh, or a selenide, R–SePh, with allyltri-*n*-butyltin in an addition-fragmentation process (20.26) (Scheme 20.10).

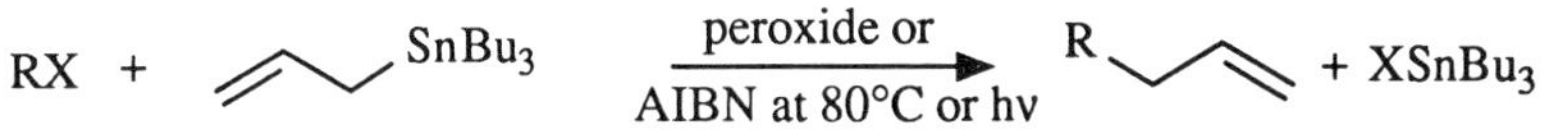

Addition-fragmentation: allylation reaction

$$RX + \underset{}{\diagdown\!\!\!\diagup\!\!\diagdown} SnBu_3 \xrightarrow[\text{AIBN at 80°C or hv}]{\text{peroxide or}} R\diagdown\!\!\!\diagup\!\!\diagdown + XSnBu_3 \quad (20.26)$$

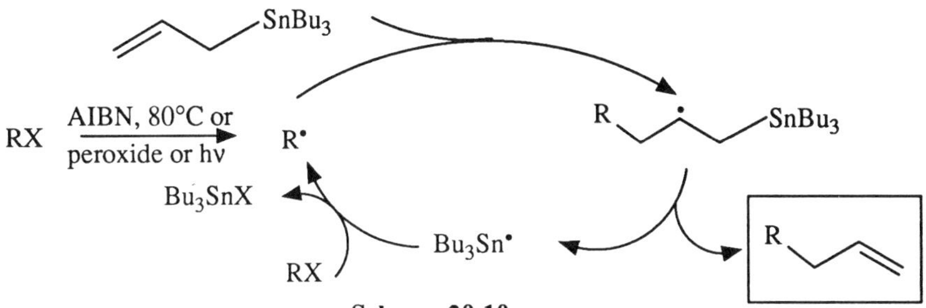

Scheme 20.10

The yields are generally acceptable, even with iodobenzene, and the stereoselectivity is often good. The allylstannane adds to the less hindered face of the molecule (Table 20.9). This reaction is compatible with the presence of other functional groups, such as acetal, ketal, ether, epoxide, lactone, ester or free OH. One can therefore use molecules of very varied structure.

21 - CYCLIZATION

Intramolecular radical additions, i.e. cyclization, have been very widely used in organic synthesis since the beginning of the 1980s. Many mechanistic studies (Chap. 13) have shown that a wide range of mono- and polycyclic products can be obtained regio- and stereoselectively: natural products, alkaloids, antibiotics, etc. Carbon, aminyl, oxyl radicals, etc. can be cyclized with different types of multiple bonds: C=C, C≡C, C=O and C≡N.

21-1 METHODOLOGY

The formation of a cyclized product by a radical process is described in Scheme 21.1.

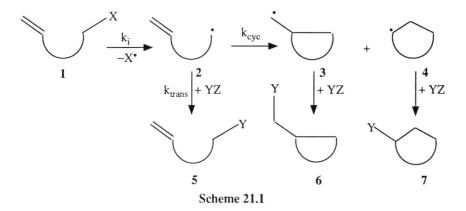

Scheme 21.1

To obtain a satisfactory yield of the cyclized product, **6** or **7**, it is necessary that the reaction rate be greater than that of transfer:

$$k_{cyc} [2] > k_{trans} [2] [YZ] \quad \Rightarrow \quad k_{cyc} > k_{trans} [YZ] \quad (21.1)$$

The nature and the concentration of YZ (Y = H or halogen) must be chosen in terms of k_{cyc}. Generally, the structure of **2** affects k_{cyc} more than k_{trans}. When k_{cyc} is relatively small one can avoid the initial radical **2** being trapped by a hydrogen-donor by using Bu_3GeH rather than Bu_3SnH, since the transfer rate constant for the former is $10 \sim 20$ times lower than that of the second. The value of k_{trans} is the same for the primary radicals, **2** and **3**, and little different for the secondary radical **4**. There is a large range of reactants, whose rate constants are known, making it possible to adapt the reaction conditions to different situations.

The choice of YZ depends also on the way in which radical **2** is generated from the precursor **1**, which itself is chosen for its ease of preparation and its ability to give **2**. The methods used are the same as those described in Chapter 20 for addition reactions.

21-2 CYCLIZATION OF CARBON RADICALS

The aim of carbon radical cyclization is the construction of 5- or 6-membered carbon rings or heterocycles which are the basic structural elements of many organic molecules.

21-2-1 Monocyclic compounds

Cyclization on a >C=C< double bond

$$\text{exo} \quad (21.2)$$

$$\text{endo} \quad (21.3)$$

The yield and the regio- and stereoselectivity depend on the substituents, R^1 to R^5, the length of the chain and the nature of X. When the double bond is δ to the radical center, cyclization goes essentially *exo* to give a 5-membered ring with *trans* stereoselectivity (21.5).

$$(21.4)$$

$$(21.5)$$

81% (*trans/cis* = 97/3)   98%

This reaction can be applied to sugars since the conditions are neutral and therefore compatible with most protecting groups. Very often, moreover, the OH groups do not have to be protected. In the following example, cyclization is stereoselective with the Z isomer but not with the E isomer, which gives an α/β ratio of 50/50.

Cyclization of carbon radicals

$$\text{(21.6)}$$

R = Ph–C=O   yield 89%   α/β = 91/9
R = H   yield 80%   α/β = 86/14

If the radical is stabilized (Chap. 13) cyclization becomes reversible and leads to the 6-membered ring *endo* compound, which is thermodynamically the more stable.

$$\xrightarrow[\text{48 h}]{\substack{\text{benzoyl peroxide}\\\text{cyclohexane at reflux}}}\quad 78\% \quad (21.7)$$

An iodide precursor is a better transfer agent than cyclohexane, which makes it possible to *trap* the kinetic product, the 5-membered ring *exo* compound.

$$\xrightarrow[\substack{\text{hv, 10 mn}\\\text{yield 86\%}}]{Bu_3Sn-SnBu_3} \quad exo/endo = 9/1 \qquad (21.8)$$

Cyclization on a –C≡C– bond

$$\xrightarrow{Bu_3SnH}\quad \approx 90\% \text{ for } R^1, R^2, R^3 = H \text{ or Me} \qquad (21.9)$$

Cyclization on a >C=O bond

Cyclization goes only on the carbon of the carbonyl double bond; it is therefore chemo- and regioselective.

$$\xrightarrow{Bu_3SnH}\quad 85\% \qquad (21.10)$$

Vinyl radicals

Vinyl radicals can be obtained by means of Bu_3SnH but, as in reaction (21.11), the ratio of 5- and 6-membered ring products depends on the concentration of the transfer agent, Bu_3SnH.

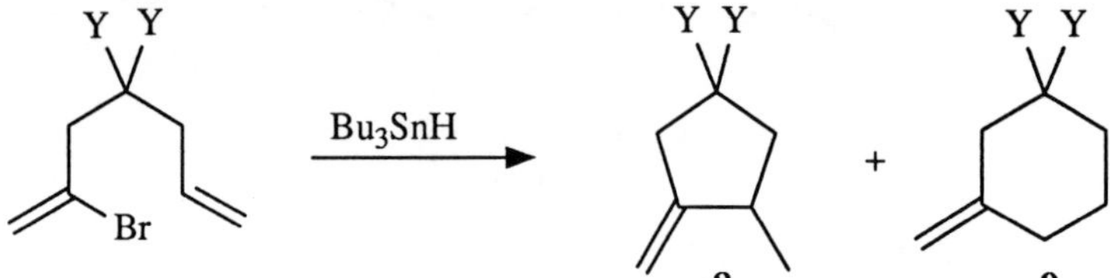
Y Y
Br
Bu_3SnH
Y Y
+
Y Y

(21.11)

8 **9**

8/**9** = 3/1 for $[Bu_3SnH] = 0.02$ M and 97/1 for $[Bu_3SnH] = 1.7$ M

When the tri-*n*-butyltin hydride concentration is low a certain amount of the 6-membered ring compound is obtained; this is the thermodynamically more stable and results from equilibrium between the radicals shown in (21.12).

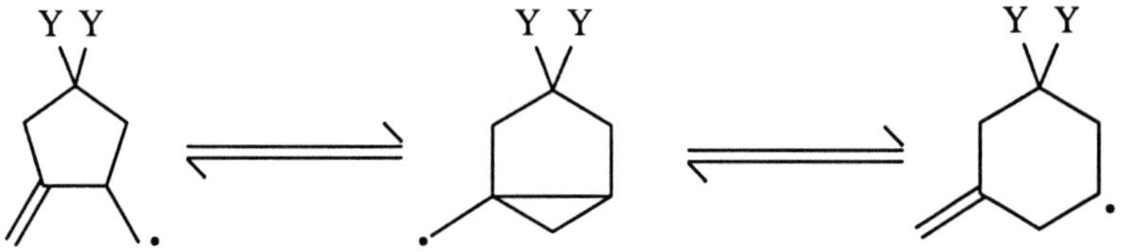
Y Y
Y Y
Y Y

(21.12)

Redox system

The use of a redox system instead of Bu_3SnH applies to halo compounds

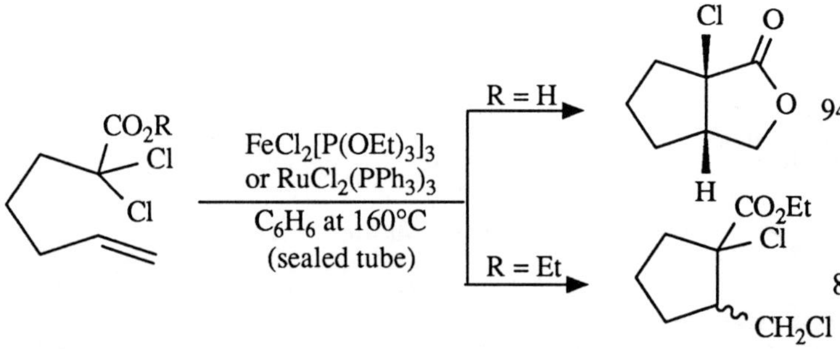
CO_2R
Cl
Cl
$FeCl_2[P(OEt)_3]_3$
or $RuCl_2(PPh_3)_3$
C_6H_6 at 160°C
(sealed tube)
R = H
R = Et
Cl
O
O
H
CO_2Et
Cl
CH_2Cl

94% (21.13)

80% (21.14)

or compounds with enolizable carbonyls.

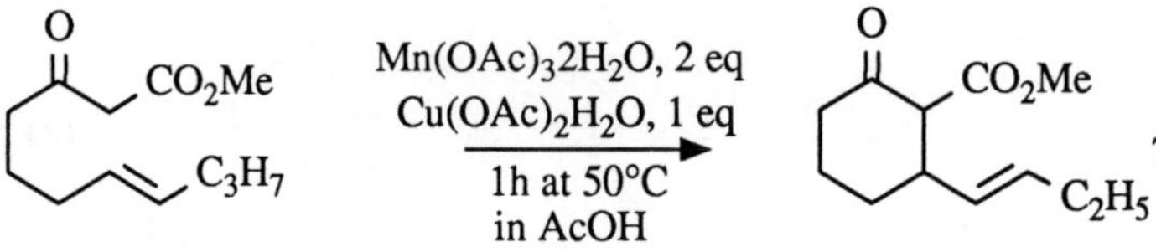
O
CO_2Me
C_3H_7
$Mn(OAc)_3 2H_2O$, 2 eq
$Cu(OAc)_2H_2O$, 1 eq
1h at 50°C
in AcOH
O
CO_2Me
C_2H_5

71% (21.15)

Macrocyclization

By working at high dilution (5 x 10^{-3} M iodide) macrocycles can be synthesized, but 10 ~ 12 carbon rings are much more difficult to obtain than those with 14 ~ 16 carbons.

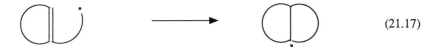
(21.16)

21-2-2 Bicyclic compounds

Two approaches can be used, depending on whether the unsaturated bond is:

– intracyclic and the radical center on a side-chain;

(21.17)

– extracyclic and the radical center on the ring.

(21.18)

Cyclization on an intracyclic double bond

The construction of 5-membered rings is the most common; for obvious geometric reasons the junction can only be *cis* (21.19)(21.20).

96% (21.19)

93% (21.20)

Lengthening the side-chain makes it possible to synthesize 6-membered rings, again *cis* (21.21).

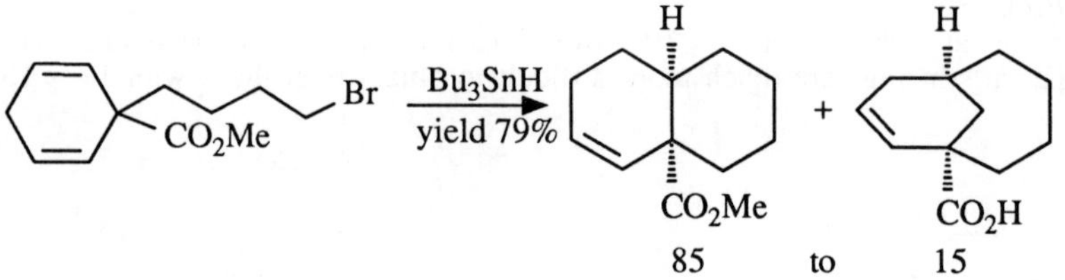

$$\text{(21.21)}$$

Cyclic allyl alcohols can be used to perform a synthesis where two or three substituents are introduced regio- and stereoselectively (Scheme 21.2). The reaction is carried out with a $Bu_3SnCl/NaBH_4$ mixture (see reaction (20.17)) to avoid trapping the radical intermediate by hydrogen transfer. Applied to cyclopentenol this makes it possible to prepare synthons which are potentially prostaglandin precursors.

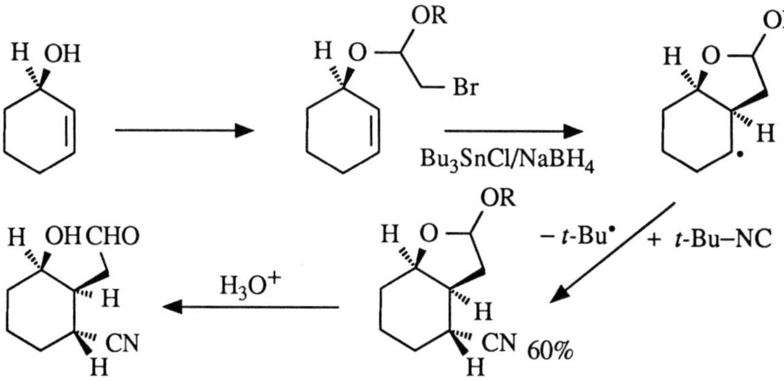

Scheme 21.2

The cyclization of a *carbonyl* leads, after β-fragmentation of the intermediate alkoxyl radical, to ring enlargement (Scheme 21.3). The presence of the ester function controls the direction of β-fragmentation of the alkoxyl radical. The yields are often over 75%, in particular when n = 0, and are better for iodide than for bromide. With n = 1 (4-membered ring) the reaction does not occur. This type of reaction is also possible with cyclohexanones.

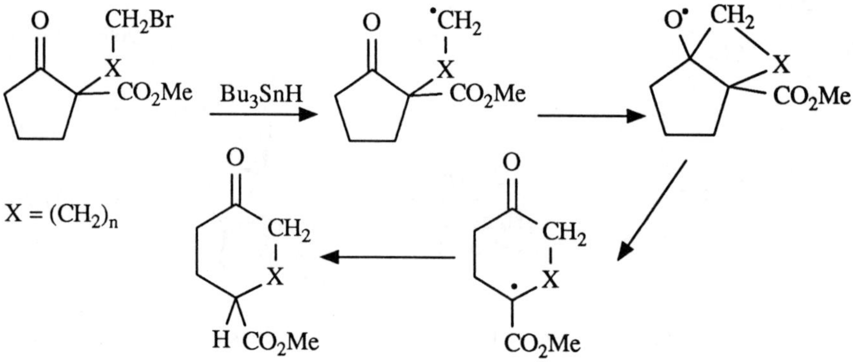

$X = (CH_2)_n$

Scheme 21.3

Cyclization on an extracyclic double or triple bond

When the radical center is part of a ring, cyclization on a δ double bond goes mainly *exo* and the junction is *cis* (21.22).

$$\text{[structure]} \xrightarrow[X = I, Br]{Bu_3SnH} \text{[structure]} \quad 60\% \quad (21.22)$$

A *trans* fusion can be obtained by fixing the stereochemistry of the substituent bearing the radical center (21.23).

$$\text{[structure]} \xrightarrow[X = I, Br]{Ph_3SnH} \text{[structure]} \quad (21.23)$$

Thus, *cis* (21.24) and *trans* (21.25) γ-lactones can be prepared by these two paths.

$$\text{[structure]} \xrightarrow[\text{yield } 63\%]{Bu_3SnH} \text{[structure]} + \text{[structure]} \quad (21.24)$$
$$\qquad\qquad\qquad\qquad\qquad 20 \quad\text{to}\quad 80$$

$$\text{[structure]} \xrightarrow[\text{yield } 90\%]{Bu_3SnH} \text{[structure]} \quad (21.25)$$

Cyclization on a triple bond in the 5 or 6 position, depending on the chain length, allows subsequent functionalization via the final double bond. The radical precursor can be an iodide (21.26), xanthate (21.27) or selenide (21.28). The corresponding ketones are obtained by oxidation.

$$\text{[structure]} \xrightarrow{(Bu_3Sn)_2} \text{[structure]} \quad 63\% \quad (21.26)$$

Many syntheses of alkaloid structures and compounds of biological interest start

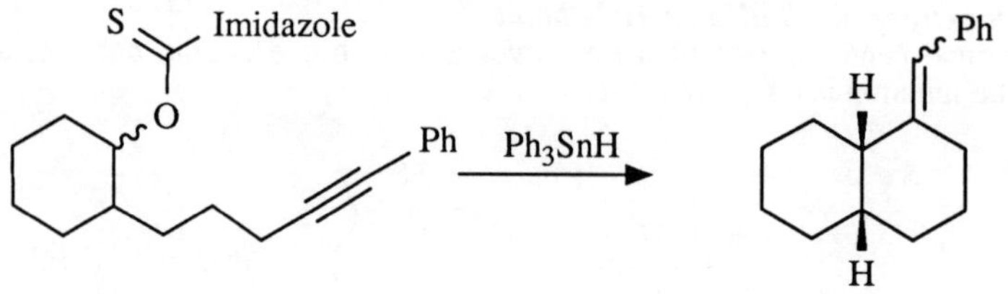

[Scheme (21.27): cyclization with Ph₃SnH, 79%]

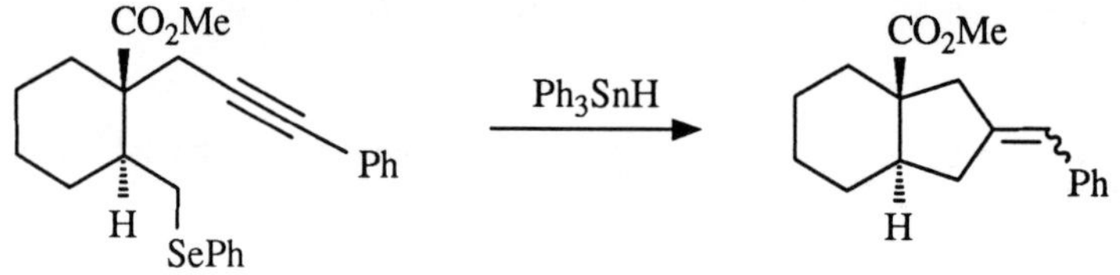

[Scheme (21.28): cyclization with Ph₃SnH, 96%]

with γ- or β-lactams. The regioselectivity of the cyclization of the N-chloromethyl-β-lactam in (21.30) depends markedly on the volume of the substituent, R, on the double bond.

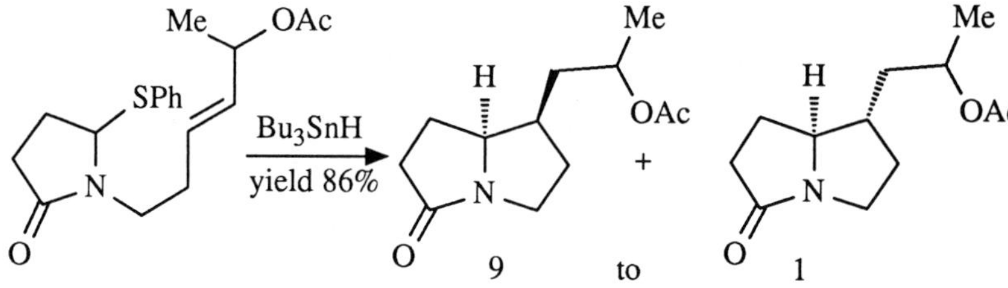

[Scheme (21.29): Bu₃SnH, yield 86%, ratio 9 to 1]

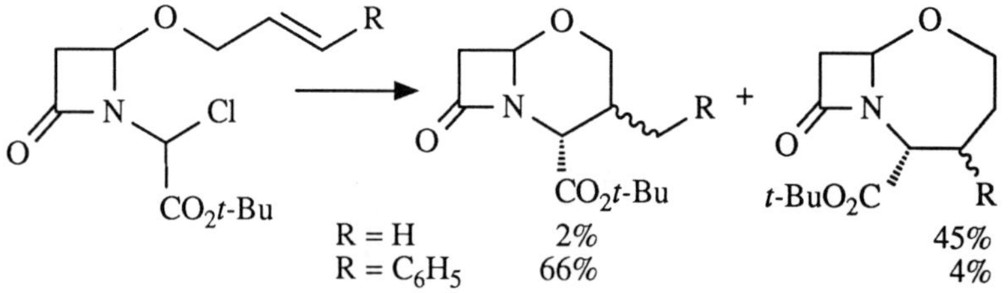

[Scheme (21.30):
R = H: 2%, 45%
R = C₆H₅: 66%, 4%]

In the sugar series cyclizations on carbonyl can be performed without any particular precautions regarding the other functional groups (21.31).

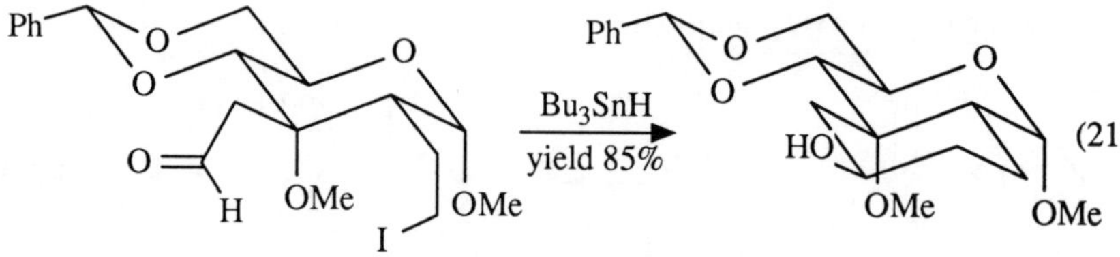

[Scheme (21.31): Bu₃SnH, yield 85%]

Bridgehead cyclization, difficult to perform by other means, can be achieved.

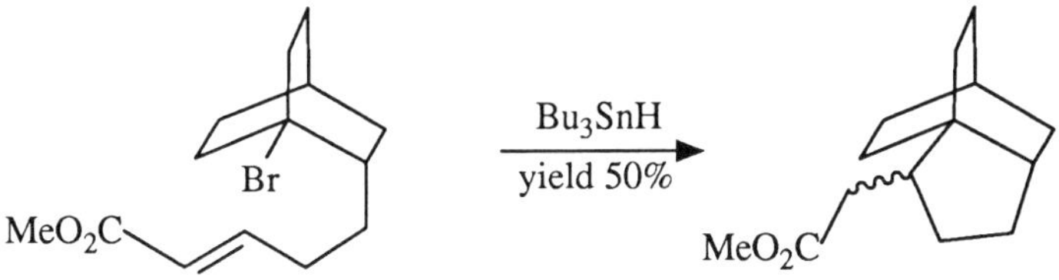

Cyclization of carbon radicals

$$\text{(21.32)}$$

Aryl radicals give indoles or substituted benzofurans in good yields.

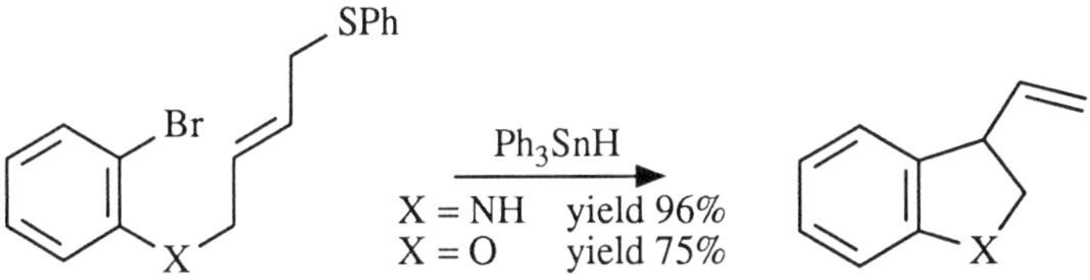

$$\text{(21.33)}$$

X = NH yield 96%
X = O   yield 75%

21-2-3 Tandem or sequential cyclization

When a molecule contains several unsaturated bonds in appropriate positions it is possible to obtain a polycyclic product by successive intramolecular reactions, often with very satisfactory regio- and stereoselectivity. These reactions are referred to as *tandem* or *sequential*. By this means capnellene (triquinane series) has been synthesized in good yield (21.35).

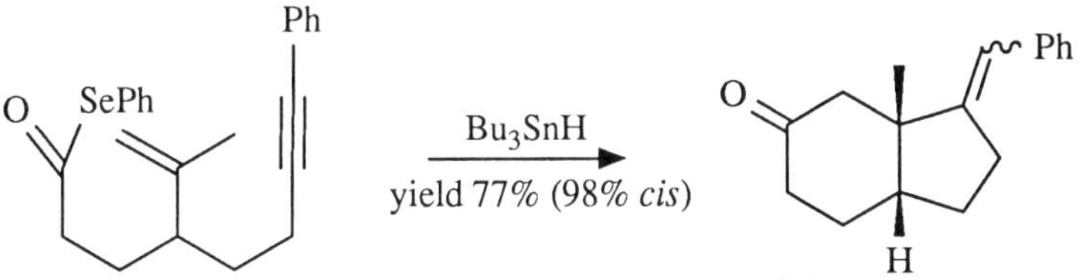

$$\text{(21.34)}$$

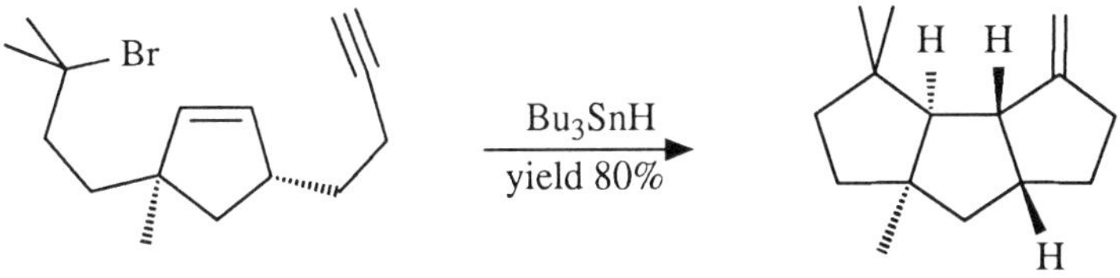

yield 77% (98% *cis*)

$$\text{(21.35)}$$

yield 80%

Cyclodecadiene, by *transannular* cyclization, gives a *cis*-decalin compound.

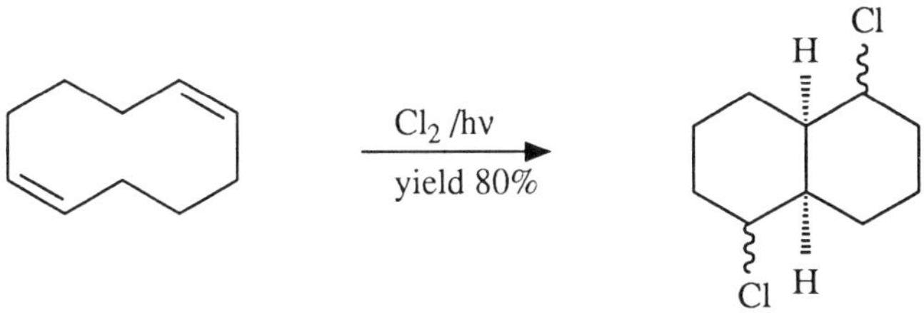

$$\text{(21.36)}$$

Cl_2 /hv yield 80%

The production of radicals from enolizable carbonyl compounds by a redox system gives good results. The last step involves oxidation of the radical either to an aromatic system (21.37)

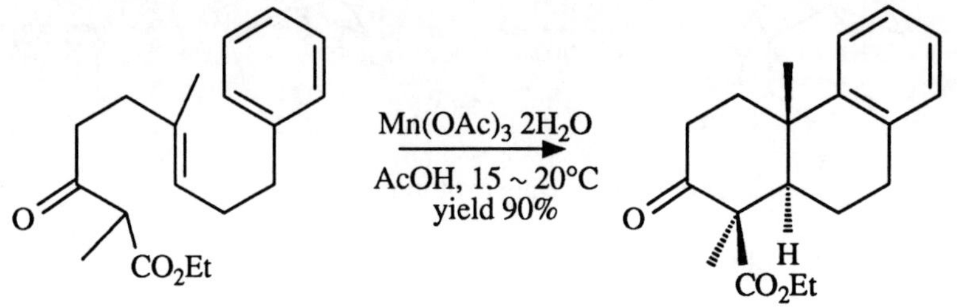

(21.37)

or to a double bond (21.38) by adding Cu(OAc)$_2$.

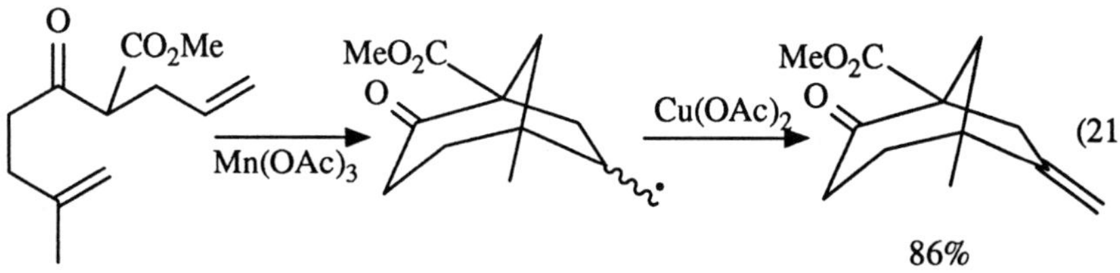

(21.38)

To illustrate the often remarkable selectivity of these sequential reactions, we shall cite the following example which gives only one product very stereoselectively despite the very different types of reaction involved: cyclization on >C=O, β-fragmentation of the alkoxyl radical and cyclization on –C≡C– (Scheme 20.4).

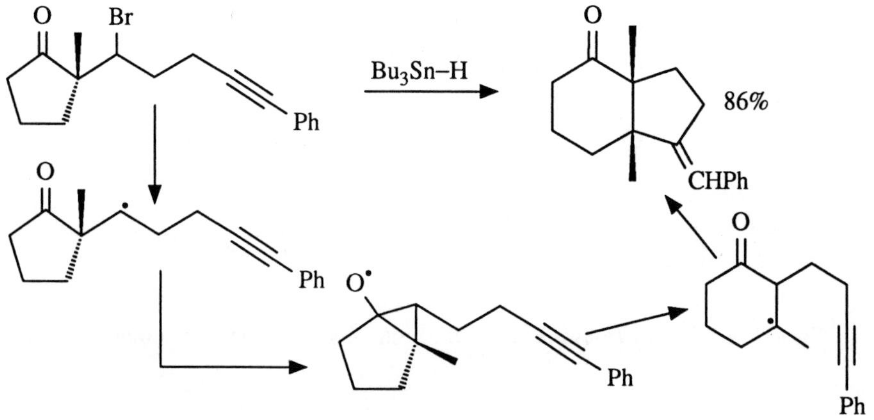

Scheme 20.4

21-3 CYCLIZATION OF HETEROATOMIC RADICALS

21-3-1 Aminyl radicals

Aminyl radicals, obtained by photolysis or redox reactions of N-chloroamines, readily cyclize, even in the neutral form, unlike intermolecular additions (Chap. 8). In this way mono- and polycyclic products, the basis of alkaloid structures, are synthesized.

$$\text{hv (AcOH/H}_2\text{O) yield 70\%} \qquad \text{TiCl}_3 \qquad \text{yield 81\%} \tag{21.39}$$

With a disubstituted double bond the reaction is very stereoselective when one uses a redox system which produces an intermediate radical-metal ion complex (21.40).

$$\tag{21.40}$$

Isomer	yield	erythro : threo
cis	79%	91 : 9
trans	86%	100 : 0

Polycyclic compounds can also be synthesized (21.41).

$$\xrightarrow{\text{AcOH/H}_2\text{O}}_{\text{TiCl}_3} \qquad 77\% \tag{21.41}$$

It is sometimes necessary to protonate the radical intermediate in order to get a good yield (21.42).

$$\xrightarrow{\text{H}_2\text{SO}_4}_{\text{hv}} \qquad 90\% \tag{21.42}$$

21-3-2 Alkoxyl and thiyl radicals

Alkoxyl radicals are easily obtained by photolysis of nitrites, but their cyclization is much less used in synthesis than that of carbon radicals. The same is true of thiyl radicals. Nevertheless, both give, in simple cases, good yields and the 5-*exo* cyclization of alkoxyl radicals is fast and regioselective.

$$\text{[alkene-O-NO]} \xrightarrow[\text{yield 70\%}]{h\nu} \text{[tetrahydrofuran with CH=NOH]} \qquad (21.43)$$

The cyclization of thiyl radicals, by simple photolysis of thiols, is easy to perform but, because the C–S bond is longer and thiyl addition is reversible, the product is generally an *endo/exo* mixture. Nonetheless, when the double bond is intracyclic the regioselectivity is very good.

$$\text{[cyclohexene-CH}_2\text{SH with R]} \xrightarrow{h\nu} \text{[bicyclic sulfide with R]} \qquad 95\% \quad (21.44)$$

21-4 COMBINATION OF INTERMOLECULAR ADDITION AND CYCLIZATION

The order of the reactions, addition then cyclization or cyclization then addition, depends on the reactants involved.

21-4-1 Addition plus cyclization

This reaction sequence, called *annellation*, can give one-to-one or one-to-two adducts, depending on the structure of the reactants (21.45).

$$\text{NC-CH=CH}_2 + \text{[Ph}_2\text{C=CH-CH}_2\text{-CH}_2\text{-I]} \xrightarrow{\text{Bu}_3\text{SnH}} \text{[cyclopentane with NC and CPh}_2\text{]} \qquad (21.45)$$

75% of a *cis/trans* mixture

If Ph is replaced by Me, the radical intermediate resulting from cyclization is more nucleophilic. It reacts rapidly with an electron-poor olefin to give a one-to-two adduct (21.46).

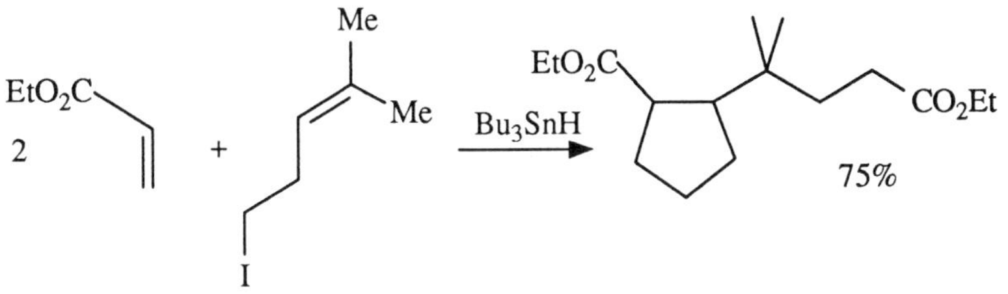

Combination of intermolecular addition and cyclization

(21.46)

Redox systems can be used with enolizable ketones for annellation. Thus, with methylenecyclopentane the *spiro* compound is obtained quantitatively.

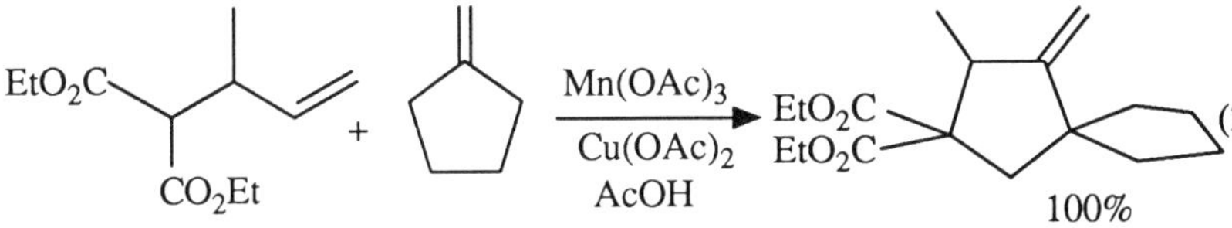

(21.47)

21-4-2 Cyclization plus addition

The following two examples show the importance of the relative concentration of the reactants on the final product yield. Cyclization of the aryl radical is very fast ($k_{5exo} \approx 5 \times 10^8$ s^{-1}). By using a slight excess of olefin with respect to Bu$_3$SnH the reaction is oriented towards radical addition to olefin rather than hydrogen transfer from Bu$_3$SnH. If the olefin concentration is too high the yield falls.

(21.48)

The cyclization (21.49) is slower than that in (21.48). To avoid trapping both the initial radical and that arising from cyclization the reaction is run with a low concentration of Bu$_3$SnH, supplied by the Bu$_3$SnCl/NaBH$_4$ system, and a high concentration of olefin to trap the cyclized radical.

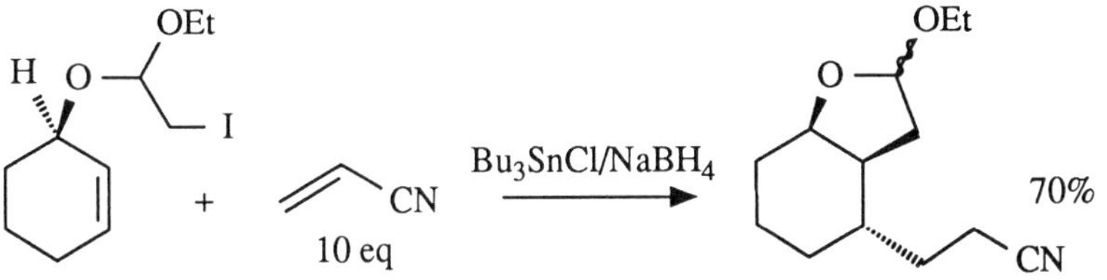

(21.49)

22 – AROMATIC SUBSTITUTION

Aromatic substitution (22.1) can be achieved in various ways:

$$\text{Y-C}_6\text{H}_4\text{-X} \longrightarrow \longrightarrow \text{Y-C}_6\text{H}_4\text{-Z} \quad (22.1)$$

– by creation of an aryl radical which then adds to an olefin (Chaps 12 and 20) (22.2);

$$\text{Y-C}_6\text{H}_4\text{-X} \longrightarrow \text{Y-C}_6\text{H}_4\cdot \xrightarrow{\;\;Z=\;\;} \text{Y-C}_6\text{H}_4\text{-}\!\!\!\sim\!\!\!\text{-Z} \quad (22.2)$$

– by addition of a radical, $Z^\bullet$, to the aromatic nucleus, then rearomatization by oxidative elimination (Chap. 14) (22.3). The reactions of neutral radicals are generally slow and unregioselective. On the other hand, good results are obtained if the reactants, radical and substrate, have very different polarities. This is the case of *amination and substitution on protonated heteroatomic bases*;

$$\text{Y-C}_6\text{H}_4\text{-H} \longrightarrow \text{Y-C}_6\text{H}_4(\cdot)(Z)(H) \xrightarrow{\text{oxidant}} \text{Y-C}_6\text{H}_4\text{-Z} + H^+ \quad (22.3)$$

– by electron transfer and reactions of radical anion intermediates, i.e. *nucleophilic radical substitution*, $S_{RN}1$ (para. 15-1-3).

22-1 AMINATION REACTIONS

Whereas neutral aminyl radicals react only slowly with an electron-rich substrate such as an aromatic system, the reaction becomes interesting if protonated aminyl radicals are used. These are obtained by redox reaction with N-chloroamines in acid.

Several factors contribute to the synthetic utility of this sort of reaction:

- ease of obtaining N-chloroamines;
- very great variety of N-chloroamines, the only limit being the steric hindrance of the substituents;
- very great variety of aromatic substrates. However, electron-attracti substituents make the reaction more difficult;
- experimental simplicity. Normally one works at ambiel $H_2SO_4/AcOH$ (85/15) mixture and the metal salt, Fe^{2+}, Ti^{3+}, Cu^+ catalytic amounts;

– the yields are high with respect to the N-chloroamine and quantitative with respect to the aromatic (Table 22.1);
  – because of polar effects the substrate selectivity is very high. This explains why aniline is unreactive and the particular selectivity of 4-nitrobiphenyl (substitution only on the unsubstituted ring);
  – preferential *para* attack (Table 22.1).

$$R_2\overset{+}{N}HCl + M^{n+} \longrightarrow R_2\overset{\bullet+}{N}H + Cl^- + M^{(n+1)+} \quad (22.4)$$

$$R_2\overset{\bullet+}{N}H + Y\text{-}C_6H_4\text{-}H \longrightarrow Y\text{-}C_6H_4\text{-}\overset{+}{N}HR_2\text{-}H \quad (22.5)$$

$$Y\text{-}C_6H_4\text{-}\overset{+}{N}HR_2\text{-}H + M^{(n+1)+} \longrightarrow Y\text{-}C_6H_4\text{-}\overset{+}{N}HR_2 + H^+ + M^{n+} \quad (22.6)$$

aromatic substrate	N-chloroamine	orientation %	yield %
benzene	dimethylamine		100[a]
benzene	piperidine		100[a]
benzene	morpholine		100[a]
phenol	piperidine	o (9%) ; p (91%)	97[b]
anisole	piperidine	o (9%) ; p (91%)	65[b]
acetanilide	piperidine	p (100%)	98[b]
acetanilide	dimethylamine	p (100%)	93[b]
naphthalene	dimethylamine	1 (97%) ; 2 (3%)	68[b]
fluorenone	dimethylamine	2 (100%)	98[b]
4-nitrobiphenyl	dimethylamine	4' (100%)	86[b]

Table 22.1 – *Amination reaction. a) Yields with respect to benzene: 60 ~ 80%.*
b) Yields with respect to N-chloroamine; those with respect to the aromatic substrate are always higher.

22-2 SUBSTITUTION ON HETEROAROMATIC BASES BY NUCLEOPHILIC RADICALS

Six-membered ring heterocycles, such as pyridine, react slowly if at all with electrophiles of the Friedel-Crafts type, R-Hal and RC(O)Cl. It is easy therefore to understand the interest of radical reactions for the synthesis of substituted nitrogen heterocycles, which are very important pharmaceutically. The high reactivity (k_{add} = 10^5 to 10^8 M^{-1}s^{-1}) of radical nucleophiles in the addition step of substitution (22.7) (Chap. 14) is due to the high electrophilicity of the protonated heteroaromatic substrate and the nucleophilicity of the radical. The more electrophilic the radical and the more X is electron-attracting, the faster the reaction. The substrate selectivity, which is due to X, is very good and the regioselectivity is high, substitution going mainly α to nitrogen. Rearomatization (22.5) is easy, even with mild oxidizing agents.

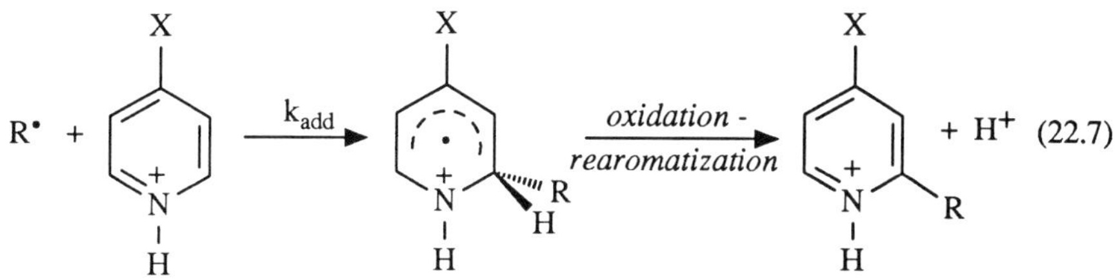

Substitution on heteroatomic bases by nucleophilic radicals   259

$$R^\bullet + \underset{\underset{H}{|}}{\underset{N^+}{\bigcirc}}^X \xrightarrow{k_{add}} \underset{\underset{H}{|}}{\underset{N^+}{\bigcirc}}^X{}_{R,H} \xrightarrow{\text{oxidation - rearomatization}} \underset{\underset{H}{|}}{\underset{N^+}{\bigcirc}}^X{}_R + H^+ \quad (22.7)$$

heteroaromatic base	radical R•	method[a]	orientation	yield[b]
alkylation				
pyridine	Et•	A	2 ; 4[c]	100
4-cyanopyridine	*i*-Pr•	A	2 ; 2,6[d]	88
quinoline	*t*-Bu•	A	2	95
benzothiazole	*t*-Bu•	A	2	75
acylation				
4-cyanopyridine	EtC•=O	C	2	60
4-cyanoquinoline	MeC•=O	C	2	93
4-cyanoquinoline	PhC•=O	C	2	86
benzothiazole	*p*-MeOPhC•=O	C	2	80
hydroxyalkylation				
4-methylquinoline	•CH$_2$OH	D	2	82
4-methylquinoline	•CH$_2$OH	B	2	98
4-methylquinoline	CH$_3$ĊHOH	D	2	70
2-methylquinoline	•CH$_2$OH	D	4	80
α-oxyalkylation				
4-cyanopyridine	dioxanyl	C	2	65
2-methylquinoline	dioxanyl	C	2	86
amidation				
4-methylquinoline	•CONH$_2$	C	2	99
isoquinoline	•CONH$_2$	C	1	100

Table 22.2 – *Substitution on heteroaromatic base by nucleophilic radicals. a) Method for producing radicals, R•. b) Yields (%) with respect to the base. c) 2/4 = 44 / 56. d) 2/2,6 = 67 / 33.*

22-2-1 Production of radicals, R•

Many methods for producing radicals, R•, can be applied to compounds of different chemical families. A general account of the various means of generating radicals was given in Chapter 9. In particular, for substitution on protonated heteroaromatic bases, R• can be produced:

– either directly by decarboxylation of a carboxylic acid (method A);
– or by reacting (22.8) a compound containing an easily removed iodine or hydrogen atom with another radical produced by various procedures (methods B-E).

$$R-H \text{ or } R-I + \Sigma^\bullet \longrightarrow R^\bullet + \Sigma-H \text{ or } \Sigma-I \quad (22.8)$$

Method A: oxidative decarboxylation of an acid
The reaction is performed with ammonium persulfate and catalytic amounts of a silver salt. This method is very practical and general.

$$Ag^+ + S_2O_8^{2-} \longrightarrow Ag^{2+} + SO_4^{\bullet -} + SO_4^{2-} \quad (22.9)$$

$$Ag^+ + SO_4^{\bullet -} \longrightarrow Ag^{2+} + SO_4^{2-} \quad (22.10)$$

$$RCO_2H + Ag^{2+} \longrightarrow R^{\bullet} + CO_2 + Ag^+ + H^+ \quad (22.11)$$

Method B: thermal decomposition of a peroxide
In the case of benzoyl peroxide both the $PhCO_2^{\bullet}$ and $Ph^{\bullet}$ radicals can take part in reaction (22.12).

$$(PhCO_2)_2 \longrightarrow 2\, PhCO_2^{\bullet} \longrightarrow 2\, Ph^{\bullet} + 2\, CO_2 \quad (22.12)$$

Method C: redox reaction on a peroxide
The system presented in (22.13) is widely used, since $RO^{\bullet}$ and $HO^{\bullet}$ are very reactive and, moreover, the intermediate cyclohexadienyl radical is rearomatized ($Fe^{3+} \rightarrow Fe^{2+}$). For this reason the reaction requires only catalytic amounts of Fe^{2+}; in other words, we have a chain process.

$$ROOH + Fe^{2+} \xrightarrow{R = H,\, t\text{-Bu}} RO^{\bullet} + HO^- + Fe^{3+} \quad (22.13)$$
$$\text{(or } RO^- + HO^{\bullet}\text{)}$$

If the radical, $R^{\bullet}$, produced by reaction (22.8), is too reactive its oxidation by Fe^{3+} (22.14) can be faster than its reaction with the protonated base or the oxidation of the cyclohexadienyl radical adduct. Thus, under these conditions, the $Me_2(OH)C^{\bullet}$ radical gives no product.

$$R^{\bullet} + Fe^{3+} \longrightarrow R^+ + Fe^{2+} \quad (22.14)$$

Method D: redox reaction on a hydroxylamine
The $^{\bullet +}NH_3$ radical obtained by this method (22.15) performs reaction (22.8). Ti^{4+} is a mild reducing agent which, nevertheless, is capable of rearomatizing the cyclohexadienyl radical.

$$^+NH_3OH + Ti^{3+} + H^+ \longrightarrow {}^{+\bullet}NH_3 + Ti^{4+} + H_2O \quad (22.15)$$

Method E: use of alkyl iodides
The conventional method for generating radicals from an alkyl iodide by means of Bu_3SnH is unsuitable here, since there is no oxidizing agent. However, $Me^{\bullet}$ radicals can be produced by means of a redox reaction, a variant of method C, from H_2O_2/Fe^{2+} in DMSO, and these react with the iodide, RI, to give the radical, $R^{\bullet}$, in a fast process.

$$H_2O_2 + Fe^{2+} \longrightarrow HO^{\bullet} + HO^- + Fe^{3+} \quad (22.16)$$

$$CH_3SOCH_3 \xrightarrow{+ HO^\bullet} CH_3 \underset{CH_3}{\overset{O^\bullet}{\underset{|}{\overset{|}{S}}}} OH \longrightarrow {}^\bullet CH_3 + CH_3SO_2H \quad (22.17)$$

$$RI + {}^\bullet CH_3 \longrightarrow R^\bullet + ICH_3 \quad (22.18)$$

22-2-2 Reaction conditions

The reactions are carried out in acid, 0.4 M trifluoroacetic acid or 0.2 M sulfuric acid, or a mixture of both, with often the addition of a cosolvent. Given the very polar character of the reaction, the polarity of the solvent has a considerable impact on the regioselectivity. Thus, for protonated pyridine the ratio for Ph$^\bullet$ radical substitution in the 2 and 4 positions is 64/31 and 70/24 in water and benzene, respectively, whereas for the more nucleophilic α-THF$^\bullet$ radical these ratios are reversed: 20/80 and 86/14, in 1:1 THF-H$_2$O and pure THF, respectively (Chap. 14). The reaction temperatures are moderate, 60-80°C for the reactions in Table 22.2 and 40-50°C for the iodide/ redox systems of Table 22.3.

The applications of these procedures concern a wide range of nucleophilic radicals produced from RH or RI to perform alkylations, arylations, hydroxyalkylations, etc. and many heteroatomic bases. The examples given in Tables 22.2 and 22.3 therefore represent a very limited sample.

heteroaromatic base	iodide	orientation	yield[c]
4-methylquinoline	isopropyl	2	88
4-methylquinoline	cyclohexyl	2	91
4-methylquinoline	tert-butyl	2	78
quinoline	tert-butyl	2	87
quinoline	cyclohexyl	2 ; 4 ; 2,4[a]	91
4-cyanopyridine	isopropyl	2 ; 2,6[b]	95
4-methylpyridine	cyclohexyl	2	99

Table 22.3 – *Substitution on heteroaromatic bases by alkyl iodide in H$_2$O/DMSO (method E). a) 23% of 2, 33% of 4 and 44% of 2,4 disubstituted product. b) 2/2,6 = 67/33. c) Yield (%) with respect to base.*

22-3 NUCLEOPHILIC RADICAL SUBSTITUTION, S$_{RN}$1

The overall S$_{RN}$1 reaction (para. 15-1-3) is written as in (22.19); it is regiospecific the nucleophile, Nu$^-$, replacing the nucleofuge at the same position (Tables 22.4 and 22.5).

$$Y\text{-Ar-}X + Nu^- \longrightarrow Y\text{-Ar-}Nu + X^- \quad (22.19)$$

substrate	anion	product	yield%
PhBr	CH₂=C(O⁻)-t-Bu	PhCH₂COt-Bu	90[a]
2,4,6-(MeO)₃C₆H₂I	CH₂=C(O⁻)-Me	2,4,6-(MeO)₃C₆H₂CH₂COMe	92[a]
2-bromopyridine	(CH₃)₂C=C(O⁻)-i-Pr	2-(CMe₂COi-Pr)pyridine	97[a]
2-bromopyridine	CH₂=C(O⁻)-Me	2-(CH₂COMe)pyridine	100[b]
PhBr	CH₂=C(O⁻)-NMe₂	PhCH₂C(O)NMe₂	72[a]
2-bromoaniline	CH₂=C(O⁻)-Me	2-methylindole	93[b]
2-bromobenzoic acid	CH₂=C(O⁻)-R	lactone (from o-CH₂COR/CO₂H)	80[a]
2-chloro-3-aminopyridine	CH₂=C(O⁻)-t-Bu	2-t-Bu-pyrrolo[3,2-b]pyridine	100[a]
2-chloro-3-(NCH₃COCH₃)pyridine	intramolecular	N-methyl-oxindole fused pyridine	82[c]
benzophenone	⁻CH(CN)(CO₂Et)	4-(CH(CN)CO₂Et)benzophenone	94[d]

Table 22.4 – *Nucleophilic radical substitution ($S_{RN}1$) in liquid ammonia with UV irradiation in the presence of: a) t-BuOK; b) KNH₂; c) in LDA/THF/ hexane with irradiation; d) by electrochemistry in DMSO.*

Reaction conditions

The most appropriate solvent is liquid NH₃, though DMF and DMSO, but not

HMPA, can also be used. Potassium *tert*-butoxide or potassium amide produces the anion, Nu⁻, by deprotonation. UV irradiation stimulates the electron transfer steps.

Substrate

A very large variety of substrates give good yields (substituted benzenes, naphthalene, anthracene, pyridine, pyrimidine, quinoline, thiophene, etc.). The nucleofuge, X, can be halogen, phosphate, phenylsulfide, triethylammonium, etc. The reaction is possible in the presence of various substituents, Y, such as alkyl, alkoxyl, phenyl, carboxylate or benzoyl, but CN and OH groups are not tolerated. The presence of an *ortho* substituent has little effect (Table 22.4, line 2) and can lead to intramolecular (Table 22.4, penultimate line) or bimolecular cyclization. Indole (Table 22.4, lines 6 and 8) or benzofuran derivatives (Table 22.4, line 7) are easily synthesized.

Anion

The yields obtained with ketone enolates (Table 22.4) are very satisfactory, while those with aldehydes and esters are relatively poor, except for malonates (Table 22.4, last line).

Enolate anions of amides give good results only if the nitrogen is disubstituted. With the acetamide anion, MeCONH⁻, there is no reaction. Results concerning selenates and thiolates are given in Table 22.5.

substrate	anion	product	yield %
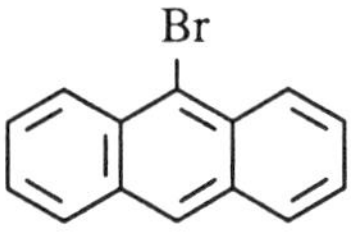 9-bromoanthracene	PhSe⁻	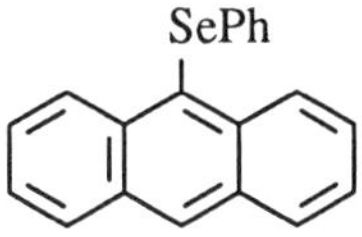 9-(phenylseleno)anthracene	72
4-cyano-2-bromopyridine	CH₃CH₂S⁻	4-cyano-2-(ethylthio)pyridine	90
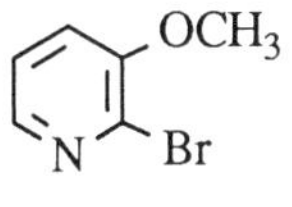 3-methoxy-2-bromopyridine	CH₃CH₂S⁻	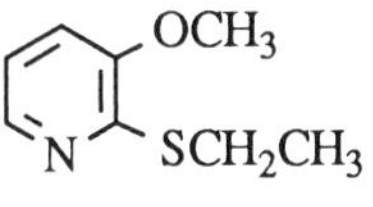 3-methoxy-2-(ethylthio)pyridine	90

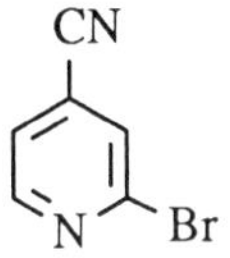

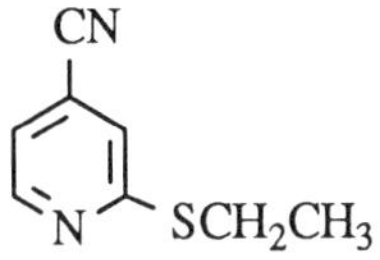

Table 22.5 – *Nucleophilic radical substitution ($S_{RN}1$); t-BuOK in liquid ammonia with irradiation.*

23 – COUPLING REACTIONS

Radical-radical reactions (23.1) are considered as a nuisance when they terminate chains (Chap. 10) but the formation of a C–C bond can be synthetically useful. The structure of the radicals and the reaction conditions are very important. The radicals are produced in various ways:

- by conventional hydrogen abstraction by means of a peroxide: *dehydrodimerization* (Scheme 23.1);
- by electrochemical electron transfer;
- by chemical electron transfer, $S_{RN}1$ (para. 15-1-3).

$$R^{1\bullet} + R^{2\bullet} \longrightarrow R^1-R^2 \qquad (23.1)$$

23-1 DEHYDRODIMERIZATION

The thermal decomposition of a peroxide, such as di-*tert*-butyl peroxide at 140°C (23.2) or di-*tert*-butyl peroxalate at 60°C (23.3) leads to $t\text{-BuO}^{\bullet}$ radicals. These radicals can abstract hydrogen from a weak C–H bond to give a thermodynamically stabilized radical (23.4). In the *absence of a solvent*, which would act as a hydrogen- or halogen-donor, a dimer is formed (Scheme 23.1).

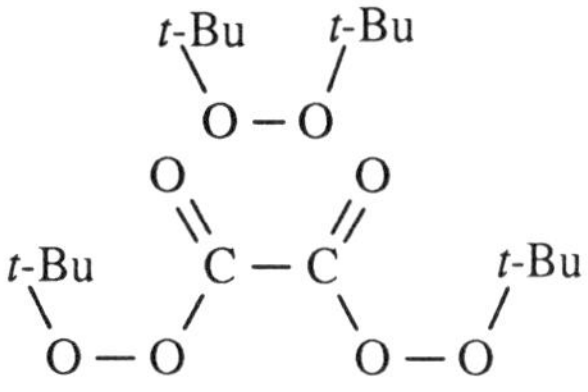

$$t\text{-Bu-O-O-}t\text{-Bu} \longrightarrow 2\ t\text{-BuO}^{\bullet} \qquad (23.2)$$

$$t\text{-Bu-O-O-C(=O)-C(=O)-O-O-}t\text{-Bu} \longrightarrow 2\ t\text{-BuO}^{\bullet} + 2\ CO_2 \qquad (23.3)$$

$$R\text{-H} + t\text{-BuO}^{\bullet} \longrightarrow R^{\bullet} + t\text{-BuOH} \qquad (23.4)$$

$$R^{\bullet} + R^{\bullet} \longrightarrow R\text{-R} \qquad (23.5)$$

Scheme 23.1

This dimerization, sometimes called *oxidative dimerization* or *oxidative coupling*, has many applications, of which we shall give a few examples (Scheme 23.2).

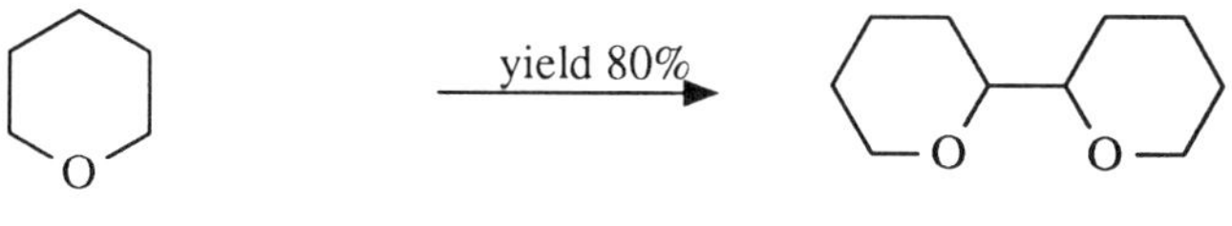

$$\text{tetrahydropyran} \xrightarrow{\text{yield 80\%}} \text{2,2'-bitetrahydropyran} \qquad (23.6)$$

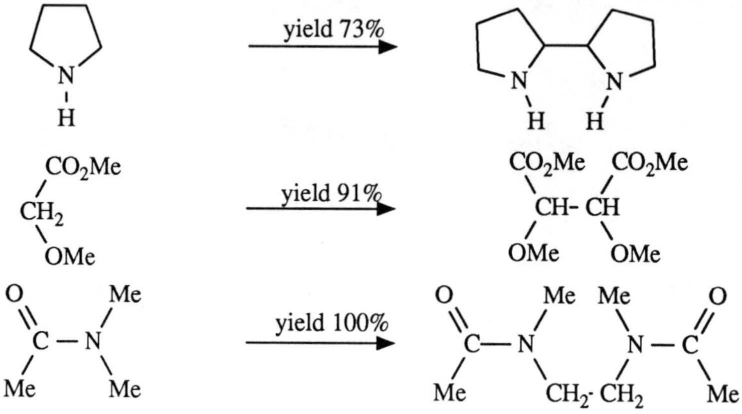

Scheme 23.2

(23.7) pyrrolidine → dimer at α-position, yield 73%

(23.8) $CH_2(CO_2Me)(OMe)$ → $(MeO_2C)(MeO)CH-CH(OMe)(CO_2Me)$, yield 91%

(23.9) $MeC(O)NMe_2$ → $MeC(O)N(Me)CH_2-CH_2N(Me)C(O)Me$, yield 100%

A captodative olefin is sufficently reactive to trap radicals by addition. The radical obtained in this way is stabilized by the captodative effect and readily dimerizes, leading to a *dimer adduct*. Captor substituents are CN, CO_2Me and donors *t*-BuS, NMe_2, OMe. The high radicophilicity of this type of double bond means that it reacts with both nucleophilic ($MeC^{\bullet}=O$, $MeCONMe_2CH_2^{\bullet}$) and electrophilic radicals ($RS^{\bullet}$ where R = Ph or alkyl). Yields of dimer adduct are about 70%.

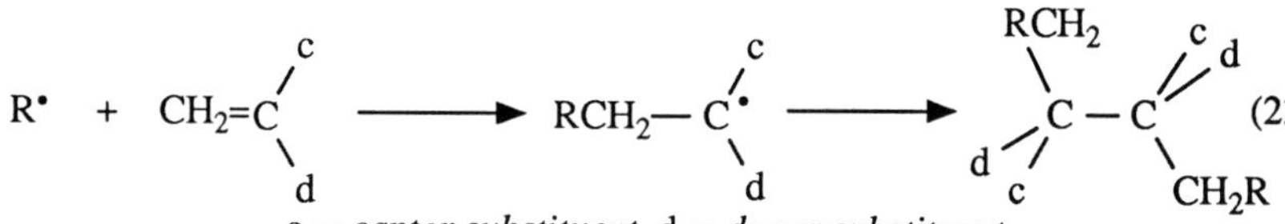

$$R^{\bullet} + CH_2=C(c)(d) \longrightarrow RCH_2-C^{\bullet}(c)(d) \longrightarrow (RCH_2)(c)(d)C-C(d)(c)(CH_2R) \quad (23.10)$$

c = *captor substituent*, d = *donor substituent*

23-2 ELECTROCHEMISTRY

In the Kolbe reaction the electrochemical oxidation of carboxylates gives radicals which then dimerize (23.11).

$$RCO_2^- \xrightarrow{-e^-} RCO_2^{\bullet} \xrightarrow{-CO_2} R^{\bullet} \xrightarrow{dimerization} R-R \quad (23.11)$$

In this way hydrocarbons are obtained from aliphatic acids (23.12) and aliphatic α,ω-diesters from hemiesters of diacids (23.13).

$$CH_3(CH_2)_nCO_2H \xrightarrow[\text{yield 75 to 93\%}]{\text{n from 0 to 18}} CH_3(CH_2)_{2n}CH_3 \quad (23.12)$$

$$CH_3O_2C(CH_2)_nCO_2H \xrightarrow[\text{yield 60 to 95\%}]{\text{n from 1 to 16}} CH_3O_2C(CH_2)_{2n}CO_2CH_3 \quad (23.13)$$

In the presence of a radicophilic olefin one obtains either a di-adduct (23.14) or a dimer adduct (23.15), depending on the structure. A good yield implies the reaction of a nucleophilic radical, R•, on a double bond which is made electron-deficient by an attractor substituent.

$$MeCO_2^- \xrightarrow[-CO_2]{-e^-} Me^{\bullet} \xrightarrow[\text{yield } 80\%]{\text{ethyl fumarate}} \underset{(meso/dl = 1/1)}{\begin{array}{c} Me \quad Me \\ CH-CH \\ CO_2Et \quad CO_2Et \end{array}} \quad (23.14)$$

$$RCO_2^- \xrightarrow[-CO_2]{-e^-} R^{\bullet} \xrightarrow{\underset{CH_2}{\overset{Me}{\underset{\|}{C-CHO}}}} \underset{\text{yield } 80\%}{\begin{array}{c} CH_2R \quad CH_2R \\ C-C \\ Me \quad Me \\ CHO \quad CHO \end{array}} \quad (23.15)$$

R = Me, Et, n-Pr

23-3 $S_{RN}1$ REACTION

The overall $S_{RN}1$ reaction (para. 15-1-3) is given by (23.16). The rate determining step is the radical-nucleophile coupling in which the C–C bond is formed (23.17).

$$Nu^- + RX \longrightarrow RNu + X^- \quad (23.16)$$

$$Nu^- + R^{\bullet} \longrightarrow RNu^{\bullet-} \quad (23.17)$$

These reactions, performed in polar aprotic solvents such as DMSO, DMF and HMPA, lead to highly branched products difficult to obtain by other means. However, the substrate, like the anion, must have an electron-attracting group, often NO_2, if the yields are to be satisfactory, i.e. above 80%. $S_{RN}1$ reactions can be applied to many systems.

23-3-1 *para*-Substituted cumyl system

The historic example (1961) is that of Kornblum (23.18). In the absence of an attractor *para* substituent (CN or NO_2, or two *meta* CF_3 groups) the yield is much poorer. The nucleofuge can be azide, sulfone, benzoate or thioether instead of Cl. The nucleophile can be PhS^-, $PhSO_2^-$, $^-CH(CO_2Me)_2$, ^-CN or N_3^-. The yields are about 90-95%.

$$\underset{CN}{\underset{|}{\text{Ar}}}\overset{Me}{\underset{Me}{C}}-Cl + Me_2\bar{C}NO_2 \xrightarrow{\text{yield } 90\%} \underset{CN}{\underset{|}{\text{Ar}}}\overset{Me}{\underset{Me}{C}}-\overset{Me}{\underset{Me}{C}}-NO_2 + Cl^- \quad (23.18)$$

23-3-2 Heterocyclic series

The $S_{RN}1$ reaction of two equivalents of nitronate anion with 1-methyl-2-chloromethyl-5-nitroimidazole **1** gives first the C–alkylation product **2**, which with excess nitronate anion leads to the vinylimidazole **3** in 88% yield. This last product, which has antiparasite and antibacterial activity, cannot be prepared by any other synthetic route.

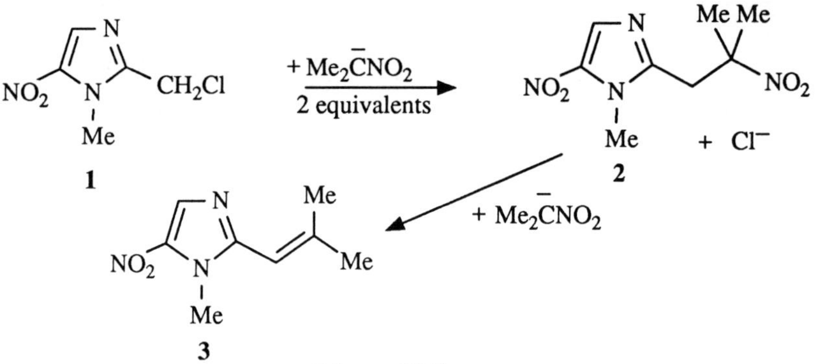

Scheme 23.3

23-3-3 Aliphatic series

The reaction also gives very good yields with aliphatics.

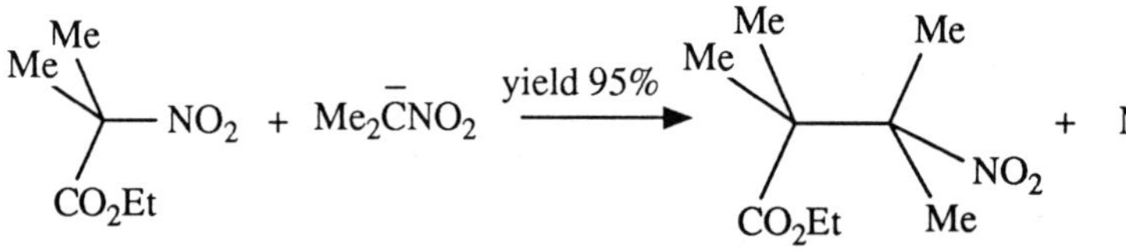

$$\text{(23.19)}$$

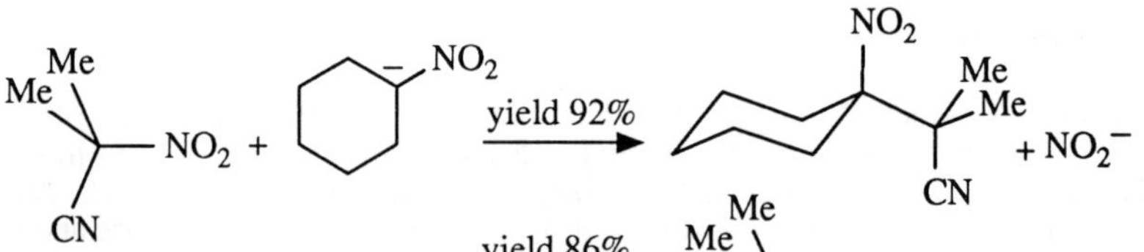

$$\text{(23.20)}$$

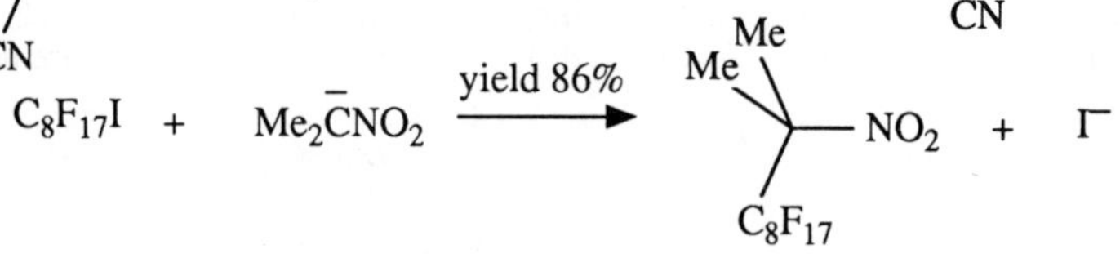

$$\text{(23.21)}$$

24 – RADICAL REACTIONS IN INDUSTRY

Radical reactions are used in the chemical industry to obtain high tonnage products, such as polymers, or smaller quantities. In fine chemistry there are also radical synthesis steps. Most industrial processes fall into one of four classes of reaction:

- addition, polymerization representing the greatest tonnage;
- photochemical chlorination;
- oxidation by atmospheric oxygen;
- oxidation-combustion.

24-1 POLYMERS

Alkenes can be polymerized cationically, anionically or by radical processes. Radical polymerization represents about 75% of the world production of all types of polymer (about 10^8 tonnes), 30% of which is in Europe. The most important polymers are polyvinyl chloride (PVC), polystyrene and polyethylene.

The reaction proceeds via a chain mechanism, described in Chapters 5 and 7, where the propagation step is a succession of radical additions to the monomer double bond. The reaction is initiated either by thermal decomposition of peroxide or azo compounds or by a redox system (Fenton-type reaction).

$$\text{initiation} \quad \Sigma-\Sigma \longrightarrow 2\,\Sigma^\bullet \quad (24.1)$$

$$\text{propagation} \quad \Sigma^\bullet + CH_2=CHR \longrightarrow \Sigma CH_2CHR \equiv \Sigma M_1^\bullet \quad (24.2)$$

$$\Sigma M_1^\bullet + CH_2=CHR \longrightarrow \Sigma M_1 CH_2 CHR \equiv \Sigma M_2^\bullet \quad (24.3)$$

$$\Sigma M_2^\bullet + CH_2=CHR \longrightarrow \Sigma M_2 CH_2 CHR \equiv \Sigma M_3^\bullet \quad (24.4)$$

$$\Sigma M_3^\bullet + CH_2=CHR \longrightarrow \text{etc.} \quad (24.5)$$

$$\text{termination} \quad \Sigma M_m^\bullet + \Sigma M_n^\bullet \longrightarrow \text{non-radical products} \quad (24.6)$$

The linear polymerization process is sometimes perturbed or modified by *transfer* reactions, that is, hydrogen abstraction from the products or the reactants by the polymer radicals. If the new radical is sufficiently reactive it can initiate a new chain and one obtains a *cross-linked* or *reticulated* polymer with properties different from those of a linear polymer. For example, the thermal reticulation of linear polydiorganosiloxanes by a peroxide leads to elastomers which can be vulcanized by heating.

Conversely, an unreactive radical tends to favor termination reactions and limits the polymerization process; in this case telomers, i.e. low molecular weight polymers, are produced.

$$RO-\underset{\underset{CH_3}{|}}{\overset{\overset{CH_3}{|}}{Si}}-O- \quad \xrightarrow{-ROH} \quad \underset{\underset{CH_3}{|}}{\overset{\overset{\overset{\bullet}{CH_2}}{|}}{Si}}-O- \quad (24.7)$$

$$2 \; \underset{\underset{CH_3}{|}}{\overset{\overset{\overset{\bullet}{CH_2}}{|}}{Si}}-O- \quad \longrightarrow \quad -O-\underset{\underset{CH_3}{|}}{Si}-CH_2-CH_2-\underset{\underset{CH_3}{|}}{Si}-O- \quad (24.8)$$

24-1-1 Polymer preparation

Polymers are prepared industrially by four different procedures.

Bulk polymerization
The peroxide or azo initiator is added to the pure monomer which is then heated.

Solution polymerization
The monomer is dissolved in a solvent to which is added the initiator. The solvent must be chosen so as to minimize chain transfer reactions.

Suspension polymerization
The monomer is suspended in water with the help of a product such as polyvinyl alcohol, gelatine or water-soluble cellulose derivatives. The peroxide initiator dissolves in the monomer droplets.

Emulsion polymerization
The sulfonate ester of a fatty acid is added to a water-monomer mixture to create a micellar medium, and polymerization takes place inside the micelles. In this case the initiator must be water-soluble and is frequently ammonium or potassium persul-

polymer	monomer	preparation mode[a]	production (kT, 1988) Europe	world
polyethylene (low density)	$CH_2=CH_2$	B: high pressure peroxides or O_2 150 ~ 300°C	5 000	15 000
PVC	$CH_2=CHCl$	B, E or S peroxide	5 000	20 000
polystyrene	$CH_2=CHPh$	S: peroxide or O_2 E: redox	2 100	7 500
fluoropolymers	$CF_2=CF_2$	E, S: under pressure	14 400	43 000
polyacrylic	$CH_2=\underset{\underset{Me}{\vert}}{C}CO_2Me$	B, S	220	700

Table 24.1 – *Main industrial polymers. a) Procedures: B, bulk; E, emulsion; S, suspension.*

fate or a hydroperoxide-ferrous salt redox system.

The main industrial polymers are listed in Table 24.1.

24-1-2 Copolymers

The polymerization of a mixture of two or more monomers gives macromolecules containing units of both, which are called *copolymers* and which cannot always be obtained by ionic processes.

The physical properties of copolymers are often different from those of the corresponding homopolymers, hence their industrial interest. Moreover, 1,2-disubstituted alkenes, which do not polymerize alone, readily copolymerize. Thus, maleic anhydride copolymerizes with styrene or stilbene.

The most common copolymers are styrene-acrylonitrile, styrene-methyl methacrylate and styrene-butadiene-acrylonitrile as well as the fluorinated copolymers, tetrafluoroethylene-perfluorovinyl ether or hexafluoropropylene or ethylene, chlorotrifluoroethylene-ethylene.

The example of the styrene-butadiene copolymer is a good illustration of the importance of radical chemistry. Before the Second World War practically all rubber objects, tires in particular, were made from natural latex, supplies of which were cut off by the conflict with Japan. In the United States industry took up the challenge and developed the synthesis of rubber by the emulsion copolymerization of styrene and butadiene with a redox initiator, consisting of a mixture of menthane or cumene hydroperoxide and a ferrous salt. Even before 1939, Germany, which did not have access to natural latex, for other reasons, developed the synthesis of rubber (Buna S) also by copolymerization of styrene and butadiene with a redox system.

24-1-3 Polymer reticulation and degradation

The mechanical properties of polymers are modified by *reticulation*, that is, the cross-linking of polymer chains. It is possible to obtain materials better in many respects than the starting materials. Almost all natural and synthetic elastomers, as well as thermoplastics, can be reticulated by means of peroxides. This transformation is performed industrially mainly on polyethylene, ethylene-propylene or ethylene-propylene-butadiene rubbers, polydimethylsiloxanes or polymethylvinylsiloxanes and ethyl-vinyl acetate copolymers. The mechanism is shown in Scheme 24.1.

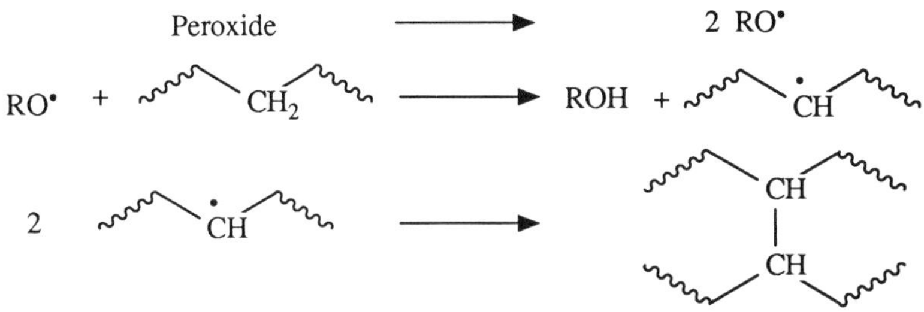

$$\text{Peroxide} \longrightarrow 2 \text{ RO}^\bullet \quad (24.9)$$

$$\text{RO}^\bullet + \sim\!\!\!\sim\!\!\text{CH}_2\!\!\sim\!\!\!\sim \longrightarrow \text{ROH} + \sim\!\!\!\sim\!\!\overset{\bullet}{\text{CH}}\!\!\sim\!\!\!\sim \quad (24.10)$$

$$2 \sim\!\!\!\sim\!\!\overset{\bullet}{\text{CH}}\!\!\sim\!\!\!\sim \longrightarrow \sim\!\!\!\sim\!\!\underset{\underset{\sim\!\sim\!\sim\!\sim\text{CH}\sim\!\sim\!\sim}{|}}{\text{CH}}\!\!\sim\!\!\!\sim \quad (24.11)$$

Scheme 24.1

Polypropylene, polybutene and polyvinyl chloride are not reticulated but, under the same conditions, are degraded by a β-scission reaction as indicated in Scheme 24.2 for polypropylene.

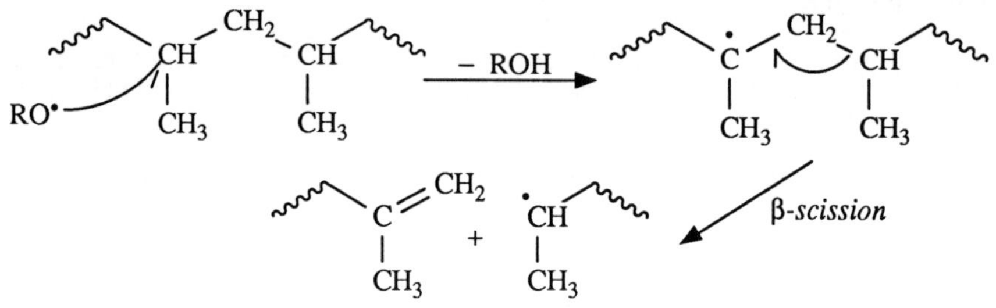

Scheme 24.2

24-2 HALOGENATION

Certain industrial halogen products are obtained by addition or substitution initiated by UV irradiation.

24-2-1 Chlorination

The chain reaction shown below is particularly efficient (Chap. 18) and is used for the synthesis of various compounds:

initiation	Cl_2 $\xrightarrow{h\nu}$	2 Cl$^\bullet$	(24.12)
propagation	Cl$^\bullet$ + RH $\longrightarrow$	HCl + R$^\bullet$	(24.13)
	R$^\bullet$ + Cl_2 $\longrightarrow$	RCl + Cl$^\bullet$	(24.14)

- higher chloromethanes from methyl chloride;
- chloroparaffins, including solid products with 70% chlorine;
- benzyl chloride from toluene;
- hexachlorocyclohexane from benzene (35 kT/annum), the γ-isomer of which is used as an insecticide (lindane).

24-2-2 Thermal reactions

90% of the 31 MT world production of dichloroethane is transformed into vinyl chloride monomer by a radical process when it is pyrolyzed at 500°C.

$$ClCH_2-CH_2Cl \longrightarrow CH_2=CH-Cl + HCl \quad (24.15)$$

Chlorination and thermal dehydrochlorination of ethylene is also a means of obtaining trichloroethylene.

24-3 NITROSATION (LACTAM)

The photochemical reaction of nitrosyl chloride on cycloalkanes leads to α,ω-aminoacids (Scheme 24.3). Cyclohexane and cyclododecane give 6-aminohexanoic and 12-aminododecanoic acids, respectively; these are precursors of nylons 6 and 12.

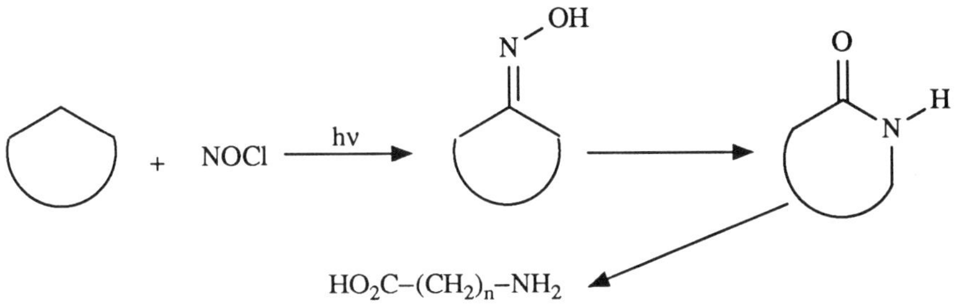

$HO_2C-(CH_2)_n-NH_2$

Scheme 24.3

24-4 OXIDATION

Certain industrial processes depend on controlled oxidation by atmospheric oxygen.

24-4-1 Hydrogen peroxide

Of the various processes for producing hydrogen peroxide, that which is the most used at present involves the oxidation of anthraquinol by air, which is resumed schematically in Scheme 24.4. The hydrogen peroxide is extracted by scrubbing the organic solution with water followed by concentrating the aqueous solution to about 70% by distillation.

Hydrogen peroxide is the perfect ecological oxidizing agent since it leaves no residue. Its principal applications are:

- bleaching of paper pulp and cotton textiles (40% concentration);

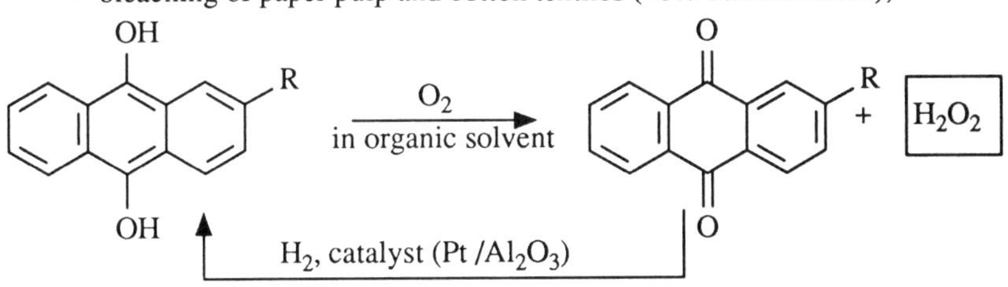

Scheme 24.4

- production of persalts for detergent powders;
- production of hydrazine (24.16);
- epoxidation of natural rubber and vegetable oils;
- cleansing of gas and liquid effluents.

$$2\ NH_3 + H_2O_2 \xrightarrow{\text{catalyst}} N_2H_4 + 2\ H_2O \qquad (24.16)$$

The world production capacity in 1991 was about 1 500 kT/annum of 100% hydrogen peroxide equivalents, about 70% of this capacity being located in Europe.

24-4-2 Phenol

Most phenol is obtained by air oxidation of cumene to its hydroperoxide, which is then transformed into phenol and acetone by acid (Europeanproduction, 1 350 kT/annum; world, 4 200 kT/annum).

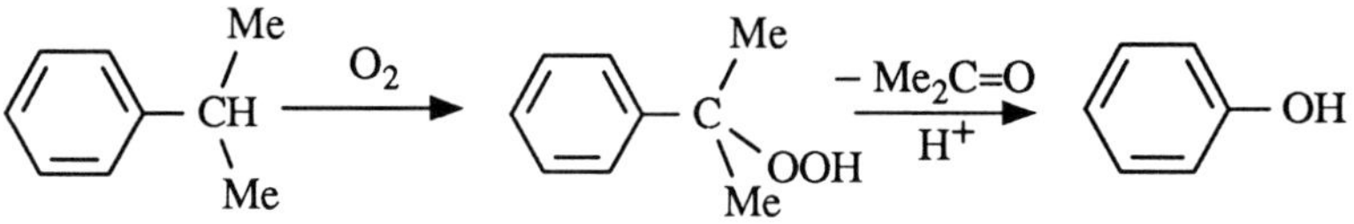

$$\text{PhCHMe}_2 \xrightarrow{O_2} \text{PhCMe}_2\text{OOH} \xrightarrow[H^+]{-Me_2C=O} \text{PhOH} \qquad (24.17)$$

24-4-3 Adipic acid

The first step in the production of adipic acid, the raw material for nylon 6, is air oxidation of cyclohexane to its hydroperoxide, which is then transformed into cyclohexanol and cyclohexanone. Nitric acid oxidation of this mixture leads to adipic acid (European and world production, 770 kT and 1 760 kT, respectively).

This apparently simple operation is much more difficult to carry out than it looks, and much effort has been put into optimizing the pressure, temperature, concentration and reactor design in order to control the reaction. Nevertheless, it is the best way of oxidizing cyclohexane cleanly. All other transition metal-catalyzed oxidation procedures give poorer results.

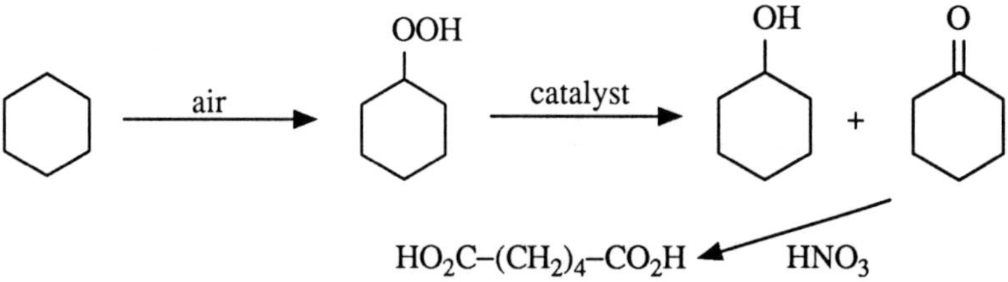

Scheme 24.5

24-4-4 Propylene oxide

Propylene oxide can be obtained by autoxidation of ethylbenzene as shown in Scheme 24.6 (European and world production, 400 kT and 1 200 kT, respectively).

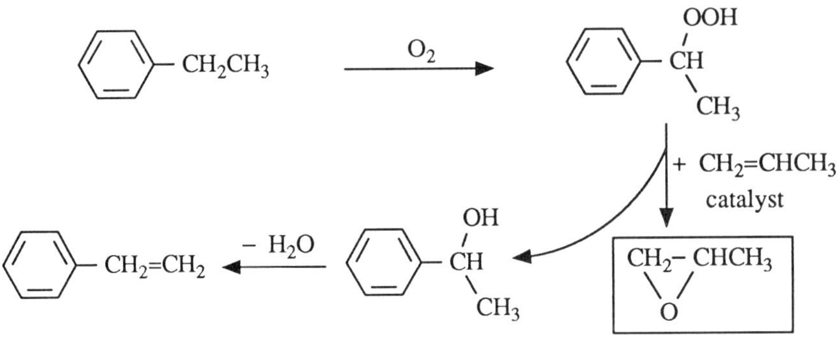

Scheme 24.6

24-5 OXIDATION-COMBUSTION

Various compounds, but particularly fossil hydrocarbons, are used as sources of energy for heating and the operation of various motors: cars, planes, etc. The overall fuel oxidation process can be represented by the thermal equation:

$$>CH_{2\ liquid} + 1.5\ O_{2\ gas} \xrightarrow{\Delta H = -156\ kcal/mol} CO_{2\ gas} + H_2O_{liquid} \quad (24.18)$$

Very few chemical reactions can claim such an energy potential. Table 24.2 gives heats of combustion for various organic compounds.

fuel	heat of combustion[a]	
	kcal/mol	kcal/g
methane	213	13.4
n-butane	688	11.9
acetylene	293.3	11.5
benzene	781.8	10.0
ethanol	327	7.17
coal		7.65
wood		4.54

Table 24.2 – *Heats of combustion. a) For reaction giving liquid H_2O.*

The conversion of chemical energy into mechanical or electrical energy is limited by the efficiency of the motor which performs this transformation (Carnot's principle).

Many studies have been performed at different temperature ranges in attempts to understand the hydrocarbon oxidation process. To simplify the problem we consider two ranges: room temperature up to 250°C and temperatures over 250°C, which we shall call "low" and "high" temperature, respectively. In both cases the reaction proceeds by a more or less complex radical chain mechanism.

The complexity is explained by the secondary reactions arising from products which are more reactive than the starting materials. Thus, hydroperoxides cause induced decomposition, since even at 100°C the homolysis of the RO–OH bond (BDE = 43 kcal/mol) is not negligible and the RO• and •OH radicals formed lead to the formation of aldehyde, then acid, then alcohol. These reactions are relatively rapid at 300°C and higher.

$$RCH_2O^{\bullet} \longrightarrow CH_2O + R^{\bullet} \quad (24.19)$$
$$RCH_2O^{\bullet} + O_2 \longrightarrow RCHO + HOO^{\bullet} \quad (24.20)$$
$$RCH_2O^{\bullet} + R^1-H \longrightarrow RCH_2OH + R^{1\bullet} \quad (24.21)$$

24-5-1 Low-temperature reactions

We meet again the mechanism described for autoxidation (Section 18-3). The overall reaction (24.22), which is 18 ~ 24 kcal/mol exothermic, is potentially autocatalytic. In this temperature range step (24.24) is irreversible; the overall reaction is slow since the activation energy of step (24.25) is fairly high: 14 kcal/mol for tertiary C–H bonds.

$$R-H + O_2 \longrightarrow R-OO-H \quad (24.22)$$

initiation $\quad R-H \longrightarrow R^{\bullet} \quad (24.23)$

propagation
$$R^{\bullet} + O_2 \longrightarrow R-OO^{\bullet} \quad (24.24)$$
$$R-OO^{\bullet} + R-H \longrightarrow R-OO-H + R^{\bullet} \quad (24.25)$$

termination $\quad 2\ R-OO^{\bullet} \longrightarrow$ non-radical products $\quad (24.26)$

24-5-2 High-temperature reactions

One of the peculiarities of the reactions of hydrocarbons with oxygen is that they have a negative temperature coefficient. For example, an equimolar mixture of *n*-pentane and oxygen at 1 atm. total pressure reacts very slowly; the rate increases from 250°C, reaches a maximum at about 350°C and decreases to become zero at about 380°C. Subsequently, between 450 and 500°C the rate increases to a point where the reaction becomes explosive. When the mixture is heated to 300°C, there is an induction period during which there is no reaction, the duration of this period depending on the nature of the hydrocarbon, and of the walls of the reaction vessel, and on the geometry of the vessel.

This phenomenon is explained by a change of mechanism. At high temperature, reaction (24.24) becomes reversible, the energy of the R–O$_2^{\bullet}$ bond being only 28 kcal/mol. One then observes the formation of an olefinic compound, via a chain mechanism:

$$R\overset{\bullet}{C}H-CH_3 + O_2 \longrightarrow RCH=CH_2 + HOO^{\bullet} \quad (24.27)$$
$$HO_2^{\bullet} + RCH-CH_3 \longrightarrow R\overset{\bullet}{C}H-CH_3 + H_2O_2 \quad (24.28)$$

It is very likely that during the induction period ROOH is formed and that, when the temperature is raised, this initiates one or other of the chain reactions.

Direct measurements have indeed shown that as little as 0.1 mol% of hydroperoxide eliminates the induction period and reduces the negative temperature effect.

The low-temperature mechanism produces hydroperoxides exothermically as the first product; this reaction, which decreases with increasing temperature, is replaced by the practically thermoneutral formation of olefin.

When the temperature is raised, H_2O_2, which is more stable than ROOH, since BDE(HO–OH) = 51 kcal/mol, can play the part of ROOH and act as a secondary radical source, but only above 480°C.

The initiation reaction remains difficult to explain (24.23).

PART IV

REFERENCES AND TABLES

25 – REFERENCES

GENERAL REVIEWS AND BOOKS

1 **Kochi JK ed**, *Free Radicals*, Wiley, New York 1973 vol 1-2
2 **Pryor WA**, *Free Radicals*, Mc Graw-Hill, New York 1966
3 **Huyser ES**, *Free Radical Chain Reactions*, Wiley-Interscience, New York 1970
4 **Pryor WA ed**, *Free Radicals in Biology*, Academic Press, New York 1976-1990 vol 1-12
5 **Nonhebel DC, Walton JC,** *Free-Radical Chemistry. Structure and Mechanism*, Cambridge University Press, 1974
6 **Nonhebel DC, Tedder JM, Walton JC,** *Radicals*, Cambridge University Press, 1977
7 **Giese B,** *Radicals in Organic Synthesis: Formation of Carbon–Carbon Bonds*, Pergamon Press, Oxford 1986
8 **Huyser ES ed**, *Methods in Free Radical Chemistry*, Marcel Dekker, New York 1969-1974 vol 1-5
9 **Williams GH ed**, *Advances in Free Radical Chemistry*, Academic and Heyden and Logos, London and New York 1965-1980 vol 1-5
10 **Ingold KU, Roberts BP,** *Free-Radical Substitution Reactions*, Wiley Interscience, New York 1971
11 **Regitz M, Giese B,** C-Radikale, *Houben-Weyl*, 4th edn, Georg Thieme Verlag, Stuttgart 1989 vol E19a
12 **Griller D, Lorand JP,** *Radical Reaction Rates in Liquids*, Landolt-Börnstein, New Series, Group II, Fischer H, Beckwith ALJ eds, Springer Verlag, Berlin 1984 vol 13a
13 **Motherwell WB, Crich D,** *Free Radical Chain Reactions in Organic Synthesis*, Academic Press, New York 1992

PART I - GENERAL CONCEPTS AND BASIC PRINCIPLES

Chapter 2 - detection and observation of free radicals

refs *1-12*
14 **Weil JA, Bolton JR, Wertz JE,** *Electron Paramagnetic Resonance: Elementary Theory and Practical Applications*, Wiley, New York 1994
15 **Kochi JK,** Configurations and Conformations of Transient Alkyl Radicals in Solution by Electron Spin Resonance Spectroscopy, *Adv Free Radical Chem* 1975 **5** 189
16 **Griller D, Ingold KU,** Electron Paramagnetic Resonance and the Art of Physical-Organic Chemistry, *Acc Chem Res* 1980 **13** 193
17 **Jantzen EG,** Spin Trapping, *Acc Chem Res* 1971 **4** 31
18 **Berliner LJ ed,** *Spin Labelling: Theory and Applications*, Academic Press, New York 1976
19 **Swartz HM, Bolton JR, Borg DC,** *Biological Applications of Electron Spin Resonance*, Wiley, New York 1972

20 **Perkins MJ,** Spin Trapping, *Adv Phys Org Chem* 1980 **17** 1
21 **Fischer H,** The Structure of Free Radicals by ESR Spectroscopy,
    ref *1* vol 2 p 435
22 **Kaptein R,** Chemically Induced Dynamic Nuclear Polarisation. Theory and
    Applications in Mechanistic Chemistry, *Adv Free Radical Chem* 1975 **5** 319
23 **Ward HR,** Chemically Induced Dynamic Nuclear Polarisation (CIDNP).
    I. The Phenomenon, Examples and Applications, *Acc Chem Res* 1972 **5** 18
24 **Ward HR,** Chemically Induced Dynamic Nuclear Polarization,
    ref *1* vol 1 p 239

Chapter 3 - radical structure

25 **Kaplan L,** The Structure and Stereochemistry of Free Radicals,
    ref *1* vol 2 p 361
26 **Norman ROC,** Structures of Organic Radicals, *Chem in Britain* 1970 **6** 66

Chapter 4 - radical stability

27 **Leroy G, Sana M, Wilante C, Nemba RM,** Bond-Dissociation Energies of
    Organic Compounds. A Tentative Rationalization Based on the Concept of
    Stabilization Energy, *J Mol Structure* 1989 **198** 159
28 **Griller D, Ingold KU,** Persistent Carbon-Centered Radicals,
    Acc Chem Res 1976 **9** 13
29 **Viehe HG, Merenyi R, Stella L, Janousek Z,** Capto-dative Substituents
    Effects in Synthesis with Radicals and Radicophiles,
    Angew Chem Int Ed Engl 1979 **18** 917
30 **Viehe HG, Janousek Z, Merenyi R, Stella L,** The Captodative Effect,
    Acc Chem Res 1985 **18** 148
31 **Sustmann R, Korth HG,** The Captodative Effect,
    Adv Phys Org Chem 1990 **26** 131
32 **O'Neal HE, Benson SW,** Thermochemistry of Free Radicals, ref *1* vol 2 p 275

Chapter 5 - elementary reactions and mechanisms

   refs *2, 5-6*

Chapter 6 - reactivity of free radicals

33 **Lefort D, Fossey J, Sorba J,** Facteurs contrôlant la réactivité des radicaux
    libres, *New J Chem 1992* **16** 219
34 **Bottoni A, Fossey J, Lefort D,** On the Role of the Orbital Interaction Concept
    in the Interpretation of Organic Free Radical Structures and Reactivities,
    Molecules in Physics, Chemistry and Biology, Maruani J ed,
    Kluwer Academic Plublishers, Dordrecht 1989 **vol 3** p 173
35 **Minisci F, Citterio A,** Polar Effects in Free-Radical Reactions in Synthetic
    Chemistry, ref *9* vol 6 p 65
36 **Giese B,** The Stereochemistry of Intermolecular Free Radical Reactions,
    Angew Chem Int Ed Engl 1989 **28** 969
37 **Tedder JM,** The Importance of Polarity, Bond Strength and Steric Effects in
    Determining the Site of Attack and the Rate of Free Radical Substitution in
    Aliphatic Compounds, *Tetrahedron* 1982 **38** 313
38 **Tedder JM,** Which Factors Determine the Reactivity and Regioselectivity of

Part I - general concepts and basic principles

Free Radical Substitution and Addition Reaction?
Angew Chem Int Ed Engl 1982 **21** 401
39 **Bonacic-Koutecky V, Koutecky J, Salem L,** A Theory of Free Radical Reactions, *J Amer Chem Soc* 1977 **99** 842
40 **Delbecq F, Ilavsky D, Nguyen Trong Anh, Lefour JM,** Etude théorique de l'orientation des additions radicalaires, *L'actualité chimique* 1986 (1-2) 49
41 **Kerr JA,** Rate Processes in the Gas Phase, ref *1* vol 1 p 1
42 **Ingold KU,** Rate Constants for Free Radical Reactions in Solution, ref *1* vol 1 p 37
43 **Russell GA,** Reactivity, Selectivity and Polar Effects in Hydrogen Atom Transfer Reactions, ref *1* vol 1 p 275
44 **Trotman-Dickenson AF,** The Abstraction of Hydrogen Atoms by Free Radicals, ref *9* vol 1 p 1
45 **Howard JA,** Absolute Rate Constants for Reactions of Oxyl Radicals, ref *9* vol 4 p 49

Chapter 7 - radical kinetics

ref *8*

Chapter 8 - radicals centered on an atom other than carbon

ref *8*
46 **Poller RC,** Free Radical Reactions of Organotin Compounds, *Rev Silicon, Germanium, Tin Compd* 1978 **3** 243
47 **Kuivila HG,** Organotin Hydrides and Organic Free Radicals, *Acc Chem Res* 1968 **1** 299
48 **Alberti A, Pedulli G,** Addition Reactions of Silyl Radicals to Unsaturated Compounds, *Rev Chem Intermed* 1987 **8** 207
49 **Tordo P,** ESR Spectra of Free Radicals Derived from Phosphines, *The Chemistry of Functional Groups, The Chemistry of Organophosphorus Compounds*, Patai S, Hartley FR eds, Wiley, New York 1990 vol 1 p 137
50 **Danen WC, Neugebauer FA,** Aminyl Free Radicals, *Angew Chem Int Ed Engl* 1975 **14** 783
51 **Mackiewicz P, Furstoss R,** Radicaux amidyle: Structure et réactivité, *Tetrahedron* 1978 **34** 3241
52 Free Radicals with Heteroatoms, Part IV of ref *1* vol 2 p 527
53 **Cadogan JIG,** Radical Reactions of Phosphorus Compounds, ref *9* vol 2 p 203
54 **Roberts BP,** The Chemistry of Phosphoranyl Radicals, ref *9* vol 6 p 225

PART II - REACTIONS: CLASSIFICATIONS AND MECHANISMS

Chapter 9 - production of free radicals

refs *1-12*
55 **Davies AG,** *Organic Peroxides*, Butterworths, London 1961
56 **Swern D ed,** *Organic Peroxides*, Wiley, New York 1971 vol 1-3
57 **Kochi JK,** *Organometallic Mechanism and Catalysis*, Academic Press, New York 1978
58 **Kochi JK,** Electron Transfer Mechanisms for Organometallic Intermediates in Catalytic Reactions, *Acc Chem Res* 1974 **7** 351

59 Minisci F, Free-Radical Additions to Olefins in the Presence of Redox Systems, *Acc Chem Res* 1975 **8** 165
60 Minisci F, Citterio A, Giordano C, Electron-Transfer Processes: Peroxydisulfate, a Useful and Versatile Reagent in Organic Chemistry, *Acc Chem Res* 1983 **16** 27
61 Koenig T, The Decomposition of Peroxides and Azoalkanes, ref *1* vol 1 p 113
62 Kochi JK, Oxidation-Reduction Reactions of Free Radicals and Metal Complexes, ref *1* vol 1 p 591
63 Gilde HG, Electrolytically Generated Radicals, ref *8* vol 3 p 1

Chapter 10 - radical-radical reactions

64 Leroy G, Sana M, A Theoretical Approach to Dehydrodimerization Reactions, *J Mol Struct (Theochem)* 1988 **179** 237
65 Gibian HJ, Corley RC, Organic Radical-Radical Reactions. Disproportionation vs Combination, *Chem Rev* 1973 **73** 441
66 Koenig T, Fischer H, "Cage" Effects, ref *1* vol 1 p 157

Chapter 11 - substitution reactions

refs *35, 37-38*
67 Davies AG, Roberts BP, Bimolecular Homolytic Substitution at a Metal Center, *Acc Chem Res* 1972 **5** 387
68 Brown HC, Midland MM, Organic Synthesis via Free-Radical Displacement Reactions of Organoboranes, *Angew Chem Int Ed Engl* 1972 **11** 692
69 Poutsma ML, Atom-Transfer and Substitution Processes, ref *1* vol 2 p 113
70 Davies AG, Roberts BP, Bimolecular Homolytic Substitution at Metal Centres, ref *1* vol 1 p 547

Chapter 12 - addition and fragmentation reactions

refs *36, 40*
71 Giese B, Formation of CC Bonds by Addition of Free Radicals to Alkenes, *Angew Chem Int Ed Engl* 1983 **22** 753
72 Tedder JM, Walton JC, The Importance of Polarity and Steric Effects in Determining the Rate and Orientation of Free Radical Addition to Olefins, *Tetrahedron* 1980 **36** 701
73 Tedder JM, Walton JC, The Kinetics and Orientation of Free-Radical Addition to Olefins, *Acc Chem Res* 1976 **9** 183
74 Arnaud R, Etude théorique des réactions d'addition radicalaire: analyse du rôle des interactions frontalières, *New J Chem* 1989 **13** 543
75 Abell PI, Addition to Multiple Bonds, ref *1* vol 2 p 63

Chapter 13 - cyclizations and rearrangements

76 Beckwith ALJ, Ingold KU, Free-Radical Rearrangements, *Rearrangements in Ground and Excited States*, de Mayo P ed, Academic Press, New York 1980 vol 1 p 161
77 Surzur JM, Radical Cyclizations by Intramolecular Additions, *Reactive Intermediates*, Abramovitch RA ed, Plenum, New York 1982 **vol 2** p 121
78 Beckwith ALJ, Schiesser CH, Regio- and Stereo-Selectivity of Alkenyl

79   Radical Ring Closure: a Theoretical Study, *Tetrahedron* 1985 **42** 3925
79   **Baldwin JE,** Rules for Ring-Closure, *J Chem Soc Chem Commun* 1976 734
80   **RajanBabu TV,** Stereochemistry of Intramolecular Free-Radical Cyclization Reaction, *Acc Chem Res* 1991 **24** 139
81   **Julia M,** Free-Radical Cyclisations, *Acc Chem Res* 1971 **4** 386
82   **Wilt JW,** Free Radical Rearrangements, ref *1* vol 1 p 333
83   **Freidlina RK, Terent'ev AB,** Rearrangement of Short-Lived Radicals in the Liquid Phase, in ref *9* vol 6 p 1

Chapter 14 - aromatic substitution

84   **Minisci F, Vismara E, Fontana F,** Recent Developments of Free-Radical Substitutions of Heteroatomic Bases, *Heterocycles* 1989 **28** 489
85   **Minisci F,** Recent Aspects of Homolytic Aromatic Substitutions, *Top Curr Chem* 1976 **62** 1
86   **Tiecco M, Testaferri L,** Homolytic Aromatic Substitution by Alkyl Radicals, *Reactives Intermediates,* Abramovitch RA ed, Plenum, New York 1983 vol 3 p 61
87   **Hey DH,** Substitution aromatique homolytique, *Bull Soc Chim (Fr)* 1968 1591
88   **Vernin G,** Substitution homolytique en série aromatique et hétéroaromatique. Récents progrès dans les réactions d'hétéroarylation, *Bull Soc Chim (Fr)* 1976 1257
89   **Perkins MJ,** Aromatic Substitutions, ref *1* vol 2 p 231
90   **Hey DH,** Arylation of Aromatic Compounds, ref *9* vol 2 p 47

Chapter 15 - reactions of charged radicals.

ref *60*

91   **Kornblum N,** Substitution Reactions which Proceed via Radical Anion Intermediates, *Angew Chem Int Ed Engl* 1975 **14** 734
92   **Bunnett JF,** Aromatic Substitution by $S_{RN}1$ Mechanism, *Acc Chem Res* 1978 **11** 413
93   **Russell GA,** Free Radical Chain Processes in Aliphatic Systems Involving an Electron Transfer Reaction, *Adv Phys Org Chem* 1987 **23** 271
94   **Savéant JM,** Catalysis of Chemical Reactions by Electrodes, *Acc Chem Res* 1980 **13** 323
95   **Savéant JM,** Single Electron Transfer and Nucleophilic Substitution, *Adv Phys Org Chem* 1990 **26** 1
96   **Norris RK,** The $S_{RN}1$ Reaction of Organic Halides, *The Chemistry of Functional Groups Supplement D: The Chemistry of Halides, Pseudo-halides and Azides,* Patai S, Rappoport Z eds, Wiley, NewYork 1983 chap 16 p 681
97   **Sawyer DT, Valentine JS,** How Super is Superoxide? *Acc Chem Res* 1981 **14** 393
98   **Bard AJ, Ledwith A, Shine HJ,** Formation, Properties and Reactions of Cation Radicals in Solution, *Adv Phys Org Chem* 1976 **13** 155
99   **Hammerich O, Parker VD,** Kinetics and Mechanisms of Reactions of Organic Cation Radicals in Solution, *Adv Phys Org Chem* 1984 **20** 55
100   **Kochi JK,** Electron Transfer and Charge Transfer: Twin Themes in Unifying the Mechanisms of Organic and Organometallic Reactions, *Angew Chem Int Ed Engl* 1988 **27** 1227
101   **Garst JF,** Electron Transfer Reactions of Organic Anions, ref *1* vol 1 p 503
102   **Sosnovsky G, Rawlinson DJ,** Free Radical Reactions in the Presence of Metal Ions. Reactions of Nitrogen Compounds, ref *9* vol 4 p 203

Chapter 16 - free radicals in biochemistry

refs 5, 7
103   **Ferradini C, Pucheault J,** Biologie de l'action des rayonnements ionisants, Masson, Paris 1983
104   **Lefort M,** Chimie des radiations des solutions aqueuses, *Actions chimiques et biologiques des radiations,* Haissinski M ed, Masson, Paris 1955
105   **Ward JF,** Radiolytic Damage to Genetic Material, *J Chem Educ* 1981 **58** 135
106   **Sies H,** Biochemistry of Oxidative Stress, *Angew Chem Int Ed Engl* 1986 **28** 1058
107   **Pryor WA,** Free Radical Pathology, *Chem Eng News* 1971 **49** 35
108   **Bland J,** Biochemical Consequences of Lipid Peroxidation, *J Chem Educ* 1978 **55** 151
109   **Mason RP, Chignell CF,** Free Radicals in Pharmacology and Toxicology - Selected Topics, *Pharmac Rev* 1982 **33** 189
110   **Aldeman RC, Roth GS eds,** *Testing the Theories of Aging,* CRC Press, Boca Raton, 1982
111   **Floyd RA ed,** *Free Radicals and Cancer,* Marcel Dekker, New York 1982
112   **Simic M, Traylor K, Ward J, von Sonntag C eds,** *Oxygen Radicals in Biology and Medicine,* Plenum, New York 1988

Chapter 17 - how to prove that a reaction is a radical process

113   **Fossey J, Sorba J, Lefort D,** Comment montrer qu'une réaction est radicalaire, *L'actualité chimique* 1990 (2) 71
114   **Newcomb M,** Radical Kinetics and Mechanistic Probe Studies, *Advances in Detailed Reaction Mechanism,* Jai Press Inc, 1991 vol 1 p 1
115   **Griller D, Ingold KU,** Free-Radical Clocks, *Acc Chem Res* 1980 **13** 317
116   **Newcomb M, Curran DP,** A Critical Evaluation of Studies Employing Alkenyl Halide "Mechanistic Probes" as Indicators of Single-Electron-Transfer-Processes, *Acc Chem Res* 1988 **21** 206
117   **Ashby EC,** Single-Electron Transfer, a Major Reaction Pathway in Organic Chemistry. An Answer to Recent Criticisms, *Acc Chem Res* 1988 **21** 414

PART III - APPLICATIONS IN SYNTHESIS

refs 7, *13, 36, 71*
118   **Davies DI, Parrott MJ,** *Free Radicals in Organic Synthesis,* Springer Verlag, Berlin 1978
119   **Sosnovsky G,** *Free Radical Reactions in Preparative Organic Chemistry,* McMillan, New York 1964
120   **Jasperse CP, Curran DP, Fevig TL,** Radical Reactions in Natural Product Synthesis, *Chem Rev* 1991 **91** 1237
121   **Curran DP,** The Design and Application of Free Radical Chain Reactions in Organic Synthesis, *Synthesis* 1988 417 and 489
122   **Curran DP,** Radical Reactions and Retrosynthetic Planning, *Synlett* 1991 63
123   **Ramaiah M,** Radical Reactions in Organic Synthesis, *Tetrahedron* 1987 **43** 3541
124   **Barton DHR, Zard SZ,** Invention of New Reactions Useful in the Chemistry of Natural Products, *Pure Appl Chem* 1986 **58** 675
125   **Barton DHR,** The Invention of Chemical Reactions: the Last Five Years, *Tetrahedron* 1992 **48** 2529

Part III - applications in synthesis

126 Barton DHR, Motherwell WB, Quelques progrès récents dans la chimie des substances naturelles, *L'actualité chimique* 1983 (10) 9

Chapter 18 - functionalization of the C-H bond

ref *35*
127 **Deno NC,** Free-Radical Chlorinations via Nitrogen Cation Radicals, ref *8* vol 3 p 135
128 **Poutsma ML,** Halogenation, ref *1* vol 2 p 159
129 **Minisci F, Citterio A,** Polar Effects in Free-Radical Reactions in Synthetic Chemistry, in ref *9* vol 6 p 65
130 **Thaler WA,** Free-Radical Brominations, ref *8* vol 2 p 121

Chapter 19 - transformation of functional groups

refs *123, 126*
131 **Neumann WP,** Tri-*n*-butyl Hydride as Reagent in Organic Synthesis, *Synthesis* 1987 665
132 **Hartwig W,** Modern Methods for the Radical Deoxygenation of Alcohols, *Tetrahedron* 1983 **39** 2609
133 **Chatgilialoglu C,** Organosilanes as Radical-Based Reducing Agents in Synthesis, *Acc Chem Res* 1992 **25** 188
134 **Crich D,** O-Acyl Thiohydroxamates: New and Versatile Sources of Alkyl Radicals for Use in Organic Synthesis, *Aldrichimica Acta* 1987 **20** 35

Chapter 20 - functionalization of multiple bonds

refs *8, 59, 60, 71*
135 **Vogel HH,** Radical Addition of Carboxylic Acids and Carboxylic Acid Derivatives to Unsaturated Compounds as a Synthetic Method, *Synthesis* 1970 99
136 **Neale RS,** Nitrogen Radicals as Synthesis Intermediates. N-Halamide Rearrangements and Additions to Unsaturated Hydrocarbons, *Synthesis* 1971 1

Chapter 21 - cyclization

refs *77-83, 120-123*
137 **Stella L,** Homolytic Cyclizations of N-Chloroalkenylamines, *Angew Chem Int Ed Engl* 1983 **22** 337

Chapter 22 - aromatic substitution

refs *84-86, 91-92*
138 **Minisci F,** Novel Applications of Free-Radical Reactions in Preparative Organic Chemistry, *Synthesis* 1973 1

Chapter 23 - coupling reactions

ref 29
139 **Schäfer HJ,** Recent Contributions of Kolbe Electrolysis to Organic Synthesis, *Top Curr Chem* 1990 **152** 91
140 **Schäfer HJ,** Anodic and Cathodic C–C Bond Formation, *Angew Chem Int Ed Engl* 1981 **20** 911

Chapter 24 - radical reactions in industry

141 The Plastics Market in Western Europe, *Kunstoffe German Plastics*, Hans Publishers, Munich 1989 **79** 6
142 **Seymour RB,** Polymers are Everywhere, *J Chem Educ* 1988 **65** 327
143 **Pryor WA,** Organic Free Radicals, *Chem Eng News* 1968 70
144 **Benson SW, Nangia PS,** Some Unresolved Problems in Oxidation and Combustion, *Acc Chem Res* 1979 **12** 223

26 – TABLES

26-1 RATE CONSTANTS

Rate constants (k) are given in $M^{-1}s^{-1}$ for bimolecular reactions and in s^{-1} for unimolecular reactions (rearrangement and fragmentation). Temperatures (T) are given in degrees Celsius (°C).

26-1-1 Hydrogen abstraction

Carbon radicals

	•CH$_3$ k	•CH$_3$ T	•CH$_2$R k	•CH$_2$R T	•CHR$_2$ k	•CHR$_2$ T	•CR$_3$ k	•CR$_3$ T
THF(H$_\alpha$)			6.0 x 10^3	25°			2.0 x 10^3	25°
CH$_3$CN	< 3 x 10^2							
HCCl$_3$							1.4 x 10^2	25°
1,4-cyclohexadiene	1.3 x 10^5	27°	4.8 x 10^5	27°			9.4 x 10^3	27°
(cy-C$_6$H$_{11}$)$_2$P–H			1.0 x 10^6	25°			2.5 x 10^3	25°
t-BuS–H			7.9 x 10^6	25°				
PhS–H			9.2 x 10^7	25°	1.1 x 10^8	25°	1.5 x 10^8	25°
Et$_3$Si–H			7.0 x 10^3	25°			3.0 x 10^3	25°
(Me$_3$Si)$_3$Si–H			3.8 x 10^5	25°	1.4 x 10^5	25°	2.6 x 10^5	25°
n-Bu$_3$Sn–H	1.2 x 10^7	25°	2.7 x 10^6	25°	1.5 x 10^6	25°	1.7 x 10^6	30°
n-Bu$_3$Ge–H	5.0 x 10^5	25°	1.0 x 10^5	25°				

Table 26.1 – *Hydrogen abstraction by methyl, primary, secondary and tertiairy carbon radicals*

	Ph• k	Ph• T	•CH$_2$Ph k	•CH$_2$Ph T	cy-C$_3$H$_5$• k	cy-C$_3$H$_5$• T
RCH$_2$–H	0.4 x 10^5	45°				
R$_2$CH–H	3.3 x 10^5	45°				
R$_3$–H	1.6 x 10^6	45°				
1,4-cyclohexadiene			1.0 x 10^2	27°		
PhCH$_2$–H	3.3 x 10^5	45°				
MeCN	1.0 x 10^5	25°				
THF(H$_\alpha$)	4.8 x 10^6	25°				
EtO–H	2.3 x 10^5	45°				
(cy-C$_6$H$_{11}$)$_2$P–H			2.5 x 10^3	25°		
PhS–H	1.9 x 10^9	25°	3.0 x 10^5	30°		
n-Bu$_3$Sn–H	5.9 x 10^8	25°	3.6 x 10^4	25°	8.5 x 10^7	25°
n-Bu$_3$Ge–H	2.6 x 10^8	25°			1.3 x 10^7	25°

Table 26.2 – *Hydrogen abstraction by phenyl, benzyl and cyclopropyl radicals*

Oxygen radicals

	HO• k T	t-BuO• k T	t-BuOO•[a] k T	PhCO$_2$• k T
cyclopentane		8.8 x 10^5 21°		
cyclohexane	6.1 x 10^9 22°	1.6 x 10^6 21°		1.4 x 10^6 22°
PhCH$_2$–H		2.3 x 10^5 21°	8.3 x 10^{-2} 30°	
1,4-cyclohexadiene	7.7 x 10^9 22°			6.6 x 10^7 22°
THF(H$_\alpha$)		8.3 x 10^6 21°		2.5 x 10^3 25°
MeCN	2.2 x 10^7 22°			≤ 1 x 10^5 22°
Me$_2$C(OH)–H		1.8 x 10^8 21°		
MeC(O)–H		1.0 x 10^7 125°		
H–CCl$_3$	10^7 22°	4.6 x 10^5 22°		
PhO–H			2.8 x 10^3 30°	
Et$_3$Si–H		5.7 x 10^6 27°		
Et$_3$Sn–H		2.0 x 10^8 25°		6.0 x 10^4 25°
Me$_2$N–H		1.1 x 10^7 25°		

Table 26.3 – *Hydrogen abstraction by oxyl radicals. a) The increase in the rate constant for phenol is due to hydrogen bonding in the the transition structure.*

26-1-2 Halogen abstraction

Carbon radicals

	•CH$_2$R k T	t-Bu• k T	Ph• k T	PhC•=O k T
CCl$_4$	7.2 x 10^3 25°	4.9 x 10^4 25°	6.0 x 10^6 25°	5.6 x 10^4 25°
Br–CCl$_3$	7.0 x 10^6 25°	2.6 x 10^8 80°	1.4 x 10^9 25°	2.2 x 10^8 25°
NBS[a]	6 x 10^2 50°			
t-Bu–Cl	6 x 10^2 50°			
t-Bu–Br	4.6 x 10^3 50°		1 x 10^6 25°	
t-Bu–I	3 x 10^6 50°		2.0 x 10^9 25°	
n-Bu–Br	6 x 10^2 50°			
i-Pr–I			1 x 10^9 25°	
PhCH$_2$SO$_2$–Cl	1.3 x 10^6 30°			
Br–CH$_2$CO$_2$Et	7 x 10^4 50°			
I–CH$_2$CO$_2$Et	2.6 x 10^7 50°			

Table 26.4 – *Halogen abstraction by carbon radicals. a) N-Bromosuccinimide.*

1,2 Transfer

•CH$_2$—C(X)(CH$_3$)(CH$_3$) $\longrightarrow$ XCH$_2$—C•(CH$_3$)(CH$_3$)

$k > 10^8$ s^{-1} for X = Cl
$k \approx 10^{11}$ s^{-1} for X = Br

Table 22.5 – *1,2 Halogen transfer*

Group 14 radicals

	$Et_3Si^\bullet$ k	T	$n\text{-}Bu_3Sn^\bullet$ k	T	$n\text{-}Bu_3Ge^\bullet$ k	T
Me–I	8.1×10^9	25°	4.3×10^9	25°		
n-Pr–Br	5.4×10^8	30°	3.2×10^7	25°	4.6×10^7	27°
n-Pr–I	4.3×10^9	30°			$\geq 3 \times 10^7$	27°
RCH_2–Cl			8.5×10^2	80°		
RCH_2–Br			5.0×10^7	80°		
RCH_2–I			2.5×10^9	80°		
$CH_2=CH(CH_2)_4$–Cl	2.6×10^5	25°	6.6×10^3	25°	1.2×10^4	25°
$CH_2=CH(CH_2)_4$–Br	5.5×10^8	25°	5.0×10^7	25°	4.6×10^7	25°
t-Bu–Cl	2.5×10^6	30°	2.7×10^4	25°	$< 5 \times 10^4$	25°
t-Bu–Br	1.1×10^9	25°	1.4×10^8	25°	8.6×10^7	25°
t-Bu–I			2.5×10^9	25°		
$PhCH_2$–Cl	1.4×10^7	25°	1.1×10^6	25°	1.3×10^6	25°
$PhCH_2$–Br	2.4×10^9	30°	1.7×10^9	25°	8.6×10^7	25°
$PhCH_2SO_2$–Cl	5.7×10^9	30°				
$p\text{-}MeOC_6H_4$–Br			2.4×10^6	25°		
$p\text{-}MeOC_6H_4$–I			8.8×10^8	25°		
CCl_4	4.6×10^9	30°				

Table 26.6 – *Halogen abstraction by $Et_3Si^\bullet$, $n\text{-}Bu_3Sn^\bullet$ and $n\text{-}Bu_3Ge^\bullet$ radicals.*

26-1-3 Substitution on a heteroatom

Intramolecular substitution

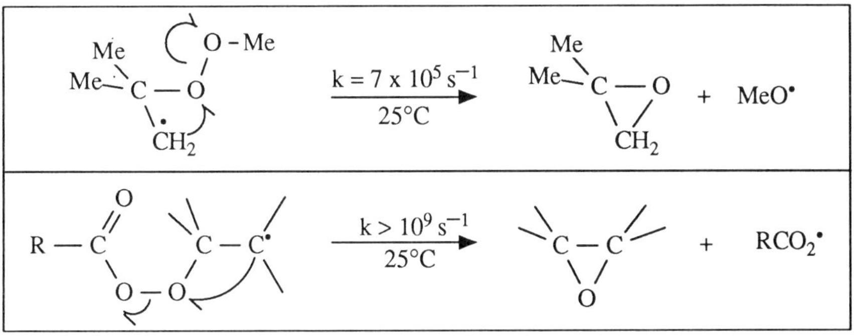

Table 26.7 – *Intramolecular substitution.*

Intermolecular substitution on S and Se

	n-Bu$_3$Sn•		n-Bu$_3$Ge•	
	k	T	k	T
n-BuOCH$_2$–SPh	1.0×10^3	25°	1.6×10^4	25°
n-BuOCH$_2$–SePh	6.0×10^6	25°	2.3×10^7	25°

Table 26.8 – *Substitution by n-Bu$_3$Sn• and n-Bu$_3$Ge• radicals.*

26-1-4 Addition to a double bond

Carbon radicals

	•CH$_2$R		t-Bu•		Ph•		•CCl$_3$	
	k	T	k	T	k	T	k	T
styrene	5.0×10^4	20°	5.9×10^7	21°	1.1×10^8	25°	3.8×10^5	22°
methyl acrylate	2.1×10^5	20°						
acrylonitrile	5.3×10^5	20°	2.4×10^6	27°				
2-methylpropene			7.4×10^2	21°				
cyclohexene					2.8×10^8	25°		
benzene					4.5×10^5	25°		
benzoquinone	2.0×10^7	69°						
CH$_2$=CHCH$_2$SnBu$_3$	3.0×10^4	80°						

Table 26.9 – *Addition of carbons radicals.*

R• + CH$_2$=CH–Z $\xrightarrow{k}$ R–CH$_2$–ĊH–Z			
R•	Z	k	T
Me•	P(O)(OEt)$_2$	2.5×10^3	–40°
Et•	P(O)(OEt)$_2$	2.6×10^3	–40°
n-Bu•	P(O)(OEt)$_2$	5.0×10^3	–40°
t-Bu•	P(O)(OEt)$_2$	5.9×10^3	–40°
n-Hex•	CN	5.9×10^5	0°
t-Bu•	CN	1.0×10^6	23°

Table 26.10 – *Rate constants for addition of R• to CH$_2$=CH–Z.*

Group 14 radicals

	$Et_3Si^\bullet$		$n\text{-}Bu_3Sn^\bullet$		$n\text{-}Bu_3Ge^\bullet$	
	k	T	k	T	k	T
acrylonitrile	1.1×10^9	27°	8.3×10^7	27°	1.8×10^8	27°
styrene	2.2×10^8	27°	1.0×10^8	27°	8.6×10^7	27°
cyclohexene	9.4×10^5	27°				
2-methylpropene	3.7×10^6	27°				
1-hexene	4.8×10^6	27°				

Table 26.11 – *Addition of $Et_3Si^\bullet$, $n\text{-}Bu_3Sn^\bullet$ and $n\text{-}Bu_3Ge^\bullet$ radicals.*

Oxygen and sulfur radicals

	$t\text{-}BuO^\bullet$		$t\text{-}BuS^\bullet$		$t\text{-}BuOO^\bullet$	
	k	T	k	T	k	T
1-octene			1.9×10^6	25°	$0.6 \sim 0.8$	118°
styrene	$\sim 10^6$	27°			1.3	30°
norbornene	1.1×10^6	27°				

	$PhCO_2^\bullet$		$PhCO_3^\bullet$		$MeCO_3^\bullet$	
	k	T	k	T	k	T
styrene	5.1×10^7	24°				
α-methylstyrene			1.8×10^5	20°		
1-octene			4.5×10^2	20°		
cis 2-butene					7.5×10^5	120°
benzene	7.8×10^7	24°				

Table 26.12 – *Addition of oxygen and sulfur radicals.*

26-1-5 Addition to aromatic substrates

$R^\bullet$	PhH	4-CNPyH$^+$
$t\text{-}Bu^\bullet$	a	6.3×10^7
$^\bullet CH_2OH$	a	$\approx 10^7$
$Ph^\bullet$	1.0×10^6	6.0×10^6
$t\text{-}BuC^\bullet{=}O$	a	$\approx 10^6$
$n\text{-}Bu^\bullet$	3.8×10^2	8.9×10^5
$PhCO_2^\bullet$	1.4×10^2	-
$HO^\bullet$	5×10^9	-
$^{\bullet+}NHR_2$	$> 10^6$	

Table 26.13 – *Addition of radicals to benzene and protonated 4-cyanopyridine. a) No reaction.*

substrate	k
PhH	3.8×10^2
PhOCH$_3$	1.3×10^3
4-Me pyridine	1.5×10^3
4-Me pyridine H$^+$	1.1×10^5
4-CN pyridine H$^+$	1.8×10^6
quinoxaline H$^+$	2.7×10^7

Table 26.14 – *Addition of primary alkyl to aromatic substrates*

26-1-6 Cyclization and ring opening

initial radical $\xrightarrow{k_c}_{k_o}$ cyclized radical		k_c	k_o
(allyl-propyl radical)	(cyclopentyl radical)	2.3×10^5	
(methyl-substituted)	(methyl-cyclopentyl)	1×10^5	
(phenyl allyl)	(indanyl)	5×10^8	
(allyloxy radical)	(tetrahydrofuranyl)	6×10^8	
(alkynyl)	(methylenecyclopentyl)	1×10^5	
(acyl-alkenyl)	(cyclopentyloxy)	8.7×10^5	4.7×10^8
(acyl-pentenyl)	(cyclohexyloxy)	1×10^6	1.1×10^7
(nitrile-alkenyl)	(iminocyclopentyl)	4×10^3	
(dimethyl-allyl)	(dimethylcyclopropyl)	1.7×10^7	3×10^8
(methyl-butenyl)	(methylcyclopropyl)	$< 3 \times 10^4$	1.7×10^9
(=NC$_3$H$_7$)	(NC$_3$H$_7$-cyclopropyl)		2.5×10^7

Table 26.15 – *Cyclization, k_c, and ring opening, k_o at 25°C.*

$\diagup\!\!=\!\!\diagdown (CH_2)_n\diagup\dot{C}H_2$	k_{exo}	k_{endo}	k_{exo}/k_{endo}	k_{-exo}
n = 1	1.8×10^4	not obs.	-	2.0×10^8
n = 2	1.0	not obs.	-	4.7×10^3
n = 3	2.3×10^5	4.1×10^3	58	-
n = 4	5.2×10^3	8.3×10^2	6	-
n = 5	< 70	1.2×10^2	< 0.6	-

Table 26.16 – *Cyclization and reopening of alkyl radicals at 25°C.*

$-X=Y$	A	B	k	E_a
$-CH=CH_2$	CMe_2	CH_2	10^7	5.7
$-C\equiv C-t\text{-}Bu$	CMe_2	CH_2	9.3×10	12.8
$-C\equiv N$	CMe_2	CH_2	0.9	16.4
$-(t\text{-}Bu)C=O$	CMe_2	CH_2	1.7×10^5	7.8
$-Ph$	CMe_2	CH_2	7.6×10^2	11.8
$-Ph$	CH_2	CH_2	5.9×10	
$-Ph$	CPh_2	CH_2	3.6×10^5	
$-Ph$	CPh_2	O	5.0×10^{10}	

Table 26.17 – *1,2 Transfer of unsaturated groups. Activation energy, E_a, in kcal/mol.*

26-1-7 Fragmentation

fragmentation			k	T
$PhCO_2^\bullet$	$\rightarrow$ $Ph^\bullet$	$+ CO_2$	2.0×10^6	25°
$p\text{-}MeOPhCO_2^\bullet$	$\rightarrow$ $p\text{-}MeOPh^\bullet$	$+ CO_2$	1.4×10^6	25°
$p\text{-}ClPhCO_2^\bullet$	$\rightarrow$ $p\text{-}ClPh^\bullet$	$+ CO_2$	0.3×10^6	25°
$MeCO_2^\bullet$	$\rightarrow$ $Me^\bullet$	$+ CO_2$	1.6×10^9	60°
$RCO_2^\bullet$	$\rightarrow$ $R^\bullet$	$+ CO_2$	$> 10^{10}$	
$PhC^\bullet=O$	$\rightarrow$ $Ph^\bullet$	$+ CO$	1.7×10^4	67°
$PhCH_2C^\bullet=O$	$\rightarrow$ $PhCH_2^\bullet$	$+ CO$	5.0×10^7	25°
$MeC^\bullet=O$	$\rightarrow$ $Me^\bullet$	$+ CO$	7.3	25°
$t\text{-}BuC^\bullet=O$	$\rightarrow$ $t\text{-}Bu^\bullet$	$+ CO$	1.3×10^5	27°
$RSO_2^\bullet$	$\rightarrow$ $R^\bullet$	$+ SO_2$	$10^4 \approx 10^5$	
$t\text{-}BuO^\bullet$	$\rightarrow$ $t\text{-}BuO^\bullet$	$+$ acetone	10^5	60°

Table 26.18 – *Fragmentation.*

26-1-8 Radical-radical reactions

radical	solvent	k	T
Me$^\bullet$	gas	$\approx 3 \times 10^{10}$	25°
Me$^\bullet$	cyclohexane	4.5×10^9	25°
$C_6H_{13}^\bullet$	cyclohexane	1.4×10^9	25°
cy-$C_6H_{11}^\bullet$	cyclohexane	1.4×10^9	25°
cy-$C_6H_{11}^\bullet$	benzene	8.3×10^8	25°
t-Bu$^\bullet$	cyclohexane	1.1×10^9	25°
PhCH$_2^\bullet$	cyclohexane	1.0×10^9	25°
PhCH$_2^\bullet$	benzene	0.9×10^9	25°
$^\bullet$CCl$_3$	CCl$_4$	2.5×10^9	25°
RCH$_2$OO$^\bullet$	benzene	$\approx 3 \times 10^8$	30°
t-BuOO$^\bullet$	cyclohexane	1.0×10^3	30°

Table 26.19 – *Radical dimerization in solution.*

recombination		k	T
Ph$^\bullet$ + O$_2$ $\longrightarrow$	PhOO$^\bullet$	4.4×10^9	27°
t-Bu$^\bullet$ + O$_2$ $\longrightarrow$	t-BuOO$^\bullet$	4.9×10^9	27°
R$^\bullet$ + >NO$^\bullet$ $\longrightarrow$	>NO–R	$10^8 \sim 10^9$	

Table 26.20 – *Recombination.*

26-1-9 Homolysis

	k	T		k	T
(PhCO$_2$–)$_2$	1.95×10^{-6}	60°	(t-BuO–O$_2$C–)$_2$	1.7×10^{-5}	35°
	3×10^{-5}	80°		1×10^{-3}	55°
(MeCO$_2$–)$_2$	0.5×10^{-5}	60°	t-BuO–Ot-Bu	1.4×10^{-8}	70°
	1.3×10^{-5}	80°	t-BuO–OH	1×10^{-5}	150°
(i-PrO–CO$_2$–)$_2$	2.3×10^{-5}	50°	PhO–OPh	5×10^{-4}	90°
PhCO$_2$–O$_2$CMe	2×10^{-5}	70°	(Me$_2$C(CN)–N=)$_2$	1.5×10^{-4}	70°
PhCO$_2$–Ot-Bu	1×10^{-4}	120°	PhS–SPh	2×10^{-8}	90°

Table 26.21 – *Rate constant for homolysis.*

26-2 IONIZATION POTENTIALS

radical	IP	radical	IP	radical	IP	radical	IP
•CH_3	9.84	Ph•	8.25	•CF_3	8.9	•CMe_2CN	8.2
1-C_4H_9•	8.02	•CH_2Ph	7.20	•CCl_3	7.8	•CH_2CO_2H	10.9
2-C_4H_9•	7.25	p-$FPhCH_2$•	8.18	MeO•	8.6	•CH_2NH_2	6.1
t-C_4H_9•	6.70	allyl (prim)	8.13	t-BuO•	12	•CH_2NMe_2	5.7
cy-C_5H_9•	7.21	Cl•	12.97	t-BuOO•	11.5	H•C=O	8.1

Table 26.22 – *Ionization potential (IP) in eV.*

26-3 BOND ENERGIES

Bond energies (BDE) are given in kcal/mol (1 cal = 4.184 Joule). The error on BDE i rarely less than 1 kcal/mol.

26-3-1 C–H and heteroatom–H bonds

R–H	BDE	$E_s(R•)$	R–H	BDE	$E_s(R•)$
CH_3–H	105	0	H–$CHMeCO_2Me$	92	13
Et–H	101	4	$BrCH_2$–H	103	2
n-Pr–H	101	4	$ClCH_2$–H	103	2
t-$BuCH_2$–H	100	5	FCH_2–H	103	2
i-Pr–H	99	6	Br_3C–H	96	9
t-Bu–H	95	10	Cl_3C–H	96	9
cy-C_3H_5–H	106	21	F_3C–H	108	–3
cy-C_4H_7–H	97	8	H–CH_2NH_2	95	10
cy-C_6H_{11}–H	99	6	H–CH_2NMe_2	88	17
H–$CH_2CH=CH_2$	86	19	H–CH_2OH	94	11
H–$CH_2CMe=CH_2$	85	20	H–CH_2OMe	93	12
H–(cyclopentadienyl)–H	73	32	H–CHMeOEt	92	13
			H_α–THF	92	13
H–$CH_2C\equiv CH$	90	15	H–CMe_2OH	91	14
H–CH_2Ph	85	20	H–$CCNMe_2$	86	19
H–CPh_3	77	28	$CH_2=CH$–H	104	1
H–CH_2CHO	92	13	Ph–H	111	–6
H–CH_2COCH_3	92	13	H–COMe	86	19
H–CH_2CO_2H	97	8	H–COPh	87	18
H–CH_2CO_2Me	99	6	H–CO_2Me	94	11

Table 26.23 – *BDE(R–H) and $E_s(R•)$ stabilisation energies (see para 4-1-2).*

•R	BDE(R–H)	E_s	E_s^*	ΔE
•CH(CHO)$_2$	99	6	26	20
•CH(NO$_2$)$_2$	99	6	18	12
•CH(t-Bu)$_2$	98	7	18	11
•CH(OCH$_3$)$_2$	91	14	24	10
•CH(CH$_3$)OCH$_3$	91	14	16	2
•CH(NH$_2$)CHO	73	31	23	–8
•CH(NH$_2$)CO$_2$H	76	29	18	–12

Table 26.24 – BDE(R–H); E_s, stabilization energy calculated from BDE; E_s^* stabilization energy calculated by additivity; ΔE, see eqn (4.7), para 4-1-2).

F–H	136	MeO–H	104	MeS–H	88
Cl–H	103	t-BuO–H	103	PhS–H	80
Br–H	88	PhO–H	86	Me$_3$Si–H	90
I–H	71	MeCO$_2$–H	106	Et$_3$Si–H	90
Me$_2$N–H	92	PhCO$_2$–H	110	(Me$_3$Si)$_3$Si–H	74
Me$_2^+$NH–H	96	HOO–H	90	n-Bu$_3$Ge–H	84
PhNH–H	88	t-BuOO–H	88	Me$_3$Sn–H	74
(cy-C$_6$H$_{11}$)$_2$P–H	74	RCO$_3$–H	93	n-Bu$_3$Sn–H	74

Table 26.25 – BDE(Σ–H).

26-3-2 C–C, C–heteroatom, C–halogen and heteroatom–heteroatom bonds

Me–Me	90	CH$_3$–NH$_2$	85	Me–OMe	83
t-BuCH$_2$–CH$_3$	87	CH$_3$–$^+$NH$_3$	112	t-Bu–OMe	83
t-Bu–t-Bu	67	PhCH$_2$–NH$_2$	74	Me–SPh	57
t-Bu$_2$CH–CHt-Bu$_2$	36	t-Bu–NO	43	Me–SiMe$_3$	76
CH$_2$=CHCH$_2$–Me	74	Ph–NO	53		

Table 26.26 – BDE(C–C) and BDE(C–Σ).

CH$_3$–Cl	83	CH$_3$–Br	71	CH$_3$–I	57
Cl$_3$C–Cl	71	Cl$_3$C–Br	56		
CH$_2$=CH–Cl	87	CF$_3$–Br	71	CF$_3$–I	56
Ph–Cl	95	Ph–Br	81	Ph–I	65
PhCH$_2$–Cl	74	PhCH$_2$–Br	60	PhCH$_2$–I	50
PhC(O)–Cl	74	Me$_3$C–Br	70	Me$_3$C–I	54

Table 26.27 – BDE(C–halogen).

F–F	38	HO–OH	51	R_2N–Cl	45
Cl–Cl	59	t-BuO–OH	44	R_2N–NO	44
Br–Br	46	t-BuO–Ot-Bu	38	n-Bu$_3$Sn–Cl	94
I–I	36	RCO$_2$–O$_2$CR	30	n-Bu$_3$Sn–Br	83
H$_2$N–NH$_2$	67	t-BuO–Cl	47	Me$_3$Sn–SnMe$_3$	63
MeHN–NHMe	65	t-BuO–Br	44	Me$_3$Si–SiMe$_3$	81
F$_2$N–NF$_2$	21	RO–NO	41		
O$_2$N–NO$_2$	14	HS–SH	66		

Table 26.28 – $BDE(\Sigma–\Sigma)$ and $BDE(\Sigma^1–\Sigma^2)$.

X	H	Cl	Br	I	Me	OH	OEt	NH$_2$
Me$_3$C–X	95	84	70	54	86	95	82	85
Me$_3$Si–X	90	112	96	77	90	128	111	100
Me$_3$Ge–X	82	116	104	63	76		107	
Me$_3$Sn–X	74	94	83	69	65	110	84	

Table 26.29 – $BDE(Me_3M–X)$.

INDEX

A

Absolute rate constants
- addition to aromatics, 168
- addition to >C=C<, 62, 63, 292-293
  - by alkyl radical, 62, 292
  - by heteroatomic radical, 293
- cyclization, 152-154, 159
- decarboxylation, 16, 217
- definition, 49
- dimerization, 180, 296
- fragmentation, 150, 295
- homolytic aromatic substitution, 160, 257
  - addition of protonated radical, 169
  - addition step, 168
- initiator decomposition, 109
- opening, 153, 154, 158, 205
- reaction with oxygen, 95
- rearrangement, 153, 161
- S_H2, 292
- S_Hi, 291
- transfer, halogen-C,
  - by alkyl radical, 71, 202, 290
  - by phenyl radical, 202, 290
- transfer, halogen-$^+NR_3$
  - by heteroatom radical, 71
- transfer, hydrogen-C
  - by alkyl radical, 289
  - by phenyl radical, 51, 289
  - by oxyl radicals, 290
- transfer, hydrogen-heteroatom
  - by alkyl radical, 51, 202, 222, 289
  - by oxyl radicals, 290
  - by phenyl radical, 202, 222, 289
  - intramolecular, 135, 162, 216

Abstraction. *See* transfer or S_H2 substitution
Acetoxyl transfer, 164
Activation energy, 49, 72, 109, 121, 127, 150, 164
Acyl radicals, 96, 149
Acyloxyl. *See* radicals, oxyl
Addition, 44, 56, 206, 229, 254
- frontier orbital, 140
- isomerizing, 45
- kinetics, 76

- mechanism, 139, 151
- on >C=C<, of radicals, 139, 151, 229, 292
- carbon, 55, 69, 130, 140, 230, 292
  - halogen, 131, 137
  - heteroatom, 85, 92, 96, 236, 238, 240, 293
  - oxyl, 96, 293
- on anion, 182
- on aromatics, 64, 86, 167
- on –C≡C– or –C≡N, 85, 147, 154, 156
- on >C=N–, 86, 148, 227
- on >C=O, 86, 114, 147, 156, 225
- on >C=S, 86, 114, 147, 225, 226, 227
- on –N=O, 228
- rate constant, 62, 63, 139, 292-293
- reactivity, 62, 139
- regioselectivity, 143
- reversibility, 139, 146
- stereoselectivity, 144

Addition/fragmentation, 99, 114, 148, 166
Adipic acid, 274
AIBN, 109, 112, 117, 221, 222, 228, 233, 239, 240
Alcohols, deoxygenation, 225
Aldehydes, autoxidation, 220
Alkaloids, 217
Alkoxyl. *See* radicals, oxyl
Alkyl. *See* radicals
Allylation, 45, 240
Allylic. *See* radicals
- substitution, 196, 214
Amidyl. *See* radicals
Aminoacids, 195
Aminoxyl. *See* nitroxide
Aminyl. *See* radicals
Annelation, 254
Antioxidant, 93, 195
Anthraquinol, 273
Arachidonic acid, 196
Aromatic alkylation. *See* reaction
Aromatic amination. *See* reaction
Aromatic substitution. *See* reaction
Arrhenius relationship, 49
Aryl halides. *See* $S_{RN}1$

– radicals, 155, 176
– transfer, 162
Autoxidation, 95, 122, 195, 218
Azoalkanes, 112, 122
Azobisisobutyronitrile. *See* AIBN

B

Barton reaction, 114, 136, 224
BDE, definition, 32
– carbon-carbon, 151 162
– carbon-halogen, 81, 201, 298
– carbon-heteroatom, 83, 86, 96, 150, 298
– carbon-hydrogen, 33, 36, 51, 54, 81, 83, 297
– halogen-halogen, 100, 212, 299
– heteroatom-heteroatom, 83, 96, 299
– heteroatom-hydrogen, 81, 83, 86, 92, 93, 202, 297
– oxygen-hydrogen, 93, 194, 298
– protonated molecules, 87, 89
Benzophenone, 106
Benzoyl peroxide. *See* peroxide
Benzoyloxyl. *See* radicals
Bicyclic compounds (cyclization), 247
Biology, 191
Bond dissociation energy. *See* BDE
Bond formation
– C–C, 139, 229, 230,
– >C=C<, 137, 147, 148, 149
– –C≡C–, 227
– –C≡N, 207
– C-metal, 240
– C–O, 155
– C–P, C–S, 155, 238, 239
– C–Si, 155
– O–B, 96
N-Bromosuccinimide. *See* NBS
tert-Butyl. *See* radical
tert-Butyl hydroperoxide, 109
tert-Butyl peroxide, 109, 112
tert-Butyl peroxybenzoate, 112
tert-Butyl peroxyoxalate, 112

C

Cage, solvent, 13
Cage recombination, 13, 111
Captodative. *See* radicals
– effect, 36
Carboxylic acids, 115, 196
Carboxy-inversion, 111

Chain,
– definition, 42
– kinetic, 73
– length, 73
– propagation, 42, 73
Charged radicals, 181
Chemically Induced Nuclear Polarization (CIDNP), 13
Chemoselectivity, 222, 245
N-Chloroamines
– addition, 236, 237
– aromatic substitution, 257
– cyclization, 88, 253
– decomposition, 106
– halogenation, 213
– intramolecular hydrogen transfer, 89, 216
– reduction, 87
CIDNP. *See* Chemically Induced Nuclear Polarization
Cobalt, 114
Combination. *See* recombination
Combustion, 275
Competition, 79
Configuration, 10, 19, 22, 130
Conformation, 10, 20, 130
Copolymerization, 55, 271
Cumene, 202, 274
Cyclization, 40, 46, 151, 215, 243
– by S_Hi, 157
– coupled with addition, 254
– heteroatom radicals, 85, 88, 253
– mechanism, 151
– on >C=C<, 151, 244, 249
– on intracyclic >C=C<, 247
– on –C≡C–, 154, 245, 249, 252
– on >C=O, 154, 156, 245, 252
– sequential, 256
– stereoselectivity, 152, 244, 249
– tandem, 251
– transannular, 251
Cyclopropyl. *See* radicals
Cytochrome, 191, 198

D

Deamination, 227
Decalyl radicals, 21, 24
Decarboxylation, 17, 18, 46, 97, 149, 223, 224, 259
Defunctionalization, 43, 84
Denitration, 228

Diazonium, 115
Dimerization. *See* recombination
Diphenylpicrylhydrazyl radical. *See* DPPH
Dismutation. *See* disproportionation
Displacement, 125, 133
Disproportionation, 41, 123, 166, 182, 187
Distonic radicals, 185
Disulfides, 133
DMSO, 261
DNA, 194
DPPH, 17, 31, 203

E

Electrochemistry, 94, 266
Electron paramagnetic resonance. *See* EPR
Electron spin resonance (ESR). *See* EPR
Electron transfer, 39, 42, 47, 94, 115, 182, 184, 186
Electronic state of radicals, 20
Electrophilicity, 53, 61, 177
Elimination, 176. *See* fragmentation
Enthalpy, 50, 71, 129, 139, 148, 158, 168
EPR, 5, 166, 186, 197
ESR. *See* EPR
Ethyl. *See* radical
Evans-Polanyi relationship, 52
Exogenous agents, 198
– halogen derivatives, 199
– nitrogen oxides, 199
– radiation, 198

F

Fenton reaction, 95, 116, 196
Flavoproteins, oxidation by, 192
Fluoromethyl radicals, 11, 22, 127
Frontier orbital, 60, 140, 171
Function, transformation of, 211, 221
Fragmentation, 41, 148, 161
– allylic substitution, 214
– carbon radicals, 86, 137, 146-149, 153-164, 225-227
– decarboxylation. *See* decarboxylation
– elimination of CO, SO_2, 149
– formation of >C=C< bond, 137, 146, 149, 150
– formation of –C≡N bond, 207, 227
– oxyl radicals, 68, 97, 147, 148, 150, 215
– phosphoranyl radicals, 93, 135
– radical anions, 47, 182

– radical cations, 185
– rate constant, 295
– ring opening, 45, 158, 164, 165

G

G-factor (Landé), 5, 7
Galvinoxyl radical, 32, 203
Geometry, radical center, 19
Germanium. *See* radicals, Ge-centered

H

Half-life, 31, 109, 123
Halide reduction, 221
Halogenation, 100, 211, 214, 232, 272
Hammond postulate, 52, 139
Heteroatomic bases, substitution, 168, 170, 174, 178, 259, 261, 268
5-Hexen-1-yl radical, 41, 46, 80, 151, 153, 205
Hofmann-Löffler-Freytag reaction, 89, 136, 215
HOMO (Highest Occupied Molecular Orbital), 57, 140, 171
Homolysis, 39, 93
Hunsdiecker reaction, 46, 223
Hydrogen. *See* transfer, see abstraction
Hydrogen peroxide, 193, 273
Hydroperoxide, 109, 195, 219, 231, 260, 274, 276
Hydroperoxyl. *See* radicals, oxyl
Hydroxylation. *See* reaction
N-Hydroxy-2-thiopyridone, 114, 147, 224, 226
Hydrides (Sn, Ge, Si, Hg), 80-81, 83-86, 113, 117, 129-132, 221-222, 233-236, 289-290, 298
– tri-*n*-butyltin, 80-81, 84, 86, 113, 117, 129, 130, 132, 222, 233-234, 289
Hyperfine coupling, 7
Hypochlorite, 78, 93, 117, 212, 214-215

I

Iminoxyl radical, 27
Induced decomposition, 110
Industry, radical reactions, 269
Inhibition, 208
Initiation. *See* reaction
Intramolecular. *See* transfer, see S_Hi
Ionization potential, radicals, 61, 297
– aromatics, 187, 297
Ions, radical. *See* radicals, anions and cations

Ipso-substitution. *See* reaction
Isomerization, 21, 131, 146, 159, 196-197, 204
Isonitrile, 207, 248

K

Kaptein's rules, 15
Karplus relationship, 11
Ketones, enolizable, 115
Kinetics, 16, 73, 208, 229
Koelsch radical, 37
Kolbe reaction, 115, 267
Kornblum reaction, 182, 267

L

Lipids, 195
LUMO (Lowest Unoccupied Molecular Orbital), 22, 57, 140, 142, 171, 182

M

Macrocyclization, 247
Mass spectroscopy, 21
McConnell relationship, 11
Mediators, 233
Mercury compounds, 114, 132, 134, 235
Mesomerism, 35
Methyl. *See* radical
Migration, 40, 135
Minisci reaction, 169, 178, 258
Monocyclic compounds (cyclization), 244

N

NAD^+-NADH, 192
N-Bromosuccinimide (NBS), 214, 244, 290
Neophyl rearrangement, 162
Net and multiplet effects (CIDNP), 15
Nitrite, 93
Nitro derivative, denitration, 228
Nitrogen, radicals centered on, 26, 86
Nitrogen oxides, 199
Nitrone, 207
Nitrosation, 273
Nitrosobenzene, 12
Nitroxide. *See* radicals, oxyl
Norbornene, 146, 206
Norbornyl radicals, 24
Nucleophilicity, 53, 61, 178

O

Orbital analysis, 22, 57, 68, 140, 171
Organometallics, 106, 113
Overall rate factor, 168
Oxidation, 275
– biological, 191-192, 195
– by metals, 42, 48, 94, 115, 177, 187, 193, 206
– chain, 196
– electrochemical, 94
– of alkyl aromatics, 115
– of amines, 88
– of carboxylic acids, 115
– of charged radicals, 182, 187
– of enolizable ketones, 115
– rearomatization, 48, 176-177, 257, 259
– to hydrogen peroxide, 273
Oxidation potential, aromatics, 187
Oxido-reduction, 39, 87, 94, 115, 246, 260
Oxygen, 95, 193, 218, 225, 273, 274
Oxyl. *See* radicals, oxyl
Ozone, 199

P

Paramagnetism, 6
Partial rate factor, 166, 172
Pascal's triangle, 9
Peracids, 21, 46, 56, 67, 77, 134, 217, 220, 223
Peresters, 109, 112, 158, 265
Peroxalates, 109
Peroxides
– benzoyl, 40, 109, 231-232, 245, 260
– diacyl, 16, 40, 109, 178
– dialkyl, 109, 112, 157-158, 231, 239
Peroxydicarbonates, 111
Peroxyl transfer, 165
Peroxyl. *See* radicals, oxyl
Persistent radicals, 12, 37, 56, 123
Perthiosulfate, 115, 260
Phenol, 274
Phosphinyl. *See* radicals, P-centered
Phosphoranyl. *See* radicals, P-centered
Photolysis, 39, 91, 100, 105
Polar effect, 52, 54, 72, 127, 142, 168, 170, 211, 213, 257
Polyhalomethanes, 45, 106, 129, 232
Polymers, 55, 269
Pre-exponential factor, 127
Propylene oxide, 274

Prostaglandins, 195
Pyridines, 167, 174, 258

Q

Quinones, 192
Quinoline, 259, 261

R

Radical
- 1-adamantyl, 179
- benzyloxyl. See radicals, oxyl
- benzoylperoxyl. See radicals, oxyl
- *tert*-butoxyl. See radicals, oxyl
- *tert*-butyl, 54, 62, 63, 66, 71, 72
- tri-*n*-butylstannyl, 21, 40, 43, 71, 80, 86, 113, 130, 134
- 2,4,6-tri-*tert*-butylphenyl, 37, 121
- trichloromethyl, 99, 120, 127, 146, 203, 292, 296-297
- ethyl, 11, 20, 23
  - chloro, 11
- trifluoromethyl, 10, 11, 22, 127, 297
- hydroperoxyl. See radicals, oxyl
- hydroxymethyl, 10
- methyl, 10, 22, 34, 62, 71, 72, 120,127, 133, 169, 289, 296-297
  - EPR spectrum, 9
- phenyloxyl. See radicals, oxyl
- phenyl, 10, 35, 51, 169, 202, 289-290, 292-293, 295-297
Radical anions, 94, 181
Radical cations, 94, 185
  - alkene, 188
  - aromatic, 187
Radical clocks, 74, 80, 159, 205
Radical coupling. See recombination
Radical mechanisms
- addition, 42, 139
- chain, 42, 139
- cyclization, 151
- homolytic aromatic substitution, 166
- non-chain, 47
Radical pairs, 13
Radical production, 83, 87, 91, 93, 98, 100, 105, 259
Radical protonation, 169, 177, 178, 182
Radical spectra
- EPR, 7, 21
- IR, 21

Radical stability, 12, 31, 37
- kinetic, 37
- thermodynamic, 31-34
Radical structure, 19, 181, 185
- EPR, 7, 21, 22
- hybridization, 22, 35
- orbital factors, 22
Radical trapping, 12, 200, 203
Radicals
- acyl, 96, 149
- alkenyl, 46, 80, 152-154, 159, 205
- alkyl, 9, 23, 34, 62-63, 66, 71, 72, 119, 130, 158, 169, 170, 289-297
  - ß-substituted, 11, 25, 136, 159
- allyl, 35
- amidyl, 26, 90
- aminyl, 21, 26, 88, 169, 177, 236, 253
  - complexed, 89
  - protonated, 88, 128, 136, 169, 177, 213, 216, 253, 258, 260
- aryl, 18, 154, 157, 180
- benzyl, 12, 35, 37, 56, 65, 120, 149, 188, 289, 295-297
- captodative, 36, 47, 266, 298
- cycloalkyl, 21, 24, 46, 69, 141, 144, 151, 158
- cyclohexadienyl, 166, 173, 202
- cyclopropyl, 21, 24, 130, 158
- destabilized, 32
- electrophilic, 53, 61, 171, 212, 214
- Ge-centered, 10, 26, 83, 129, 221, 240, 291-293
- halogen, 72, 78, 100, 127, 133
- hydroxyalkyl, 10, 134, 231, 259
- ionization potential. See ionization potential
- methyl-substituted, 22, 172
- N-centered, 26, 86, 203, 237
- nucleophilic, 53, 61, 178, 218, 258
- oxyl, 28, 93, 108, 150, 155, 161, 165, 193, 254, 270, 276
  - acyloxyl, 29, 39, 41, 46, 56, 67, 93, 96, 110, 149, 176, 290-291, 293, 295
  - acylperoxyl, 68, 93, 96, 293
  - alkoxyl, 28, 40, 55, 68, 79, 94, 150, 155, 161, 254, 270, 290-291, 293-295, 297
  - aryloxyl, 94
  - benzoyloxyl, 39, 93, 96, 110, 290, 293, 295
  - benzoylperoxyl, 96, 293
  - *tert*-butoxyl, 57, 92, 108, 112, 127,

133, 135, 148, 214, 265, 290, 293, 295, 297
- hydroperoxyl, 13, 93, 194
- hydroxyl, 10, 13, 40, 93, 193, 223, 290, 293
- iminoxyl, 27
- nitroxide, 27, 32, 90, 204
- peroxyl, 93-95, 122, 165, 196, 219, 225, 276, 290, 296-297
- phenyloxyl, 93
- phenylperoxyl, 296
- P-centered, 27, 91, 238
  - phosphinyl, 27
  - phosphonyl, 28, 130
  - phosphoranyl, 28, 97, 155
- persistent. See persistent
- short-lived, 12, 17, 32
- stable, 17, 31, 203, 207
- S-centered, 29, 97, 238
  - sulfinyl, 30, 97
  - sulfonyl, 29, 97, 238
  - thiyl, 29, 97, 130, 246, 254
- Si-centered, 10, 26, 83, 86, 113, 129, 233, 240, 291, 293
- Sn-centered, 10, 26, 40, 71, 83, 86, 113, 129, 221, 233, 234, 241, 291-293
- vinyl, 10, 25, 37, 147, 246
Radiolysis, 21, 106
Rate
- absolute, constant. See absolute rate contants
- determining step, 42, 72
- overall factor, 168
- partial factor, 166, 172
- relative, constant. See relative rate constants
Reactions
- addition. See addition
- allylation, 45, 240
- amination, 177, 257
- arylation, 177
- chlorination by t-BuOCl, 72
- competitive, 79
- cyclization. See cyclization
- decarboxylation. See decarboxylation
- degradation, 46
- dehydrodimerization, 265
- disproportionation. See disproportionation
- elementary, 39
- elimination. See fragmentation
- halogenation, 100, 211, 272
- homolytic aromatic substitution. See substitution

- hydroxylation, 179, 198, 217
- initiation, 42, 73, 106, 177, 183, 218, 227, 230
- isomerization, 21, 197, 204
- rearrangement, 41, 151, 161, 204
- ring opening, 41, 45, 153, 154, 158, 205
- S_H2. See substitution
- $S_{RN}1$, 46, 182, 261, 267
- substitution. See substitution
- transfer. See transfer
Reactivity, 49, 177
- enthalpy control, 50, 71, 148
- frontier orbitals, 57
- introduction, 49
- orbital analysis, 57, 61
- orbital interactions, 57
- polar effect. See polar effect
- stereoelectronic control, 69, 132, 159
- steric control. See Steric effect
Rearrangement. See reaction
Recombination, 41, 47, 111, 120, 122, 166, 182, 203, 207, 269, 296
Redox. See oxido-reduction
Reduction, 92, 94, 116, 221
Regioselectivity
- addition, 143, 147, 206
- aromatic substitution, 167, 169, 172, 175, 179, 259
- cyclization, 152, 156, 206, 244, 253
- hydrogen transfer, 55, 215, 216
- opening, 159
Relative rate constants
- addition, 62, 141
- aromatic substitution, 169, 170, 259
- cyclization, 153
- disproportionation/recombination, 124
- hydrogen transfer, 51, 127, 128, 211
Reversibility, 99, 100, 121, 146, 168, 178
Ring opening. See reactions

S

Samarium iodide, 116
Scavengers, 12, 203, 225
Scission. See fragmentation
SCD. See state correlation diagram
Selectivity, 127, 169, 212
Sequential cyclization, 251
S_H1. See substitution
S_H2. See substitution
Short-lived. See radicals

Silicon. *See* radicals, Si-centered
Solvent effect, 120, 128, 175, 211, 212, 216, 261
SOMO (Singly Occupied Molecular Orbital), 10, 22, 57, 140, 171
Spin, 5, 13
– polarization, 10
Spin marking. *See* radical trapping
Spin trapping. *See* radical trapping
$S_{RN}1$ reaction, 46, 182, 261, 267
State correlation diagram, 70-72
Steady state, 74
Stereoselectivity, 121, 130, 132, 144, 149, 154
Steric effect, 56, 121, 123, 126, 130, 141, 142, 213
Sulfinyl. *See* radicals, S-centered
Sulfonyl. *See* radicals, S-centered
Substituent effect, 140, 142, 153, 172
Substitution. *See* also transfer
– allyl, 214
– homolytic aromatic, 44, 47, 64, 166, 169, 257
    – addition step, 167
    – alkylation, 176
    – intramolecular, 180
    – ipso-substitution, 179
    – mechanism, 166
    – oxidation step, 176, 178
    – regioselectivity, 172, 257, 259
– S_{Hi} (intramolecular), 96, 126, 157, 177
– S_H2, 125, 133
    – concerted, 133
    – on heteroatoms, 133
    – on monovalent atoms, 126
    – on multivalent atoms, 133
    – on peroxide bonds, 56, 67, 133, 157, 217
Superoxide anion, 13, 185, 193

T

T radical, 207
Tandem cyclization, 251
TEMPO radical, 207
Termination, 42, 73-75
Tetroxides, 95, 122
Thermolysis, 100, 107
– peracids, 77
Thiohydroxamic esters, 114, 224, 226
Thiols, 106, 201, 226, 290
Thiyl. *See* radicals S-centered
Tin. *See* radicals Sn-centered
Tocopherol, 195
Transfer. *See* also substitution
– halogen, 81, 84, 92, 101, 113, 128, 136, 201, 215, 224, 233, 291
– hydrogen, 50, 54, 66, 68, 81, 84, 91, 108, 110, 132, 135, 201, 204, 225, 226, 289
    – allylic, 96
– intramolecular, 40, 52, 89, 90, 95, 126, 135, 204, 215
    – 1,2 halogen, 137
Transition state, 49, 70, 139, 167, 170, 177
Tri-*n*-butylstannyl. *See* radical

U

Ubiquinone, 192
Unsaturated fatty acids, 195, 196
UV spectroscopy, 17

V

Vinyl transfer, 163
Vinyl. *See* radicals
Vitamins (C and E), 195

X

Xanthates, 113, 226

Z

Zeeman energy, 5

MASSON Éditeur
120, boulevard Saint-Germain
75280 Paris Cedex 06
Dépôt légal : juillet 1995

SNEL S.A.
Rue Saint-Vincent 12 - 4020 Liège
juillet 1995